SpringerWienNewYork

CISM COURSES AND LECTURES

The series presents lecture notes, monographs, edited works and proceedings in the field of Mechanics, Engineering, Computer Science and Applied Mathematics.
Purpose of the series is to make known in the international scientific and technical community results obtained in some of the activities organized by CISM, the International Centre for Mechanical Sciences.

INTERNATIONAL CENTRE FOR MECHANICAL SCIENCES

COURSES AND LECTURES - No. 535

VARIATIONAL MODELS AND METHODS IN SOLID AND FLUID MECHANICS

EDITED BY

FRANCESCO DELL'ISOLA
UNIVERSITÀ "LA SAPIENZA", ROMA, ITALY

SERGEY GAVRILYUK
POLYTECH MARSEILLE, FRANCE

SpringerWienNewYork

This volume contains 63 illustrations

SPIN 80124210

All contributions have been typeset by the authors.

ISBN 978-3-7091-0982-3 SpringerWienNewYork

PREFACE

For this would be agreed by all:
that Nature does nothing in vain
nor labours in vain

Olympiodorus, Commentary on Aristotle's Meteora translated by Ivor Thomas in the Greek Mathematica Works Loeb Classical Library

La nature, dans la production de
ses effets, agit toujours par les
voies les plus simples

Pierre de Fermat

The CISM course C-1006 "Variational models and methods in solid and fluid mechanics" was held July 12-16, 2010 in Udine, Italy. There were about forty five participants from different european countries. The papers included in this volume correspond to the content of five mini-courses of 6 hours each which have been delivered during this week.

Variational formulation of the governing equations of solid and fluid mechanics is a classical but a very challenging topic. Variational methods give an efficient and elegant way to formulate and solve mathematical problems that are of interest for scientists and engineers. This formulation allows for an easier justification of the well-posedness of mathematical problems, the study of stability of particular solutions, a simpler implementation of numerical methods. Often, mechanical problems are more naturally posed by means of a variational method. Hamilton's principle of stationary (or least) action is the conceptual basis of practically all models in physics. The variational formulation is also useful for obtaining simpler approximate asymptotical models as done in the theory of homogeneization. In many problems of mechanics and physics, the functionals being minimized depend on parameters which can be considered as random variables. Variational structure of such problems always brings considerable simplifications in their study.

In this course, three fundamental aspects of the variational formulation of mechanics will be presented: physical, mathematical and applicative ones.

The first aspect concerns the investigation of the nature of real physical problems with the aim of finding the best variational formulation suitable to those problems. A deep knowledge of the physical problems is needed to determine the Lagrangian of the system and the nature of the variations of its motions which may be considered admissible. Actually one could say that all knowledge which is available about a system is resumed by the choice of:

- *a configuration space used to describe mathematically the system*
- *a set of admissible motions used to describe the different ways in which the system may evolve*
- *a Lagrangian functional which once minimized supplies evolution equations and boundary conditions*

The second aspect is the study of the well-posedeness of those mathematical problems which need to be solved in order to draw previsions from the formulated models. It is relatively simple to conjecture properties to be required to the Lagrangian functional in order to be assured the well-posedness of the corresponding evolution system. Much more complex is to get such results of well-posedness studying some evolution equations which are obtained with euristic schemes different from those based on Hamilton's principle. In fact always, when one needs to study mathematically a set of evolution equations, the first move is to try to put them in a variational form. It is then advisable and wiser to try to use a variational principle at the beginning of the formulation of a mathematical model.

The third aspect is related to the direct application of variational analysis to solve real engineering problems. Variational principles supply very powerful tools for getting qualitative previsions about the behaviour of the studied systems, but also for formulating effective numerical methods to get quantitative previsions.

The following problems have been presented and studied during the course :

- *Rayleigh-Hamilton's Principle for establishing governing equations and boundary conditions for second gradient models for heterogeneous deformable bodies ;*

- *A variational approach to multiphase flow problems and description of diffuse solid-fluid interfaces;*
- *New variational models of brittle fracture mechanics and some related problems ;*
- *The methods of stochastic calculus of variations and their applications to the homogenization problems and modeling of microstructures and their evolution ;*
- *Dynamical problems in damping generation and control in the situations where the energy initially conferred to a system undergoes a principle of irreversible energy confinement into a small region ;*

We are extremely grateful to all participants of the course for creating a nice atmosphere for scientific discussions. We would like also to express our thanks to the CISM staff for their assistance in the running of the course.

Francesco dell'Isola, University of Rome "La Sapienza"

Sergey Gavrilyuk, University of Aix-Marseille

CONTENTS

Variational principles are a powerful tool also for formulating field theories

Francesco dell'Isola [*‡] and Luca Placidi [†‡]

[*] University of Roma "La Sapienza"

[†] International Telematic University "Uninettuno"

[‡] International Research Centre on "Mathematics & Mechanics of Complex Systems" M& MOCS

Abstract Variational principles and calculus of variations have always been an important tool for formulating mathematical models for physical phenomena. Variational methods give an efficient and elegant way to formulate and solve mathematical problems that are of interest for scientists and engineers and are the main tool for the axiomatization of physical theories.

1 Introduction and historical background

1.1 Metrodoron and his followers

The ideas we want to evocate in this lecture are very old and were put forward already in the hellenistic period: for a detailed discussion about this point the reader is referred to the beautiful book by Lucio Russo (2003). In that book it is established that "modern" science actually was born in the hellenistic era, when Metrodoron lived. Metrodoron was a pupil of a famous greek philosopher, Epicurus, and, in our opinion, the following Metrodoron's sentence is a statement (the first?) belonging to the modern philosophy of science:

«Μέμνεσο ὅτι θνητὸς ὢν τῇ φύσει καὶ λαβὼν χρόνον ὡρισμένον ἀνέβης τοῖς περὶ φύσεως διαλογισμοῖς ἐπὶ τὴν ἀπειρίαν καὶ τὸν αἰῶνα καὶ κατεῖδες "τὰ τ' ἐόντα τά τ' ἐσσόμενα πρό τ' ἐόντα"».

Metrodoron,

"Always remember that you were born mortal and such is your nature and you were given a limited time: but by means of your reasonings about Nature you could rise to infinity and to eternity

and you indeed contemplate ***"the things that were, and that were to be, and that had been before"***". Metrodoron

Gnomologium Epicureum Vaticanum X (fr.37 Alfred Körte, Metrodori Epicurei Fragmenta, " Jahrbücher für classiche Philologie", Suppl. XVII, 1890, p. 557).

This dictum, following Körte, comes from a lost letter or book by Metrodoron (the Epicurean philosopher) addressed to Menestratos who was presumably one of his pupils. The words quoted in bold are a citation from Iliad, I 70 (the translation into English of the sentence in boldface is ours; except for this citation the translation has been taken from Homer by Murray (1924), see the ref. (14)).

In different words, Metrodoron states that by using (the right!) equations you can forecast future behavior of physical systems.

1.2 Why Variational Principles and Calculus of Variation?

In recent time, a lost Archimedes' book (19) has been rediscovered. Some authors claim that Archimedes seems to have solved, in this book and using a variational principle, the technological problem of finding the optimal shape of a boat. Archimedes seems to have chosen, as optimality criterion, that the vertical position must be a "very" stable configuration (see Rorres (2004)). In the book of Russo (21) it is demonstrated in even a more convincing way that many optimization techniques were well-known in hellenistic science. In particular Russo proves that the problem of the determination of the regular polygon having maximal area has been solved in that period. Thus, the use of a variational principle and optimization methods to solve technological problems is less recent than it is usually believed. In general, variational formulation of the governing equations of solid and fluid mechanics is a classical but very challenging topic. This kind of formulation allows for an easier proof of the well-posedness of mathematical problems, for an easier investigation of the study of stability of particular solutions, and for a simpler implementation of numerical methods. Often (but one who believes in Russo's vision about the birth of science could say instead "always"), mechanical problems are more naturally posed by means of variational methods. Hamilton's principle of stationary (or least) Action is the conceptual basis of practically all models in physics. The variational formulation is also useful for obtaining simpler approximate asymptotical models as it is done in the theory of homogenization.

We want simply to state here that the Principle of Virtual Works and the Principle of Least Action have roots much deeper than many scientists believe (see Vailati, 1897). Although many histories of science claim differently, most likely the majority of physical theories were first formulated in terms of these Principles, and only subsequently they were reconsidered from other points of view. In our opinion the Principle of Least Action, which supplies a "geometric" version of mechanics was indeed the tool used by the true founders of mechanics (i.e. the scientists of the hellenistic period) to establish it. As argued also by Colonnetti (5) and Netz and Noel (19)) surely also Archimedes and ancient greek scientists were accepting such a point of view.

The epigones of the hellenistic science, who were not able to understand the delicate mathematical arguments used by the first scientists, however could understand the minimality conditions obtained by their "maitres" (i.e. the conditions corresponding to those which we call now Euler-Lagrange equations and boundary conditions) and could grasp the "physical" arguments used to interpret them. This phenomenon is perfectly clear to everyone who is ready to consider carefully -for instance- the evolution of the theory of Euler-Bernoulli Beam (a useful reference about this point is the book of Benvenuto (1981)). Euler postulated a Principle of Least Action for the Elastica, and gets the celebrated equilibrium differential equation and boundary conditions for the equilibrium of beams by using the calculation procedure due to Lagrange (which is the departing idea of Calculus of Variations). Then Navier prepared his lectures for the Ecole Polytechnique and resumed the results obtained by Euler deciding to "spare" to the (engineering) students the difficulties of the calculus of variations. He started directly from the equilibrium equation, obtained by means of an "ad hoc" principle of balance of force and couple, and imposed boundary conditions based on "physical" assumptions. As a consequence, for a long while, generations of engineers believed that the beam equations were to be obtained in this way. Only when numerical simulations became popular, then they (actually, some of them) became aware of variational "principles". However these principle were proven as theorems starting from "balance postulates" and were considered simply as a mathematical (rather abstruse) tool and not as a fundamental heuristic concept. And this attitude is not changed even when it became clear that every serious advancement of mechanical science has been obtained using variational principles. Indeed the so called "physical sense" (a gift that many claim to posses but which nobody can claim to be able to master or to teach) is not very useful to postulate the right "balance principles" when one is in "terra incognita". For instance, when Lagrange and Sophie Germain wanted to find the plate equations they needed to employ a variational principle (and they could find the (right!) natural boundary conditions). Again when Cosserat brothers wanted to improve Cauchy Continuum Mechanics they "rediscovered" the right technique: i.e. the Principle of Least Action. Also Quantum Mechanics has been developed starting from a Variational Principle (see e.g. the references of Feynman (11), Lagrange (15) and Lanczos (16)).

Therefore an important warning is due to young researcher: refrain from trying to extend available models by means of "ad hoc" adaptations of available theories: always look for the right Action functional to be minimized!

1.3 The problem of including dissipation

One useful tool for handling complicated situations is used in Continuum Mechanics by Paul Germain when formulating second gradient theories: the Principle of Virtual Powers. Again, as remarked always in the history of the development of ideas, when this history can be reconstructed, the effective way to be used to proceed is that which starts from a Principle of Least Action, eventually generalized into a Principle of Virtual Powers. For a long time the opponents to Second Gradient Theories argued about its lack of consistency, due to the difficulties they claim to find in "getting" boundary conditions. This is a really odd statement. Indeed variational principles easily produce mathematically correct boundary conditions. So maybe what those opponents want to say is that as they are not so clever as Navier, they do not manage to interpret physically the boundary conditions found via a (correct and meaningful) variational principle. Of course if one refuses to use the Principle of Least Action he can find very difficult the job of determining some set of boundary conditions which are compatible with the (independently postulated!) bulk evolution equations. If instead one accepts the Archimedean (the reader will allow us to dream, without definitive evidence that such was the point of view of Archimedes) approach to mechanics then all these problems of well-posedness of mathematical models completely disappear.

Variational Principles always produce intrinsically well-posed mathematical problems, if the Action functional is well behaving. Of course passing from Lagragian systems (the evolution of which are governed by a Least Action functional) to non-Lagragian systems (for which such a functional may not exist) maybe difficult. This problem is related (but is not completely equivalent) to the problem of modelling dissipative phenomena. It is often stated that dissipation cannot be described by means of a Least Action Principle. This is not exactly true, as it is possible to find some Action functionals for a large class of dissipative systems. However it is true that not every conceived system can be regarded as a Lagragian one. This point is delicate and will be only evocated here. In general a non-Lagragian system can be regarded as Lagragian in two different ways: i) because it is an "approximation" of a Lagrangian system (see the case of Cattaneo equation for heat propagation), and this approximation leads to "cancel" the lacking part of the "true" Action Functional ii) because the considered system is simply a subsystem of a larger one which is truly Lagrangian. The assumption that variational principle can be used only for non-dissipative systems is contradicted by, e.g., the work presented in this book by Prof. Frankfort (12), where you find variational principles modelling dissipative systems. Indeed it is often stated that a limit of the modelling procedure based on variational principles consists in their impossibility of encompassing "nonconservative" phenomena. We do not believe that this is the case: however in order to avoid to be involved in a problem which is very difficult to treat, when dealing with dissipative systems, we will assume a slightly different point of view, usually attributed to Hamilton and Rayleigh.

2 Finding a mathematical model for natural phenomena

2.1 Principle of Least Action

We want to discuss here about the problem of finding a mathematical model for natural phenomena. We start with an epistemological Principle:

> "The Principle of Least Action tells us how to construct a mathematical model to be used for describing the world and for predicting the evolution of the phenomena occurring in it".

In the following modeling scheme, we give the right heuristic strategy to be used for finding an effective model using the Principle of Least Action. The recipe includes the following ingredients:

1. Establish the right kinematics needed to describe the physical phenomena of interest, i.e. the kinematical descriptors modeling the state of considered physical systems.
2. Establish the set of admissible motions for the system under description, i.e. establish the correct model for the admissible evolution of the system.
3. Employ the "*physical intuition*" to find the right *Action functional* to be minimized, i.e. modeling what Nature wants to minimize.

We start by finding the kinematical descriptors, because of the need of modeling the states of the considered system. Then we introduce motion, in such a way we model the evolution of the system to be described. Finally we ask Nature what is the quantity to minimize. Keeping this quantity in mind, we introduce the Action functional. To start with, it is necessary to focus the attention on a specific class of systems and on phenomena occurring to them. A configuration is the mathematical object used to model the state of considered systems: the set of possible configurations will be denoted by $\mathcal{C}$. The motion is the mathematical model describing the evolution of considered systems: it is a $\mathcal{C}$-valued function defined on time interval (t_0, t_f); the set of all admissible motions will be denoted by $\mathcal{M}$. The Action is a real-valued function, defined on $\mathcal{M}$, which models the "preferences" of nature.

Finally, to use the Principle of Least Action one needs three steps further,

4. Find the Euler-Lagrange conditions which are consequence of the postulated Least Action Principle
5. Interpret these condition on a physical ground
6. Determine, in terms of the postulated Action functional, the numerical integration scheme to be used to get the previsions needed to drive, by means of our theory, our experimental, technological or engineering activity.

2.2 The Rayleigh-Hamilton principle

When postulating an extended Rayleigh-Hamilton principle, the point 4 of the previously presented heuristic strategy will be further divided into two steps as follows:

4a. Once the quantities which expend power on the considered velocity fields are known in terms of postulated Action, introduce a suitable definite positive Rayleigh dissipation functional

4b. Equate the first variation of Action functional to the Rayleigh dissipation functional and get the evolution equations (including boundary conditions) which govern the motion of the system

Although in the literature the choice of including a Rayleigh-Hamilton principle in the class of variational principles is sometimes considered inappropriate, we will follow what seems to us the preference of the majority of the authors: therefore we do call "variational" also the strategy which we just described, not limiting the use of this adjective to the models using exclusively the Least Action Principle.

2.3 La Cinèmatique d'Abord !

According to Metrodoron, mathematical and physical objects are two different concepts. Indeed, equations are necessary for modeling physical systems but they refer to mathematical objects. When one solves the equations formulated in the framework of his model then he has to transform the obtained equations into previsions valid for the physical system he is studying. A good modeling procedure uses mathematics for finding the motion which minimizes Action. If this mathematics gives a reasonable forecasting of the observed evolution, then the model is valid. However, not everything is described by a given model. A model is always focused on a set of phenomena.

The set of phenomena that are focused by a model is established by the kinematics:

La Cinèmatique d'Abord !

In the previous scheme it is clear that the most "fundamental" step concerns the choice of the set of configurations used for characterizing the "accessible" states of the system. When constructing a mathematical model using the discussed epistemological principle, one must start with a precise and clear determination of the set C. The second step concerns the determination of admissible motions which clearly depend on the evolutionary phenomena one wants to model. A correct modeling process always starts specifying "admissible" kinematics.

2.4 La Nature agit toujours par les voies les plus simples

After having specified the admissible kinematics, one can wonder about the *desire* of Nature. The *utility* of Nature is a real-valued function defined on M. Following Maupertuis, we will call Action this "utility". Also Nature must consider which is the contingent situation: not all admissible motions are accessible by a physical system under given specific conditions. Therefore we must specify a subset M_A of the set M: the set of accessible motions. The *real* motion will be chosen by the system minimizing the Action in the subset M_A. Indeed:

La Nature agit toujours par les voies les plus simples.

2.5 Two possible choices for the set of admissible and accessible motions

In the famous textbook of Arnold (1) the author, following the tradition, does not "try" to explain Maupertuis' Principle of Least Action. We instead dare to try to deal with this. In the process of minimization of the Action, we need to specify the set of motions among which we look for minima. The choice of Lagrange is that of isochronous motions. Two motions are isochronous when they both start, at the given instant t_0, from a given configuration C_0 and arrive, at the same instant t_f, at the same final configuration C_f. On the other hand, the choice of Maupertuis is to focus on the set of motions with a "fixed energy content" and which are starting from the same configuration C_0 and ending (the instants of start and stop are not specified!) at the configuration C_f. In the set of admissible motions an "energy" functional must then be introduced: i.e. a functional which associates an energy content to any motion and any time instant t. The set of accessible motions is constituted by all motions from C_0 to C_f which have a constant energy content. The choice of Maupertuis, if not suitably modified, seems to limit the range of applicability of variational principles to non dissipative phenomena.

2.6 Further famous quotes

Many books in Calculus of Variations and/or Variational Principles, see e.g. that of Lanczos (1970), start with a preface, introduction or introductory chapter dealing with historical prolegomena and sometimes end with a philosophical chapter. In presenting this lecture notes, we did not dare to break with tradition.

> "For this would be agreed by all: that Nature does nothing in vain nor labours in vain". Olympiodorus, Commentary on Aristotle's ™Meteora translated by Ivor Thomas in the Greek Mathematical Works Loeb Classical Library
>
> "La nature, dans la production de ses effets, agit toujours par les voies les plus simples". Pierre de Fermat.

Now, the problem is:

What is utility?

3 In other words: How to find "Real Motions"?

Up to now no mathematical structure has been assumed for M_A. Indeed, Action functional is simply a real-valued map defined on M_A. "Practical" problems require the *calculation* of *real motions* by means of introduced model. Following Lagrange (15), we introduce a particular class of Action functionals in terms of a Lagrangian Action density function: so constructing in a particular way Action functional to obtain so called "Lagrangian functionals".

We need to introduce a topological structure in M_A, i.e. we need to clearly define what we mean when we say that "two motions are close". If we want to find minima of a real-valued function, then we need to estimate derivatives and equate these derivatives to zero. Action is a function defined in the set of motions (not real numbers!). Thus, we need

- to understand what is an infinitesimal variation of motion,
- to find a differential of a functional and
- to estimate the order of infinitesimal of its remainder.

In other words, we need to learn how to find a first order Taylor expansion for a Lagrangian functional by establishing the meaning of the expressions :

- Infinitesimal variation of motion.
- Differential of a functional.
- Order of infinitesimals for remainders.

This implies the need of Frechèt and Gateaux derivatives in manifolds with charts in Banach spaces. This is the right mathematical frame for studying this subject. However, Lagrange did not know that he was using such a mathematical frame and did not know anything about Frechèt and Gateaux derivatives. Thus, in this notes we try to go around the related mathematical difficulties and follow the original approach of Lagrange.

The motion minimizing Action will be searched among the motions for which the first variation of Action vanishes.

For Lagrangian functionals this condition is equivalent to a partial differential equation which is called Euler-Lagrange condition relative to the given Action functional. This procedure generalizes the corresponding one used for real-valued functions of several real variables. One serious problem with papers that start from balance equations and "play" with forces is that they do not "find" boundary conditions. In these references ((7; 8; 9)) one can find examples of modelling procedures in which one finds simultaneously bulk and boundary conditions.

From an historical point of view, in the theory of beam we deal with contact actions (normal and shear forces and momenta) because Navier has written lecture notes for l'Ecole Polytechique, trying to produce a text for students that was as simple as possible. He wrote final equations and explain not only bulk but also boundary conditions with the aid of "physical sense". However, it is very difficult in general to **find** evolution equations and boundary conditions with physical sense. On the other hand, variational principles give boundary conditions automatically and without the help of any physical sense.

Thus, Variational Principles allow Science to unveil Nature and for unveiling Nature you need a Lagrangian functional.

4 Lagrangian Action Functionals: technical details

We follow Landau and Lifshitz (1977) and Moiseiwitsch (1966).

Let $\Psi_\sigma(x_\mu)$ be any set of n tensor fields defined on $\mathbb{R}^m$, (σ being a multi-index and $\mu = 1, 2, ..., m$). We define the Lagrangian density as:

$$\mathfrak{L}\left(x_\mu, \Psi_\sigma, \frac{\partial \Psi_\sigma}{\partial x_\mu}\right). \tag{1}$$

We can then introduce the Action functional as

$$\mathfrak{A} = \int_T \mathfrak{L}\left(x_\mu, \Psi_\sigma, \frac{\partial \Psi_\sigma}{\partial x_\mu}\right) \tag{2}$$

Where T is a hyper-volume in the $m-$th dimensional space determined by the coordinates x_μ. When we will want to derive the theory of second gradient materials, this approach will not be appropriate, because we would need to add the dependence on the second gradient of Ψ_σ in (1).

4.1 Variation of the Action Functional

We now consider small variations $\varepsilon\eta_\sigma(x_\mu)$ of the considered fields $\Psi_\sigma(x_\mu)$:

$$\tilde{\Psi}_\sigma(x_\mu) = \Psi_\sigma(x_\mu) + \varepsilon\eta_\sigma(x_\mu), \tag{3}$$

where the $\eta_\sigma(x_\mu)$ are any set of linearly independent functions of the x_μ which vanish on the part $\partial_d T$ ($\partial_d T \subseteq \partial T$) of the boundary ∂T of the hyper-volume T, on which the kinematical condition are prescribed. The variation of the Action functional can then be computed as:

$$\Delta\mathfrak{A} = \int_T \mathfrak{L}\left(x_\mu, \tilde{\Psi}_\sigma, \frac{\partial \tilde{\Psi}_\sigma}{\partial x_\mu}\right) - \int_T \mathfrak{L}\left(x_\mu, \Psi_\sigma, \frac{\partial \Psi_\sigma}{\partial x_\mu}\right), \tag{4}$$

where T is a hyper-volume in the m-th dimensional space determined by the x_μ. The computation of the variation of the Action functional now proceeds as follows:

$$\Delta\mathfrak{A} = \int_T \mathfrak{L}\left(x_\mu, \Psi_\sigma + \varepsilon\eta_\sigma, \frac{\partial \Psi_\sigma}{\partial x_\mu} + \varepsilon\frac{\partial \eta_\sigma}{\partial x_\mu}\right) - \int_T \mathfrak{L}\left(x_\mu, \Psi_\sigma, \frac{\partial \Psi_\sigma}{\partial x_\mu}\right) + O\left(\varepsilon^2\right) \tag{5}$$

which, with a slight abuse of notations, can be written at the first order in ε as:

$$\delta\mathfrak{A} = \varepsilon \int_T \sum\nolimits_\sigma \left(\frac{\partial \mathfrak{L}}{\partial \Psi_\sigma}\eta_\sigma + \sum_{\mu=1}^{m} \frac{\partial \mathfrak{L}}{\partial\left(\partial \Psi_\sigma / \partial x_\mu\right)} \frac{\partial \eta_\sigma}{\partial x_\mu}\right) \tag{6}$$

Integrating by parts and recalling that η_σ vanish on $\partial_d T$ it is easy to get:

$$\delta\mathfrak{A} = \varepsilon \int_T \sum\nolimits_\sigma \eta_\sigma \left(\frac{\partial \mathfrak{L}}{\partial \Psi_\sigma} - \sum_{\mu=1}^{m} \frac{\partial}{\partial x_\mu} \left(\frac{\partial \mathfrak{L}}{\partial \left(\partial \Psi_\sigma / \partial x_\mu \right)} \right) \right) \tag{7}$$
$$+\varepsilon \int_{\partial T/\partial_d T} \sum\nolimits_\sigma \eta_\sigma \sum_{\mu=1}^{m} \frac{\partial \mathfrak{L}}{\partial \left(\partial \Psi_\sigma / \partial x_\mu \right)} N_\mu,$$

where $\partial T/\partial_d T$ is the difference between ∂T and $\partial_d T$ and N_μ is the external unit normal of $\partial T/\partial_d T$. Imposing $\delta\mathfrak{A} = 0$, the arbitrariness of η_σ gives, for any σ:

$$\frac{\partial \mathfrak{L}}{\partial \Psi_\sigma} - \sum_{\mu=1}^{m} \frac{\partial}{\partial x_\mu} \left(\frac{\partial \mathfrak{L}}{\partial \left(\partial \Psi_\sigma / \partial x_\mu \right)} \right) = 0, \quad \forall x_\mu \in T, \tag{8}$$

$$\sum_{\mu=1}^{m} \frac{\partial \mathfrak{L}}{\partial \left(\partial \Psi_\sigma / \partial x_\mu \right)} N_\mu = 0, \quad \forall x_\mu \in \partial T/\partial_d T. \tag{9}$$

In the case of a discontinuity material surface Σ (with unit normal N_μ) the (9) have to be completed by

$$\sum_{\mu=1}^{m} [| \frac{\partial \mathfrak{L}}{\partial \left(\partial \Psi_\sigma / \partial x_\mu \right)} |] N_\mu = 0, \quad \forall x_\mu \in \Sigma, \tag{10}$$

where $[| \, (\cdot) \, |]$ is the jump of $(\cdot)$ across the surface Σ. These equations are known as the Euler-Lagrange equations corresponding to the considered Lagrangian density.

4.2 The Space-Time Case ($m = 4$)

Let us now consider the particular case in which $m = 4$. This case corresponds, for instance, to the case $x_\mu = (x_1, x_2, x_3, t)$. We have that $\eta_\sigma(x_\mu)$ are any set of linearly independent functions of the x_μ which vanish on the boundary of time type domain,

$$\eta_\sigma(x_1, x_2, x_3, t_0) = \eta_\sigma(x_1, x_2, x_3, t_1) = 0$$

and on the part $\partial_d V$ of the boundary ∂V of the volume V, on which the kinematical conditions are prescribed,

$$\eta_\sigma(x_1, x_2, x_3, t) = 0, \quad \forall (x_1, x_2, x_3) \in \partial_d V, \quad \forall t \in [t_0, t_1] \,.$$

It is easy to show that in this particular case eq. (7) yields

$$\delta\mathfrak{A} = \varepsilon \int_{t_0}^{t_1} dt \int_V \sum\nolimits_\sigma \eta_\sigma [\frac{\partial \mathfrak{L}}{\partial \Psi_\sigma} - \sum_{k=1}^{3} \frac{\partial}{\partial x_k} \left(\frac{\partial \mathfrak{L}}{\partial \left(\partial \Psi_\sigma / \partial x_k\right)} \right) - \frac{\partial}{\partial t} \left(\frac{\partial \mathfrak{L}}{\partial \left(\partial \Psi_\sigma / \partial t\right)} \right)] + \varepsilon \int_{\partial V / \partial_d V} \sum\nolimits_\sigma \eta_\sigma \sum_{k=1}^{3} \frac{\partial \mathfrak{L}}{\partial \left(\partial \Psi_\sigma / \partial x_k\right)} N_k + \int_\Sigma \sum\nolimits_\sigma \left[\left| \eta_\sigma \sum_{k=1}^{3} \frac{\partial \mathfrak{L}}{\partial \left(\partial \Psi_\sigma / \partial x_k\right)} \right|\right] N_k \tag{11}$$

The stationarity $\delta\mathfrak{A} = 0$ of the Action implies, for any $\sigma = 1, 2, ..., n$,

$$\frac{\partial \mathfrak{L}}{\partial \Psi_\sigma} - \sum_{k=1}^{3} \frac{\partial}{\partial x_k} \left(\frac{\partial \mathfrak{L}}{\partial \left(\partial \Psi_\sigma / \partial x_k\right)} \right) - \frac{\partial}{\partial t} \left(\frac{\partial \mathfrak{L}}{\partial \left(\partial \Psi_\sigma / \partial t\right)} \right) = 0, \quad \forall x_k \in V, \tag{12}$$

$$\sum_{k=1}^{3} \frac{\partial \mathfrak{L}}{\partial \left(\partial \Psi_\sigma / \partial x_k\right)} N_k = 0, \quad \forall x_\mu \in \partial V / \partial_d V, \tag{13}$$

$$\sum_{k=1}^{3} \left[\left| \frac{\partial \mathfrak{L}}{\partial \left(\partial \Psi_\sigma / \partial x_k\right)} \right|\right] N_k = 0, \quad \forall x_\mu \in \Sigma. \tag{14}$$

Which are the standard Euler-Lagrange equations. We will see in the next chapters of this book how to generalize (14) when Σ can move freely.

5 Principle of Virtual Power and Principle of Least Action

The principle of least Action, when formulated for Action functionals admitting first differentials, can be regarded as a particular form of the principle of virtual powers. Indeed, if

$$\mathfrak{A} = \mathfrak{A}^{int} + \mathfrak{A}^{ext} + \mathfrak{A}^{ine} \tag{15}$$

then

$$\delta\mathfrak{A} = 0 \quad \Longleftrightarrow \quad \delta\mathfrak{A}^{int} + \delta\mathfrak{A}^{ext} + \delta\mathfrak{A}^{ine} = 0. \tag{16}$$

Identifying

$$\delta\mathfrak{A}^{int} = \mathfrak{P}^{int} \quad \delta\mathfrak{A}^{ext} = \mathfrak{P}^{ext}, \quad \delta\mathfrak{A}^{ine} = \mathfrak{P}^{ine}, \tag{17}$$

we get

$$\mathfrak{P}^{int} + \mathfrak{P}^{ext} + \mathfrak{P}^{ine} = 0. \tag{18}$$

Which is the standard form of principle of virtual powers.

Is the principle of virtual power more general than principle of least Action? First answer: the principle of virtual powers involves differentials which are not exact, in general. Therefore, once fixed the kinematics, the principle of virtual power is actually more general. In both $\mathfrak{P}^{int}$ and $\mathfrak{P}^{ext}$ one can include dissipative terms, which cannot, in general, be derived from an Action functional. However, it is not clear if, suitably extending the space of configurations and the set of admissible motions, one can introduce an Action functional also for systems which, in a restricted kinematics, appear as dissipative. Controversies in the literature about this subject are not yet solved.

6 Hamilton-Rayleigh Approach

We propose to use the Hamilton-Rayleigh compromise. We introduce an Action functional and a Dissipation Rayleigh functional and, by means of them, we formulate the Principle of Virtual Work. Rayleigh dissipation functional $\mathfrak{R}$ is defined as a linear functional on the set of velocities, not on the set of motions as $\mathfrak{A}$. Therefore, $\delta\mathfrak{R}$ is defined as a linear functional of the variation $\delta\dot{m}$. The principle of virtual works formulated following Hamilton-Rayleigh takes the form: (the lack of the upper dot on RHS is not a mistake!)

$$\delta\mathfrak{A}\left(\delta m\right) = \mathfrak{R}\left(\delta m\right). \tag{19}$$

7 Conclusions

We recall an ancient and useful recipe for building theories for describing effectively physical phenomena:

"In Nomina est Natura Rerum". Anonymous

This statement (passed to us by the middle age tradition) is formulated for defending mathematical formalism. This sentence claims that it is impossible to talk about any mathematical model without using the appropriate language. So, for instance, it is impossible to say clearly what is the first variation of Action using simply "words" from natural language, i.e. without writing integrals on T and, to proceed, we need to give "precisely" names to things. Therefore, to specify precisely how our models are constructed we need to introduce symbols and formulas.

However, we can also say that

"Nomina sunt Consequentia Rerum." Iustinianus, Institutiones Liber II,7,3

This because we are not blindly building our mathematical model. We get informations about physics and from these informations we actually formulate our models.

We can finally state that the "old" method of basing the formulation of mathematical models on the variational approach works: indeed it works very well.

Bibliography

[1] Arnold, V.I., Mathematical Methods of Classical Mechanics, Springer, 2nd edition, May 16, 1989.

[2] Benvenuto, E., La scienza delle costruzioni e il suo sviluppo storico, Sansoni, Firenze, 1981.

[3] Berdichevsky V., Variational Principles of Continuum Mechanics, Springer, 2009.

[4] Bourdin, B., Francfort, G.A., Marigo, J.-J., The variational approach to fracture, J. Elasticity, 91, 1-3, 1-148, 2008. (The paper also appeared as a Springer book: ISBN: 978-1-4020-6394-7).

[5] Colonnetti, G., Scienza delle costruzioni, Torino, Edizioni scientifiche Einaudi, 3rd ed., 1953-57.

[6] Cosserat, E., Cosserat, F., Sur la Théorie des Corps Déformables, Herman, Paris, 1909.

[7] dell'Isola, F., Seppecher, P., The relationship between edge contact forces, double force and interstitial working allowed by the principle of virtual power, Comptes Rendus de l'Academie de Sciences - Serie IIb: Mecanique, Physique, Chimie, Astronomie, vol. 321, pp. 303-308, 1995.

[8] dell'Isola, F., Seppecher, P., Edge Contact Forces and Quasi-Balanced Power, Meccanica, vol. 32, pp. 33-52, 1997.

[9] dell'Isola, F., Madeo, A., Seppecher, P., Boundary Conditions in Porous Media: A Variational Approach, Int. Journal of Solids and Structures, Vol. 46, 3150-3164, 2009.

[10] dell'Isola, F., Madeo, A., Seppecher, P., "Beyond Euler-Cauchy continua". In this book.

[11] Feynman, R.P., Leighton, R., Sands, M., The Feynman Lectures on Physics: The Definitive and Extended Edition, (Feynman, Leighton, Sands), 3 vv., Addison Wesley, Reading (MA) 2nd ed., 2005.

[12] Francfort, G.A., The chapter in this book by Prof. Gilles Frankfort.

[13] Germain, P., La méthode des puissances virtuelles en mécanique des milieux continus. Premi'ere partie. Théorie du second gradient, J. Mécanique, 12, 235-274, 1973.

[14] Homer. The Iliad. Translated by Murray, A. T., Loeb Classical Library Volumes1. Cambridge, MA, Harvard University Press; London,William Heinemann Ltd., 1924.

[15] Lagrange, J.L., Mécanique Analytique, Editions Jaques Gabay, Sceaux, 1788.

[16] Lanczos, C., The Variational Principles of Mechanics. Toronto: University of Toronto, 1970.

[17] Landau, L.D., Lifshitz, E.M., Quantum Mechanics: Non-Relativistic Theory. Vol. 3 (3rd ed.), Pergamon Press, 1977.

[18] Moiseiwitsch, B.L., Variational Principles, Dover, 1966.

[19] Netz, R., Noel., W., The Archimedes Codex: How a Medieval Prayer Book Is Revealing the True Genius of Antiquity's Greatest Scientist, Da Capo press, 2009. See also the website http://www.archimedespalimpsest.org/

[20] Rorres, C., Completing Book II of Archimedes's On Floating Bodies, The mathematical intelligencer, 26,3. Pages 32-42, 2004.

[21] Russo, L., The Forgotten Revolution, Springer Verlag, 2003.

[22] Vailati, G., Il principio dei lavori virtuali da Aristotele a Erone d'Alessandria, Scritti (Bologna, Forni, 1987), vol. II, pp. 113-128. Atti della R. Accademia delle Scienze di Torino, vol. XXXII, adunanza del 13 giugno 1897, quaderno IG (091) 75 I - III. 1987.

Beyond Euler-Cauchy Continua: The structure of contact actions in N-th gradient generalized continua: a generalization of the Cauchy tetrahedron argument.

Francesco dell'Isola * + and Pierre Seppecher †+ and Angela Madeo ‡+

* Università di Roma "La Sapienza" Dipartimento di Ingegneria Strutturale e Geotecnica, Laboratorio Strutture e Materiali Intelligenti. email: francesco.dellisola@uniroma1.it

† Institut de Mathématiques de Toulon U.F.R. des Sciences et Techniques,Université de Toulon et du Var

‡ Université de Lyon, Institut National des Sciences Appliquées (INSA) GCU et LGCIE

+ International Research Centre on "Mathematics & Mechanics of Complex Systems" M& MOCS, Cisterna di Latina.

This work is dedicated to Professor Antonio Romano in occasion of his 70-th birthday.

Abstract The most general and elegant axiomatic framework on which continuum mechanics can be based starts from the Principle of Virtual Works (or Virtual Power). This Principle, which was most likely used already at the very beginning of the development of mechanics (see e.g. Benvenuto (1981), Vailati (1897), Colonnetti (1953), Russo (2003)), became after D'Alembert the main tool for an efficient formulation of physical theories. Also in continuum mechanics it has been adopted soon (see e.g. Benvenuto (1981), Salençon (1988), Germain (1973), Berdichevsky (2009), Maugin (1980), Forest (2006)). Indeed the Principle of Virtual Works becomes applicable in continuum mechanics once one recognizes that to estimate the work expended on regular virtual displacement fields of a continuous body one needs a distribution (in the sense of Schwartz). Indeed in the present paper we prove, also by using concepts from differential geometry of embedded Riemanniam manifolds, that the Representation Theorem for Distributions allows for an effective characterization of the contact actions which may arise in $N-th$ order strain-gradient multipolar continua (as defined by Green and Rivlin (1964)), by univocally distinguishing them in actions (forces and $n-th$ order forces) concentrated on contact surfaces, lines (edges) and points (wedges). The used approach reconsiders the results found in the pioneering papers by Green and Rivlin (1964)-(1965) , Toupin (1962), Mindlin (1964)-(1965) and Casal (1961) as systematized, for second gradient models, by Paul Germain (1973). Finally, by recalling the results found in dell'Isola and Seppecher (1995)-(1997), we indicate how Euler-Cauchy approach to contact actions and the celebrated tetrahedron argument may be adapted to $N-th$ order strain-gradient multipolar continua.

1 Introduction

In a forthcoming review paper the authors will try to describe why, how and when many theories were conceived to go beyond the conceptual framework established for continuum mechanics in the Euler and Cauchy era. In this introduction are formulated only few comments about some papers which seem to be the starting point of the most modern studies in continuum mechanics. In this field -among the many available in the literature- the textbooks which we have found more instructive are those of Paul Germain and Jean Salençon. In them, without any loss of mathematical rigour, what nowadays seems the most effective approach to the axiomatization of mechanics is presented to the students of the École polytechnique. This approach is based on the Principle of Virtual Works. Paul Germain research paper on second gradient continua (1972) shows how fruitful is the aforementioned approach. Most likely one of the most illuminating paper in modern continuum mechanics is due to Green and Rivlin (1964): it is entitled "Simple force and stress multipoles" and formulates what the authors call "multipolar continuum mechanics". Indeed Green and Rivlin start there the foundation of what has been later called also the theory of generalized continua. They also address the problem of establishing simultaneously the bulk evolution equations and the correct boundary conditions for generalized continua: equations and boundary conditions which they find by postulating the Principle of Virtual Work. In this aspect their theory is perfectly orthodox with respect to the paradigm put forward, many years earlier, by Cosserat brothers, in their fundamental textbooks (1908)-(1909). Of great importance for understanding the relationship between Cosserat continua and higher order gradient continua studied by Green and Rivlin is the short but very clear paper by Bleustein (1967), where the boundary conditions found by Toupin in a previous work are interpreted also from a physical point of view. We must also cite here the papers by Mindlin, who also contributed greatly to the development of important generalizations of Euler-Cauchy continuum models. In particular in Mindlin (1965) is started the study of third gradient continua, which is developed in a great extent.

However in all cited papers the Cauchy cuts considered are very regular: therefore the cited authors refrain from the consideration of contact actions concentrated on edges and wedges. Instead Germain considers Cauchy cuts in which the normal can suffer discontinuities of the first kind: he therefore needs to consider contact actions concentrated on edges. However Germain limits his treatment to second gradient continua: in his theory there are not wedge contact actions. Also in Toupin's strain-gradient theory (1962) the consideration is limited to second gradient continua. In cited paper Toupin limits himself to the consideration of a particular class of second gradient continua: those in which only a particular class of contact double-forces (using the nomenclature by Germain) can be exerted: the class constituted by "couple-stresses".

Those which are called by many authors (see e.g. Maugin (2010) and Forest (2005)-(2006)) generalized continua actually strictly include higher gradient continua which we consider here. In generalized continua together with the placement field one can introduce many other kinematical descriptors, which are other fields defined in the material configuration of considered continuum. The first example of such a set further kinematical descriptors is given by Cosserats who add to the placement field also a field of "changes of attitude", i.e. a field of rotations, which describe a large class of "microscopically structured" continua. As clarified by Bleustein (1967) Toupin's continua are a class of Cosserat continua in which an internal constraint has been introduced. In the following sections, while commenting some papers recently published in the field, it is discussed how the approach used by Germain can be reconciled with an approach which parallels more strictly the one used, for first gradient continua, by Cauchy.

It is not easy (but this analysis will be attempted in the aforementioned review paper) to explain why the foundation of continuum mechanics "à la Cauchy" has been considered "more physically grounded" than the axiomatization based on the Principle of Virtual Powers. In the present work we prove that .at least for higher gradient continuum theories, the two approaches are completely equivalent.

Variational Principles and Calculus of Variations have always been an important tool in formulating mathematical models for physical phenomena. Among many others the textbook by Berdichevsky (2009) clearly shows that this statement holds, in particular, for Continuum Mechanics.

We are sure that the Principle of Virtual Works and the Principle of Least Action have roots much deeper than many scientists believe. (see e.g. Vailati (1897)). One can conjecture that the majority of physical theories were first formulated in terms of these Principles, and only subsequently reconsidered from other points of view. The Principle of Least Action, which supplies a "geometric" (see Russo (2003)) version of mechanics, is likely to be indeed the tool used by the true founders of mechanics (i.e. the scientists of the Hellenistic period) to establish it. As conjectured also by Colonnetti (1953) and Rorres (2004) Archimedes himself was basing his mechanical investigations on the Principle of Virtual Works. More recently, as already stated above in a more detailed way, Green, Rivlin, Toupin, Mindlin, Casal, and Germain formalized various versions of the theory of generalized continua basing them on the Principle of Virtual Powers: however the most illuminating treatises in this subject remain those due to Cosserat brothers (1908)-(1909).

For a long time some opponents to second gradient theories argued about its "lack of consistency", due to the difficulties in "interpreting" boundary conditions. However it has to be remarked that if one refuses to use the Principle of Virtual Powers he can find very difficult the job of finding some set of boundary conditions which are compatible with the (independently postulated!) bulk evolution equations. Actually it happens that many epigones, after having initially refused to use this principle also in continuum mechanics, have later rephrased with different notations many of the results already available in the literature.

If instead one accepts the *D'Alembertian* approach to mechanics all these problems of well-posedness of mathematical models completely disappear.

2 Second and Higher Gradient Continuum Theories

In the last fifty years it has been widely recognized that in order to describe a wealth of physical phenomena it is needed to introduce mechanical theories which take into account contact actions more complex than those considered in the format given by Cauchy to continuum mechanics. Some well-known contributions in this regard are given in the papers listed in the references by Toupin, Mindlin, Green, Rivlin, Maugin, Forest, Germain, Suiker, Sokolowski, Triantafyllidis among many others.

More recently it has been recognized that second or even higher gradient models are needed when continuum models are introduced for describing systems in which strong inhomogeneities of physical properties are present at eventually different length scales (see e.g. Abu et al. (2008), Alibert et al. (2003), Polizzotto (2007), Pideri and Seppecher (1997), Trianttafyllidis et al. (1986)-(1998), Yang and Misra (2010), Yang et al. (2011)), and may be of great importance also in continuum systems in which some "microscopical" degrees of freedom can "capture" a relevant amont of deformation energy (see e.g. Carcaterra (2005) or Carcaterra et al. (2006)).

Actually, immediately after the development of the Cauchy format of continuum mechanics, a first relevant generalization in the aforementioned direction was conceived by Eugène and François Cosserat, but their efforts were not continued until late in XX century. Cosserat described continuum bodies in which contact actions were to be modelled not only by means of surface forces, but also by means of surface couples. The conceptual differences between Cauchy-type continuum mechanics and Cosserat-type continuum mechanics were relevant, and the second one could not be obtained by means of simple modifications of the first one. The remarkable mathematical difficulties confronted by Cosserat rendered their work difficult to be accepted, and for a long period their results were nearly completely ignored. This circumstance can be easily understood: the structure of Cosserat contact actions is complex. Indeed in Cosserat continua one needs, together with Cauchy stress tensor also a Couple stress tensor, for representing contact Couples.

2.1 A first method for extending Cauchy model for continuous bodies

In order to develop continuum mechanics by going beyond the Cauchy format it is possible to use at least two different approaches.

The most simple of them, used also by Cosserats, starts by postulating how the power expended by internal actions in a body depends on the "virtual" velocity field and its gradients. Starting from this postulate one can deduce, by means of a successive application of the theorem of divergence, i.e. by means of several iterative integrations by parts, which are the contact actions which can be exerted at the boundary of the considered body. Hence, this method starts from the notion of stress tensors and deduces from it the concept of contact actions. It is based on the D'Alembert Principle of Virtual Work and has been resumed by Green and Rivlin, Mindlin, Casal and subsequently by Paul Germain, in his enlightening papers (1972-1973). This Principle is undoubtedly a great tool in Mechanics which has not been improved since its original first and "standard" formulation, differently to what stated in Fried and Gurtin (2006)-(2008) and in Podio-Guidugli (2009). It is not clear why these last authors consider as "non-standard" a formulation of the Principle of Virtual Powers which can be found stated "word-for-word" for instance in the textbooks of Jean Salençon..

Indeed other authors (e.g. the paper by Degiovanni, Marzocchi, Musesti (1999)-(2010) in the references) stated that:

In particular, the approach by means of the theory of distributions, mentioned by Germain himself but not fully developed, is here adopted from the beginning. Clearly, in order to obtain deeper results such as the Cauchy

Stress Theorem, some extra regularity has to be assumed. Note that a power depends in general from two variables, the velocity field and the subbody. So it is a bit more complex than a mere distribution.

In the same spirit in dell'Isola and Seppecher (1995)-(1997) the starting assumptions concerning contact actions are: i) for every subbody of considered body the power expended by contact actions on a generic velocity field is a distribution (i.e. a linear and continuous functional on velocity fields) ii) the power expended by contact actions is quasi-balanced (generalizing the assumption used in Noll and Virga (1990)). Then in aforementioned papers by using different polynomial test velocity fields and different families of subbodies, the Cauchy construction for stress tensors is obtained.

The works of Green and Rivlin, Mindlin and Germain have been taken up again and again, (e.g. in Fried and Gurtin (2006)-(2008)) often rephrasing them without introducing any notable amelioration and often second gradient continua are somehow confused with Cosserat continua.

Paul Germain, following a tradition set in France by André Lichnerowicz, uses the original version (and more efficient) absolute notation due to Levi-Civita. This version, at least in this context, is the most adapted, as many objects of different tensorial order are to be simultaneously handled. Sometimes those who are refraining from using the most sophisticated version of Levi-Civita absolute Calculus are lead to refer to the needed stress tensors and the related contact actions indistinctly using the names "hyperstresses" and "hypertractions". On the contrary Germain (following the spirit of Green and Rivlin) tries to convey through the nomenclature chosen the physical meaning to be attached to the new mathematical objects which he is introducing: for instance he calls "double forces" the actions which are expending powers on the velocity gradient in the directions which are normal to the surfaces of Cauchy cuts. Germain then decomposes these "double forces" into "couples" and "symmetric double forces" recognizing (following Bleustein) that couples were already introduced by Cosserats. Germain's notation supports the mechanical and physical intuition contrarily to what does a generic nomenclature based on some "hyper" prefixes.

2.2 A second method for extending Cauchy model and its relationship with the first

The second method starts by postulating the type of contact action which can be exerted on the boundary of every "regular" part of a body and then proceeds by proving a "representation" theorem for the considered class of contact actions: the existence of stress tensors is then deduced from the postulated form of contact actions with the addition of a "balance-type" postulate, based on physical grounds. In other words: to the "constitutive" assumption adopted for characterizing the class of contact action under consideration one must add a Principle of Balance: the contact actions have to be balanced by a bulk action. This is the method followed by Cauchy which is often considered as the only firm foundation of Continuum Mechanics. The important contribution due to Noll and Virga (1990) is to have introduced the assumption of "quasi-balance" for powers, which generalized, in the most suitable way, the Euler-Cauchy Postulate used in Cauchy continuum mechanics.

The mathematical difficulties presented by this second method

As remarked explicitly in dell'Isola and Seppecher (1995)-(1997) and in Degiovanni, Mazzocchi and Musesti (1999)-(2010) the mathematical difficulty to be confronted in order to establish a firm foundation for this second method relies on the dual dependence of power functional on velocity fields and on subbodies of the considered continuum. It is obvious, starting from physical plausibility considerations, that power functionals must be regarded as distributions on the set of test functions represented by the admissible velocity fields (see e.g. the textbooks of Salençon and Germain).

A fundamental results due to Schwartz allows for representing distributions (with compact support) as finite sums of derivatives of measures Schwartz (1973). When (as it is important for considering contact actions) the distribution is concentrated on a smooth submanifold of three-dimensional Euclidean space, then the derivatives to be considered are only those "normal" or "transversal" to the submanifold itself. Unfortunately in Schwartz it is not considered a representation theorem for families of distributions "attached to" the family of measurable subset of a given measurable set.

Some of the efforts of Degiovanni, Marzocchi and Musesti (1999)-(2010), Lucchesi, Šilhavý and Zani (2008) are directed, with remarkable results, to the search of such a generalized Schwartz representation theorem and to the formulation of weaker versions of Gauss divergence Theorem.

Indeed it is also of relevance the problem arising when one must define generalized "stresses" having a flux which allows for the representation of contact action and a divergence to be used for formulating bulk "local" form of balance laws. This problem has been also addressed with some interesting results (see Lucchesi, Silhavý and Zani (2008), Degiovanni, Marzocchi e Musesti (1999), Šilhavý (1985)-(1991).

2.3 The two methods can be reconciled.

During a long period the first method of the two previously described has been rejected by many researchers and it is lucky for advancement of science that its power has been, in the last decade, finally nearly unanimously accepted.

Moreover the two methods can be reconciled.

Indeed the equivalence of the two methods has been explicitly established by Cauchy him-self and made precise by Noll, for First Gradient Theories.

The same equivalence has been proven for the so called Second Gradient Theories, i.e. for theories in which the internal power is a second order distribution: this results has been obtained in the sequence of papers Noll and Virga (1990), dell'Isola and Seppecher (1995)-(1997). In these last two papers the relationship between the concept of contact line force and surface double forces was proven and also a representation formula relating the two kind of forces was obtained.

3 Some commentaries about a recent paper on second gradient continua

Unfortunately it seems that the fundamental connection between the two methods (and the available proof of the existence of this connection at least for those materials which were called by Germain *second gradient materials*) seems still not well understood in part of the mechanics community, while it has been considered as established by others (see e.g. in the references the works by Maugin, Markus and Forest).

- For instance, one can read in Podio-Guidugli Vianello (2010) that:

Although here we do not deal with this difficult issue directly, in Sect. 3, the bulk of this article,we do provide a full set of representation formulae not only, as is relatively easy, for tractions and hypertractions in terms of stresses and hyperstresses (see definition (26) for diffused tractions and hypertractions, and definition (27) for tractions concentrated on edges), but also, conversely, for stresses and hyperstresses in terms of diffused and concentrated tractions and hypertractions (see definitions (28, 29), and (34). Such representation formulae generalize the corresponding formulae for simple (!first-gradient) materials, that we derive in our preparatory Sect. 2. Since we work in a nonvariational setting, our results apply whatever the material response. The PVP we use includes edge tractions, both internal and external; without them, it would not be possible to arrive at the complete representation formula for the hyperstress in terms of hypertractions we construct in Sect. 3.5.

However it is not possible to see any difference between the results listed in the previous statement and those obtained by Casal and then by Germain. The quoted results are also available in Seppecher (1987)-(1989), Casal and Gouin (1985) and even recalled in the textbooks by Forest. An interesting application of second gradient theories to the mechanics of porous media are proposed e.g. in Collin et al. (2006) or in Sciarra, dell'Isola and Coussy (2007), where many results listed in the commented paper are explicitly exploited.

- In the commented one then reads:

An interesting feature of second-gradient materials is that, if bodies and subbodies having non everywhere smooth boundary are considered, then edge forces, i.e., line distributions of hypertractions are to be expected (and, if a dependence on gradients higher than two is allowed, one has to deal also with vertex forces, as exemplified by Podio-Guidugli [15]). To our knowledge, a rigorous interaction theory accommodating such a nonstandard behavior remains to be constructed; interesting attempts in this direction have been carried out by Forte and Vianello [3], Noll and Virga [14], and Dell'Isola and Seppecher [1].

This statement is not accurate: indeed the efforts leading to in the papers Noll and Virga (1990), dell'Isola and Seppecher (1995)-(1997), as recognized for instance in the works by Maugin and Forest cited in the references, actually constructs the searched rigorous interaction theory.

- Again in the commented paper the following statement can be found:

Finally, in Sect. 4, we provide a new proof of the following not very well-known fact in the theory of second-gradient materials: if edge tractions are constitutively presumed null on whatever edge, then the hyperstress takes a very special form whose information content is carried by a vector field. We surmise that inability to develop edge interactions be characteristic of certain second-gradient fluids, an issue that we take up in a forthcoming article [17], continuing a line of thought proposed by Podio-Guidugli [16].

Indeed, this results, rather obvious, is obtained in dell'Isola and Seppecher (1997), Remark 3, page. 48 and systematically exploited in the application of second gradient theory presented in Sciarra et al. (2001)-(2008) or in Madeo et al. (2008). Some interesting consideration about this point are already available in Seppecher (1987) together with some consideration about third gradient fluids. This result is well-known: for instance remark that Equation (35) on page. 173 in the commented paper exactly is equal to Equation (18) page. 6612 in Sciarra et al. (2007) or to Equation (13) pag.107 in Sciarra et al. (2008).

- Finally always in the commented paper one reads

Relations (7) and (8) are also arrived at when, as is customary, only tractions on body parts are introduced, because stress is constructed à la Cauchy as a consequence of balance of tetrahedron-shaped parts. The Cauchy construction is the pillar on top of which the standard theory of diffuse (i.e., absolutely continuous with respect to the area measure) contact interactions stands. For complex (i.e., nonsimple) material bodies, a Cauchy-like construction has been attempted often, but not achieved so far, to our knowledge.

On the contrary in the Conclusions of the paper dell'Isola Seppecher (1997) one can read:

The most important concepts introduced in this paper are:

(i) the concept of quasi-balanced power of contact force distribution and

(ii) that of prescribed shapes.

They allowed us to develop a system of axioms "à la Cauchy' for continua in which edge contact forces are present.

It is not clear if the authors of the commented paper are completely aware of this last statement and of the assumptions and theorems presented in the just cited paper. In our opinion, the demanded Cauchy-like construction for second gradient materials is supplied there.

3.1 Concluding remarks

The connection between internal power and the power expended by external actions has not been yet completely established for a generic $N-$th Gradient Theory, although interesting and useful considerations can be found in the papers by Green, Rivlin, Mindlin, Di Carlo and Tatone and Podio-Guidugli cited in the references.

In the following sections it will be shown how the work started in dell'Isola Seppecher (1997) can be continued. The aim in these lecture notes will be to give a firm framework to those researchers which need to deal with more complex contact actions (for instance "wedge forces"), wish to refrain from using the Principle of Virtual Power and instead prefer to adopt an approach based on "contact interactions" rather than on "virtual power expended on virtual velocity fields".

Indeed the ideas presented in the just mentioned paper can be extended rather easily to treat the case of all types of contact distributions: more precisely the Cauchy tetrahedron argument can be generalized to prove that all types of mechanical contact actions can be represented in terms of a generalized stress tensors.

To our knowledge the results which we present are novel. It is however difficult to establish how many of them were already obtained in the literature: in general while trying to write a scientific work and in particular when revisiting Cauchy's, Green's, Rivlin's and Germain's results it is advisable to try to avoid the attitude of the novelist Pierre Menard (see Borges):

> (...) *Menard's visible work can be easily enumerated. Having examined with care his personal files, I find that they contain the following items:* (a list of works follows) *I turn now to his other work: the subterranean, the interminably heroic, the peerless. And—such are the capacities of man!—the unfinished. This work, perhaps the most significant of our time, consists of the ninth and thirty-eighth chapters of the first part of Don Quixote and a fragment of chapter twenty-two. I know such an affirmation seems an absurdity; to justify this "absurdity" is the primordial object of this note.* (....) *He did not want to compose another Quixote —which is easy— but the Quixote itself. Needless to say, he never contemplated a mechanical transcription of the original; he did not propose to copy it. His admirable intention was to produce a few pages which would coincide—word for word and line for line—with those of Miguel de Cervantes. "My intent is no more than astonishing,"* (...) *To be, in some way, Cervantes and reach the Quixote seemed less arduous to him—and, consequently, less interesting—than to go on being Pierre Menard and reach the Quixote through the experiences of Pierre Menard.* (.....) *"My undertaking is not difficult, essentially," I read in another part of his letter. "I should only have to be immortal to carry it out."*

4 Resumé of some results in Differential Geometry of Riemannian manifolds embedded in the Euclidean Space.

The reference configuration of the continuum which we want to consider is a regular region C^* embedded in E^3. Its boundary ∂C^* is assumed to be a piecewise regular orientable and rectifiable surface.

Following the ideas already expressed in Cosserat (1908) and formalized e.g. by Germain (1972) or Salençon (1988-2005) we will assume that the Principle of Virtual Powers holds at least for every "regular" subbody of the considered body. In this section we specify what we mean with "regular" subbody and in particular we specify how regular must be its topological boundary.

5 Gaussian Geometry of Piecewise Regular Surfaces in E^3

The family of subbodies which we will consider in the Principle of Virtual Powers will have a topological boundary which is a piecewise regular surface, as defined in the following section. These surfaces represent a particular class of "shapes" as introduced in dell'Isola and Seppecher (1995)-(1997).

5.1 Piecewise regular surfaces embedded in E^3.

In the present paper we will use the following nomenclature:

Definition 5.1. A surface S is a piecewise regular (orientable and rectifiable) surface embedded in $E^3$when there exist a finite set

$$\{\gamma_i \subset S, i = 1, ...N\}$$

of C^1 curves (called edges) and a finite set of points (called wedges)

$$\{W_i \in S, i = 1, ..., M\}$$

such that, once introduced the notation (which is reminiscent of the one usually encountered in the formulation of Poincaré theorem for exterior forms)

$$Support(\partial S) := \left(\bigcup_{i=1}^{N} \gamma_i \right); \tag{1}$$

$$Support(\partial\partial S) := \left(\bigcup_{i=1}^{M} W_i \right), \tag{2}$$

the conditions appearing in the following list are verified.

LIST OF CONDITIONS

- for every $p \in S - (\partial S \cup \partial\partial S)$ (i.e. for every regular point belonging to S) there exists a neighborhood in S which is locally (C^2-)diffeomorphic to $\mathbb{R}^2$: we call any such local diffeomorphism

  $$r : \mathbb{R}^2 \to S - (\partial S \cup \partial\partial S)$$

  an internal chart of S, We assume that for every internal chart r the set $r(\mathbb{R}^2)$ is a rectifiable surface;

- for every $p \in S - (\partial S \cup \partial\partial S)$ there exists a translation vector in E^3 , denoted with the symbol $N(p)$, which is orthogonal to every tangent vector to S,
- for every $p \in \partial S - \partial\partial S$ there exist two diffeomorphisms (called also "border charts")

$$r^{\pm} : [0, \infty[\times \mathbb{R} \to I^{\pm} \subset S \tag{3}$$

such that

$$r^{\pm}(0,0) = p; \qquad r^{\pm}(0,\mathbb{R}) = \partial S \cap I^{\pm} \tag{4}$$

$$; (\forall y \in]0, \infty[\times \mathbb{R}) \left(r^{\pm}(y) \in S - (\partial S \cup \partial\partial S) \right) \tag{5}$$

and both the following limits exist

$$\lim_{x \to (0,0)} N(r^{\pm}(x)) := N^{\pm}(p).$$

Therefore every curve γ_i can be regarded as the border of two regular surfaces $S^{\pm}$ one on the side + the other on the side − with respect to γ_i . We will denote the unit outward pointing normal vector to γ_i in the tangent plane to $S^{\pm}$ respectively with the symbol $\nu^{\pm}$.
- for every curve γ_i (the length of which is denoted by l_i) there exists a global parametric C^1 representation r_i

$$r_i : s \in [0, l_i] \mapsto p \in S$$

such that

$$\left\| \frac{dr_i}{ds} \right\| = 1, \ \frac{dr_i}{ds} \cdot N^{\pm} = 0.$$

We will assume that

$$\frac{dr_i}{ds} \times N^{\pm} = \pm \nu^{\pm}.$$

- for every $W_j \in \partial\partial S$ there exists at least one curve γ_i such that one of the two following conditions holds

$$r_i(0) = W_j \ \text{ or } \ r_i(l_i) = W_j.$$

Roughly speaking piecewise regular orientable and rectifiable surfaces, in the particular conventional sense specified above, are surfaces where the normal vector is defined in all points except those belonging to a finite set of wedges and to a finite set of regular curves (the edges of the surface). Along these curves a tangent vector is always defined together with both the normals of the two subsurfaces concurring on the edge.

Definition 5.2. We call **face** of S every connected component of the set $S - (\partial S \cup \partial\partial S)$

5.2 Local parametrizations for regular curves and surfaces. Local curvilinear coordinate systems in E^3 adapted to surfaces and surface edges.

We start recalling some basic definitions from differential geometry

Definition 5.3. A parametrization of a regular curve is a C^1 one-to-one function $r :]a_1, b_1[\to E^3$ such that

$$(\forall p \in r\left(]a_1, b_1[\right)) \left(t(p) := \frac{dr}{dx^1}\left(r^{-1}\left(p\right)\right) \neq 0 \right).$$

The vector $t(p)$ is the tangent vector to the curve in p.

Definition 5.4. At every point p of a regular curve γ we can define the following projection operators

$$P_{\gamma,p} := t(p) \otimes t(p); \quad Q_{\gamma,p} := N_1(p) \otimes N_1(p) + N_2(p) \otimes N_2(p)$$

where $N_1(p)$ and $N_2(p)$ form an independent set of vectors both orthogonal to $t(p)$.When this will not cause confusion we will skip one or both the indices of the introduced projectors. $P_{\gamma,p}$ will be called the projector in the tangent bundle (or line) of γ at point p, while $Q_{\gamma,p}$will be called the projector in the orthogonal bundle (or plane) of γ at point p.

Definition 5.5. A map

$$r : \left(x^1, x^2\right) \in]a_1, b_1[\times]a_2, b_2[\subset \mathbb{R}^2 \mapsto r\left(x^1, x^2\right) \in E^3$$

is called a local parametrization for the regular surface S in the neighborhood of p if r is a C^1 diffeomorphism between $]a_1, b_1[\times]a_2, b_2[$ and $r\left(]a_1, b_1[\times]a_2, b_2[\right) =: I_p$ such that

$$r\left(]a_1, b_1[\times]a_2, b_2[\right) =: I_p \subset S; \tag{6}$$

$$r^{-1}(p) \in]a_1, b_1[\times]a_2, b_2[\tag{7}$$

Once a local parametrization for the regular surface S is introduced then a set of coordinate curves on S is established, together with a field of bases for the tangent planes to S.

We will consider in the following the fields of vectors induced by a local parametrization

$$(\forall \alpha \in \{1, 2\}) \left(\forall q \in I_p\right) \left(a_\alpha(q) := \frac{\partial r}{\partial x_\alpha}\left(r^{-1}\left(q\right)\right) \right).$$

The couple $\{a_1(q), a_2(q)\}$ is a basis of the tangent plane to S in the point q. The set of points

$$r\left(\{x^1\} \times]a_2, b_2[\right), r\left(]a_1, b_1[\times \{x^2\}\right)$$

are called the coordinate x^2 and x^1 curves. Remark that the vectors a_α are tangent to the coordinate x_α curves. The C^1 normal (to S) unit vector field N can be calculated by the formula

$$N(q) = \frac{a_1(q) \times a_2(q)}{\|a_1(q) \times a_2(q)\|}$$

Definition 5.6. Let the couple $\{a_1(p), a_2(p)\}$ denote an othonormal basis of the tangent plane to S in the point p. At every point p of a regular surface S we can define the projector operators

$$P_{S,p} := a_1(p) \otimes a_1(p) + a_2(p) \otimes a_2(p); \tag{8}$$

$$Q_{S,p} := N(p) \otimes N(p). \tag{9}$$

When this will not cause confusion we will skip one or both the indices of the introduced projectors. $P_{S,p}$ will be called the projector in the tangent bundle (or plane) of S at point p, while $Q_{S,p}$will be called the projector in the ortogonal bundle (or line) of S at point p.

Remark 5.7. Because of the previous definitions, regular curves and surfaces are respectively one dimensional and two dimensional manifolds embedded in the three dimensional Euclidean space.

The scalar fields

$$g_{\alpha\beta} : q \in I_p \mapsto a_\alpha(q) \cdot a_\beta(q)$$

represent the components of a tensor field which is called the **Riemannian metric** induced on S by the inner product in E^3.

Remark 5.8. A given regular curve or surface can be endowed with the structure of Riemannian manifold simply using the inner product of the Euclidean space in which they are embedded. Indeed for any couple v and w of vectors in their tangent bundle one can calculate their inner product simply by regarding them as vectors in E^3.

Although in the Euclidean space E^3 the Cartesian system of coordinates, using at every point the same vector basis to represent displacement vectors, is in general sufficient, in the present context one needs to introduce the following

Definition 5.9. Local **curvilinear coordinate systems in** E^3. A map

$$\varphi : \left(x^1, x^2, x^3\right) \in \left]a_1, b_1\right[\times \left]a_2, b_2\right[\times \left]a_3, b_3\right[\subset \mathbb{R}^3 \mapsto \varphi\left(x^1, x^2, x^3\right) \in E^3$$

is called a local chart in E^3 in the neighborhood I_p of $p \in E^3$ determining a local curvilinear coordinate system when it is a diffeomorphism between

$$\left]a_1, b_1\right[\times \left]a_2, b_2\right[\times \left]a_3, b_3\right[\text{ and } \varphi\left(\left]a_1, b_1\right[\times \left]a_2, b_2\right[\times \left]a_3, b_3\right[\right) =: I_p$$

A $i-th$ ($i \in \{1,2,3\}$) coordinate curve is obtained fixing in the function φ all arguments except the x_i variable, and the tangent vectors of such coordinate curves are denoted as follows

$$(\forall q \in I_p)\left(a_i(q) := \frac{\partial \varphi}{\partial x_i}\left(\varphi^{-1}(q)\right)\right).$$

Obviously for every $q \in I_p$ the set $\{a_i(q), i \in \{1,2,3\}\}$ is a basis of the vector space of translations in E^3.Therefore the curvilinear coordinate system which has been introduced generates a field of bases in all I_p.

Definition 5.10. In I_p we can introduce the following scalar fields

$$(\forall q \in I_p)\left(g_{ij}(q) := a_i(q) \cdot a_j(q)\right)$$

which are the components, in considered curvilinear coordinate system, of the Riemann metric in the Euclidean field E^3.

Let us consider a piecewise regular surface: as we will see in what follows, it is possible to introduce in the neighborhood of p charts (i.e. curvilinear coordinate systems) in the Euclidean space E^3 which are "adapted" to i) the surface in the neighborhood of regular point $p \in S - (\partial S \cup \partial\partial S)$ ii) to the edge of the surface in the neighborhood of a point which is not a wedge.

Local curvilinear coordinate system in E^3 adapted to S in the neighborhood of a regular point.

The following Lemma is a consequence of the inverse function Theorem and is the basis of an important part of Gaussian differential geometry (for a proof see e.g. Kosinski (1986), dell'Isola and Kosinski (1989).

Lemma 5.11. *Let r be a local parametrization of S in the neighborhood of the regular point $p \in S - (\partial S \cup \partial\partial S)$. For a suitably small positive ε the map φ defined by*

$$\begin{aligned} \varphi : \left(x^1, x^2, x^3\right) &\in \left]a_1, b_1\right[\times \left]a_2, b_2\right[\times \left]-\varepsilon, +\varepsilon\right[\subset \mathbb{R}^3 \\ &\mapsto r\left(x^1, x^2\right) + x^3 N\left(r^{-1}\left(x^1, x^2\right)\right) \in E^3 \end{aligned} \tag{10}$$

actually is an invertible function and a diffeomorphism.

Definition 5.12. When the map φ considered in the previous Lemma is a diffeomorphism it is the called the chart adapted to S induced by the parametrization r.

When this will not be cause of confusion we will admit an abuse of notation and we will use the symbols $N\left(x^1,x^2\right), g_{\alpha\beta}\left(x^1,x^2\right)$ and $a_\alpha\left(x^1,x^2\right)$ instead of the symbols $N\left(r^{-1}\left(x^1,x^2\right)\right), g_{\alpha\beta}\left(r^{-1}\left(x^1,x^2\right)\right)$ and $a_\alpha\left(r^{-1}\left(x^1,x^2\right)\right)$. The same abuse of notation will be repeated for all the fields a_i and g_{ij} not specifying the composition with the function φ^{-1}.

The class of charts we have now introduced was first introduced by Gauss (see e.g. Spivak (1979)).

We are now able to extend in the neighborhood in E^3 of a regular point $p \in S$ the fields of projectors in the tangent and in the orthogonal bundles:

Definition 5.13. Let us consider a chart φ adapted to the surface S in the neighborhood of a point p. For every $x_3 \in \left]-\varepsilon,\varepsilon\right[$ we can consider the (regular) surface S_{x_3} which is defined by the following local parametrization

$$r_{x_3} := r + x_3 N.$$

On every surface S_{x_3} it is easy to introduce the tangent and orthogonal projectors, which for $x_3 = 0$ reduce to the projectors introduced already for $S = S_0$.We will denote these fields of projectors, also when defined in the opens set

$$\varphi\left(\left]a_1,b_1\right[\times\left]a_2,b_2\right[\times\left]-\varepsilon,+\varepsilon\right[\right)\subset E^3$$

with the same symbols $P_{S,p}$ and $Q_{S,p}$.

Local curvilinear coordinate system in E^3 adapted to an edge of S in the neighborhood of a point which is not a wedge.

Let r be a local parametrization of an edge γ of S in the neighborhood of a point $p \in \partial S - \partial\partial S$. When it is a diffeomorphism, we will call the map φ defined by

$$\begin{aligned}\varphi : \left(x^1,x^2,x^3\right) &\in \left]a_1,b_1\right[\times\left]-\varepsilon,+\varepsilon\right[\times\left]-\varepsilon,+\varepsilon\right[\subset\mathbb{R}^3 \\ &\mapsto r\left(x^1\right)+x^2N^{\pm}(r^{-1}\left(x^1\right))+x^3\nu^{\pm}\left(r^{-1}\left(x^1\right)\right)\in E^3\end{aligned} \tag{11}$$

a chart adapted to S at the considered edge as induced by the edge parametrization r on the side $+$ or $-$ (respectively) depending on the consistent choice adopted.

Adapting the argument used in the proof of the Lemma of the previous subsection, and using the assumed regularity hypotheses about S and its edges, it can be proven again that, when the value of ε is chosen to be positive and suitably small, φ actually is a diffeomorphism and can be used as a chart in E^3.

Definition 5.14. Let us consider a regular point of the edge $\gamma$and a chart adapted to S at γ. For every $\left(x^2, x^3\right) \in \left]-\varepsilon, +\varepsilon\right[\times \left]-\varepsilon, +\varepsilon\right[$ we can consider the regular curves parametrized by the function

$$r^{\pm}_{x^2,x^3} := r + x^2 N^{\pm} + x^3 \nu^{\pm}.$$

For each of these curves we can define the projection on the orthogonal and tangent bundles, thus obviously extending in the neighborhood of γ the already introduced projection fields $P_{\gamma,p}$ and $Q_{\gamma,p}$.

6 Gauss Divergence Theorem for embedded Riemannian manifolds

We choose a global othonormal basis $(e_i, i = 1, 2, 3)$ for the vector field of displacements in E^3. All tensor fields, unless differently specified, will be represented by means of the components with respect this basis. In this section we consider an embedded Riemannian manifold M in E^3. This manifold can be therefore a regular curve or surface. Because we were able to construct the so-called Gaussian coordinate systems adapted to considered manifolds, then in a whole neighborhood of these manifolds it is possible to introduce the projection operator fields P and Q. For reducing the complication of the calculation which we will perform in what follows we do not use directly the adapted curvilinear coordinates: instead, after having established the existence of the fields P and Q in the neighborhood of M,we introduce a global Cartesian coordinate system and represent all fields in it. This technical choice is exactly the same one which allowed to Germain the generalization, for second gradient materials, of the results found by Green, Rivlin, Toupin and Mindlin.

It is easy to prove the following:

Lemma 6.1. *If on each manifold M, P denotes the projection on the tangent bundle : we have*

$$\begin{aligned} \delta_i^j = P_i^j + Q_i^j, \quad P_i^j P_j^k = P_i^k, \\ Q_i^j Q_j^k = Q_i^k, \quad P_i^j Q_j^k = 0. \end{aligned} \tag{12}$$

The unit external normal to M on its border is denoted ν; it belongs to the tangent space to M.

Using these notations the divergence theorem reads (see e.g. Spivak (1979))

Theorem 6.2. *For any vector field W defined in the vicinity of M*

$$\int_M (P^i_j W^j)_{,k} P^k_i = \int_{\partial M} W^i P^k_i \nu_k \tag{13}$$

This theorem together with relation

$$Q^i_{j,k} P^k_i = -Q^i_j P^k_{i,k}$$

implies that

Corollary 6.3. *For any vector field W defined in a neighborhood of M*

$$\int_M (W^i)_{,k} P^k_i = \int_M (P^i_j W^j)_{,k} P^k_i + (Q^i_j W^j)_{,k} P^k_i = \tag{14}$$

$$= \int_M W^j Q^i_{j,k} P^k_i + \int_{\partial M} W^i P^k_i \nu_k = \tag{15}$$

$$= -\int_M W^j Q^i_j P^k_{i,k} + \int_{\partial M} W^i P^k_i \nu_k. \tag{16}$$

7 Power expended by internal or external actions

Once we fix a subbody B of a given continuous body C and consider the set $\mathcal{A}$ of all admissible velocity fields for B it is natural to admit that in $\mathcal{A}$ are included all "test functions" (i.e. infinitely differentiable functions) having compact support.

It is also natural (as done e.g. by Salençon (1988-2005) or Germain (1973)) to assume that the power expended by internal or external actions (with respect to B) is a linear and continuous functional when defined in the set of test functions (with respect to Frechèt topology).

In other word we accept the following

Postulate (Power)

The power expended by actions exerted on or in a subbody B is a distribution (in the sense of Schwartz).

It is clear that, once the previous postulate is accepted, the following Theorems and Definitions, due to Schwartz (1973), become really relevant in continuum mechanics:

Theorem 7.1. *Every distribution having compact support K can be, in infinitely equivalent ways, represented as the sum of a finite number of derivatives (in the weak sense) of continuous functions all having their support included in a neighborhood of K.*

Definition 7.2. A distribution is said to have order smaller than or equal to k if one can represent it as the sum of derivatives of continuous functions all having order smaller than or equal to k.

Theorem 7.3. *Every distribution having support included in a regular embedded submanifold M can be uniquely decomposed as a finite sum of transversal derivatives of extensions of distributions defined on M.*

7.1 Representations of distributions which are $N-th$ order derivatives of absolutely continuous measures concentrated on submanifolds of R^H with border

In what follows we will need to consider distributions having support concentrated on a regular submanifold M embedded in R^H which are exactly the $N-th$ order derivatives of measures which are absolutely continuous with respect to the corresponding Hausdorff measure.

More specifically we will be interested to consider distributions having the form

$$\mathcal{P}(V) = \int_M T \cdot \nabla^N V \tag{17}$$

where T is a suitably integrable (with respect to the Hausdorff measure of M) N -times contravariant tensor field.

We have now to get a Lemma which is essential when one wants to study the structure of contact actions in $n-th$ order continua.

Lemma 7.4. *Let us consider a Riemannian "regular" embedded manifold M. Let T be a symmetric tensor field defined in the neighborhood of M. Let us introduce the following "projected" field*

$$\mathbb{P}(T)^{j_1 \dots j_{N-1} l} = \left(\sum_{\alpha=0}^{N-1} C_\alpha^N T^{i_1 i_2 \dots i_N} Q_{i_1}^{j_1} .. Q_{i_\alpha}^{j_\alpha} P_{i_{\alpha+1}}^{j_{\alpha+1}} .. P_{i_{N-1}}^{j_{N-1}} \right) P_{i_N}^l \tag{18}$$

where C_α^N denotes the appropriate binomial coefficient. Then the following formula of integration by parts holds:

$$\int_M T^{i_1 i_2 \dots i_N} V_{,i_1 i_2 \dots i_N} = \tag{19}$$

$$= \int_M T^{i_1 i_2 \dots i_N} Q_{i_1}^{j_1} Q_{i_2}^{j_2} .. Q_{i_N}^{j_N} V_{,j_1 j_2 \dots j_N} + \tag{20}$$

$$- \int_M \left(\left(\mathbb{P}(T)^{j_1 \dots j_{N-1} l} \right)_{,j_N} P_l^{j_N} \right) V_{,j_1 j_2 \dots j_{N-1}} + \tag{21}$$

$$+ \int_{\partial M} \mathbb{P}(T)^{j_1 \dots j_{N-1} l} V_{,j_1 j_2 \dots j_{N-1}} P_l^{j_N} \nu_{j_N} \tag{22}$$

Proof. We start decomposing identity tensor field in terms of tangent and orthogonal projector fields adapted to M

$$\int_M T^{i_1 i_2 \dots i_N} V_{,i_1 i_2 \dots i_N} = \tag{23}$$

$$= \int_M T^{i_1 i_2 \dots i_N} V_{,j_1 j_2 \dots j_N} (P_{i_1}^{j_1} + Q_{i_1}^{j_1})(P_{i_2}^{j_2} + Q_{i_2}^{j_2}) \dots (P_{i_N}^{j_N} + Q_{i_N}^{j_N}). \tag{24}$$

Subsequently we use the symmetry of T and orthogonality of P and Q projectors thus obtaining with simple algebra

$$\int_M T^{i_1 i_2 \dots i_N} V_{,i_1 i_2 \dots i_N} = \tag{25}$$

$$= \int_M \sum_{\alpha=0}^{N} C_\alpha^N \left(T^{i_1 i_2 \dots i_N} Q_{i_1}^{j_1} .. Q_{i_\alpha}^{j_\alpha} V_{,j_1 j_2 \dots j_N} P_{i_{\alpha+1}}^{j_{\alpha+1}} .. P_{i_N}^{j_N} \right) = \tag{26}$$

$$= \int_M T^{i_1 i_2 \dots i_N} Q_{i_1}^{j_1} Q_{i_2}^{j_2} .. Q_{i_N}^{j_N} V_{,j_1 j_2 \dots j_N} + \tag{27}$$

$$+ \int_M \sum_{\alpha=0}^{N-1} C_\alpha^N \left(T^{i_1 i_2 \dots i_N} Q_{i_1}^{j_1} .. Q_{i_\alpha}^{j_\alpha} V_{,j_1 j_2 \dots j_N} P_{i_{\alpha+1}}^{j_{\alpha+1}} .. P_{i_N}^{j_N} \right). = \tag{28}$$

$$= \int_M T^{i_1 i_2 \dots i_N} Q_{i_1}^{j_1} Q_{i_2}^{j_2} .. Q_{i_N}^{j_N} V_{,j_1 j_2 \dots j_N} + \tag{29}$$

$$+ \int_M \left(\sum_{\alpha=0}^{N-1} C_\alpha^N T^{i_1 i_2 \dots i_N} Q_{i_1}^{j_1} .. Q_{i_\alpha}^{j_\alpha} P_{i_{\alpha+1}}^{j_{\alpha+1}} .. P_{i_{N-1}}^{j_{N-1}} \right) V_{,j_1 j_2 \dots j_{N-1} j_N} P_{i_N}^{j_N}. \tag{30}$$

We finally introduce the definition of the projected field $\mathbb{P}(T)$, use Leibnitz differentiation rule and apply Divergence Theorem on M

$$\int_M T^{i_1 i_2 \ldots i_N} V_{,i_1 i_2 \ldots i_N} = \tag{31}$$

$$= \int_M T^{i_1 i_2 \ldots i_N} Q^{j_1}_{i_1} Q^{j_2}_{i_2} .. Q^{j_N}_{i_N} V_{,j_1 j_2 \ldots j_N} + \tag{32}$$

$$+ \int_M \mathbb{P}(T)^{j_1 \ldots j_{N-1} l} V_{,j_1 j_2 \ldots j_{N-1} j_N} P^{j_N}_l = \tag{33}$$

$$= \int_M T^{i_1 i_2 \ldots i_N} Q^{j_1}_{i_1} Q^{j_2}_{i_2} .. Q^{j_N}_{i_N} V_{,j_1 j_2 \ldots j_N} + \tag{34}$$

$$- \int_M \left(\mathbb{P}(T)^{j_1 \ldots j_{N-1} l}\right)_{,j_N} P^{j_N}_l V_{,j_1 j_2 \ldots j_{N-1}} + \tag{35}$$

$$+ \int_M \left(\mathbb{P}(T)^{j_1 \ldots j_{N-1} l} V_{,j_1 j_2 \ldots j_{N-1}}\right)_{,j_N} P^{j_N}_l = \tag{36}$$

$$= \int_M T^{i_1 i_2 \ldots i_N} Q^{j_1}_{i_1} Q^{j_2}_{i_2} .. Q^{j_N}_{i_N} V_{,j_1 j_2 \ldots j_N} + \tag{37}$$

$$- \int_M \left(\left(\mathbb{P}(T)^{j_1 \ldots j_{N-1} l}\right)_{,j_N} P^{j_N}_l \right) V_{,j_1 j_2 \ldots j_{N-1}} + \tag{38}$$

$$+ \int_{\partial M} \mathbb{P}(T)^{j_1 \ldots j_{N-1} l} V_{,j_1 j_2 \ldots j_{N-1}} P^{j_N}_l \nu_{j_N} \tag{39}$$

so concluding the proof. □

Remark 7.5. If T is a $N-th$ order tensor field then also $\mathbb{P}(T)$ is a $N-th$ order tensor field.

Definition 7.6. Let S be a $N-th$ order tensor field having $S^{i_1 i_2 \ldots i_N}$ as component fields . We denote by $div_M S$ the tensor field having the components $\left(\left(S^{j_1 \ldots j_{N-1} l}\right)_{,j_N} P^{j_N}_l \right)$ .Obviously $div_M S$ is a tensor field of order $N-1$.

Definition 7.7. We will denote $S_\perp$ the tensor field (equally of order N) having the following components $S^{i_1 i_2 \ldots i_N} Q^{j_1}_{i_1} Q^{j_2}_{i_2} .. Q^{j_N}_{i_N}$.

Using the notation introduced in the previous remark and denoting with a dot the complete saturation of contravariant with covariant indices, the formula obtained in the previous Lemma reads as follows

$$\int_M T \cdot \nabla^N V = \int_M T_\perp \cdot \left(\nabla^N V\right)_\perp + \tag{40}$$

$$-\int_M (div_M \mathbb{P}(T)) \cdot \nabla^{N-1} V + \tag{41}$$

$$+\int_{\partial M} \mathbb{P}(T) \cdot \left(\nabla^{N-1} V \otimes \nu\right) \tag{42}$$

We can now apply a second time the Lemma thus obtaining

$$\int_M T \cdot \nabla^N V = \int_M T_\perp \cdot \left(\nabla^N V\right)_\perp + \int_{\partial M} \mathbb{P}(T) \cdot \left(\nabla^{N-1} V \otimes \nu\right) \tag{43}$$

$$-\int_M (div_M \mathbb{P}(T)) \cdot \nabla^{N-1} V = \tag{44}$$

$$= \left(\int_M T_\perp \cdot \left(\nabla^N V\right)_\perp\right) + \int_{\partial M} \mathbb{P}(T) \cdot \left(\nabla^{N-1} V \otimes \nu\right) + \tag{45}$$

$$-\left(\int_M (div_M \mathbb{P}(T))_\perp \cdot \left(\nabla^{N-1} V\right)_\perp\right) + \tag{46}$$

$$+\left(\int_M div_M \left(\mathbb{P}\left(div_M \mathbb{P}(T)\right)\right) \cdot \nabla^{N-2} V\right) + \tag{47}$$

$$-\int_{\partial M} \mathbb{P}\left(div_M \mathbb{P}(T)\right) \cdot \left(\nabla^{N-2} V \otimes \nu\right). \tag{48}$$

Applying exactly N times the Lemma we get the following

Corollary 7.8. *Under the same assumptions of the previous lemma and having introduced the conventions*

$$(div_M \mathbb{P})^0 T = T, \qquad (div_M \mathbb{P})^\alpha T = \underbrace{div_M \mathbb{P}(div_M \mathbb{P}......(div_M \mathbb{P}(T))}_{\alpha \ times}$$

the following equality holds if T is a symmetric tensor field $K-$ times contravariant :

$$\int_M T \cdot \nabla^K V = \sum_{\alpha=0}^{K-1} (-1)^\alpha \int_M ((div_M \mathbb{P})^\alpha T)_\perp \cdot \left(\nabla^{K-\alpha} V\right)_\perp + \tag{49}$$

$$+(-1)^K \int_M \left((div_M \mathbb{P})^K T\right) V + \tag{50}$$

$$+\sum_{\alpha=0}^{K-1} (-1)^\alpha \int_{\partial M} \mathbb{P}((div_M \mathbb{P})^\alpha T) \cdot \left(\nabla^{K-1-\alpha} V \otimes \nu\right) \tag{51}$$

An alternative expression for the previous representation formula is (the lower dot indicates the saturation of the last index of the tensor at its left with the first index of the tensor at his right) obtained by renominating the indices:

$$\int_M T\cdot\nabla^K V=\sum_{J=1}^{K}(-1)^{K-J}\int_M\left((div_M\mathbb{P})^{K-J}T\right)_{\perp}\cdot\left(\nabla^J V\right)_{\perp}+ \tag{52}$$

$$+(-1)^K\int_M\left((div_M\mathbb{P})^K T\right)V+ \tag{53}$$

$$+\sum_{L=0}^{K-1}(-1)^{K-1-L}\int_{\partial M}\mathbb{P}((div_M\mathbb{P})^{K-1-L}T).\nu\cdot\nabla^L V \tag{54}$$

The previous corollary allows us to find the unique representation in terms of transversal derivatives of the distribution of the type specified by equation 17. The existence of this representation is stated in Theorem 7.3.

8 Principle of Virtual Powers applied to $N-th$ gradient continua.

On a continuous body B external world can exert actions. While deformation processes occur in the same body its subbodies interact because of internal actions. We call "internal" the power expended on an admissible velocity by internal actions, and "external" the power expended on an admissible velocity by external actions.

The following definitions are clearly inspired by Schwartz representation theorem: however it has to be remarked that such definitions were put forward by Green and Rivlin who, most likely, were unaware of Schwartz results.

Definition 8.1. We call $N-th$ gradient continuum a continuous body B for which the internal power $\mathcal{P}^{int}$ is a distribution of order smaller or equal to N. In formulas (the dot representing the saturation of contravariant with covariant indices)

$$\mathcal{P}^{int}(B,V)=\sum_{\Lambda=0}^{N}\int_B T_\Lambda\cdot\nabla^\Lambda V. \tag{55}$$

Following Green and Rivlin we will call the $\Lambda-$ times contravariant tensors T_Λ "Λ-th order multipolar stress".

Remark 8.2. The previous definition is clearly a "constitutive" assumption, specifying the "type" of internal actions which are considered "possible" inside the body. The Theorem 7.1, once one accepts the Postulate 7, proves that for a given body the internal power must be of finite order.

The energy transfer per unit time from the external world to the body B can also be calculated by means of a suitable distribution calculated on the admissible velocity field V.

We assume that in order to distinguish long range actions from contact actions the power of external actions $\mathcal{P}^{ext}$ exerted on the body B must be represented by means of the theorem 7.3: long range external actions exerted on B can be represented a distribution which is a integrable function with respect threedimensional Lebesgue measure, while contact actions are concentrated on the contact surface of B and are distributions which can be transverse distributional derivatives on the regular part of this surface and on its edges and wedges: these distributions will be represented by means of integrals calculated with Hausdorff measures of dimension two or one or by means of sums of Dirac Deltas.

We are now ready to add a second (for a beautiful presentation of the ideas inspiring this axiom the reader is referred to Salençon(1988) or to Cosserat (1908-1909))

Postulate (Principle of Virtual Powers or Power Balance)

For every subbody S of a given body and for every test velocity field V the following equality holds

$$\mathcal{P}^{int}(B,V) = \mathcal{P}^{ext}(B,V). \tag{56}$$

9 Contact actions in $N-th$ order strain-gradient multipolar continua: $N-th$ order forces and stress multipoles.

The Principle of Virtual Powers clearly implies that the admissible external contact actions which a $N-th$ gradient continuum can "sustain" belong to a particular subset of the set of distributions concentrated on ∂B, $\partial\partial B$ and $\partial\partial\partial B$. The argument we develop here follows the same spirit as the papers cited in the references by Green, Rivlin and Germain. Also of relevance are the considerations in Seppecher (1987)-(1989).

In the following we want specify the aforementioned class.

Lemma 9.1. *Let B be a $N-th$ gradient continuum. Let us assume that the topological boundary of B is a piecewise regular surface as defined in 5.1. If the equality 56 holds then for the part of $\mathcal{P}^{ext}$ having support on ∂B, $\partial\partial B$ and $\partial\partial\partial B$ (i.e. for the external actions which are contact actions concentrated on the topological boundary of B) the following properties hold i) the part of $\mathcal{P}^{ext}$ having support on ∂B and absolutely continuous with respect to H^2 measure is a distribution of order smaller or equal to $N-1$, ii) the part of $\mathcal{P}^{ext}$ having support on $\partial\partial B$ and absolutely continuous with respect to H^1 measure is a distribution of order smaller or equal to $N-2$, iii) the part of $\mathcal{P}^{ext}$ having support on $\partial\partial\partial B$ is a distribution concentrated on points, constituted by derivatives of Dirac Deltas of order smaller or equal to $N-2$.*

Proof. The proof is easily obtained repeatedly applying to the addends of 55 the Corollary 7.8 to the embedded Riemannian manifolds which constitutes the regular parts of the topological boundary of B and its edges. □

Because of the previous Lemma and the Corollary 7.8 the following representation form for $\mathcal{P}^{ext}$ holds

Lemma 9.2. *The external actions which can be sustained by a $N-th$ gradient continuum can only expend powers of the type*

$$\mathcal{P}^{ext}(B,V) = \int_{\mathcal{B}} EV + \int_{\mathcal{S}_B} F_0 V + \sum_{\Delta=1}^{N-1} \int_{\mathcal{S}_B} F_\Delta \cdot \left(\nabla^\Delta V\right)_\perp + \tag{57}$$

$$+ \int_{\mathcal{E}_B} G_0 \cdot V + \sum_{\Delta=1}^{N-2} \int_{\mathcal{E}_B} G_\Delta \cdot \left(\nabla^\Delta V\right)_\perp + \sum_{\Delta=0}^{N-3} \int_{\mathcal{W}_B} H_\Delta \cdot \nabla^\Delta V \tag{58}$$

where we used the notations $\mathcal{S}_B :=$ Support(∂B), $\mathcal{E}_B :=$ Support$(\partial\partial B)$ and $\mathcal{W}_B :=$ Support$(\partial\partial B)$

Remark 9.3. It is needed here (for more details see e.g. Arnold (1979)) to precise the relationship between e.g. $\partial\partial B$ and its support. If an edge is the border of two different regular surfaces belonging to ∂B then $\partial\partial B$ is the union of the two borders of these concurring surfaces, with a sign depending on the relative orientation of the curve, which constitutes the support of the two different borders, and the concurring surface. These definitions are those used generally in the theory of integration of differential forms defined on differential manifolds.

Definition 9.4. The fields F_Δ, G_Δ and H_Δ are called by Green and Rivlin "contact $n-$ forces" per unit surface, per unit line or concentrated on points.

Remark 9.5. In all previous considerations we have used the following obvious but important facts: i) when applying Corollary 7.8 to the body B there is no transverse direction in the embedding, ii) when the same corollary was applied to regular parts of the topological boundary of B and to its edges then the transverse directions in the corresponding embeddings are the normal to the surfaces or the planes orthogonal to the tangents to the edges, iii) in wedges all directions are transverse.

We want now to find how to generalize Cauchy representation Theorem for contact force densities in terms of Cauchy stress to multipolar $N-$ th gradient continua basing it again on the Postulate 8 and on the Corollary 7.8.

We start from the formula 55

$$\mathcal{P}^{int}(B,V) = \sum_{\Lambda=0}^{N} \int_B T_\Lambda \cdot \nabla^\Lambda V. \tag{59}$$

and apply to each of its addends the formula obtained in the Corollary 7.8 when first identifying M with B, the projection operator $\mathbb{P}$ with the identity and the outer normal ν with the normal n to the surface ∂B, we get

$$\mathcal{P}^{int}(B,V) = \sum_{\Lambda=0}^{N} \int_B T_\Lambda \cdot \nabla^\Lambda V = \tag{60}$$

$$= \int_B \left(\sum_{\Lambda=0}^{N} (-1)^\Lambda \, div^\Lambda T_\Lambda \right) V + \tag{61}$$

$$+ \sum_{\Lambda=0}^{N} \sum_{L=0}^{\Lambda-1} \int_{\partial B} ((-1)^{\Lambda-1-L} \, (div)^{\Lambda-1-L} \, T_\Lambda).n \cdot \nabla^L V \tag{62}$$

We now simply manipulate, using associativity of summation, the obtained expression for getting the coefficients of every power of ∇V appearing in it. Therefore the equality

$$\mathcal{P}^{int}(B,V) = \sum_{\Lambda=0}^{N} \int_B T_\Lambda \cdot \nabla^\Lambda V \tag{63}$$

becomes

$$\mathcal{P}^{int}(B,V)=\int_B\left(\sum_{\Lambda=0}^{N}(-1)^{\Lambda}\,div^{\Lambda}T_{\Lambda}\right)V+ \tag{64}$$

$$\sum_{L=0}^{N-1}\int_{\partial B}\left(\sum_{\Lambda=L+1}^{N}(-1)^{\Lambda-1-L}\,div^{\Lambda-1-L}T_{\Lambda}.n\right)\cdot\nabla^{L}V \tag{65}$$

A simple inspection of equations 63 and 65 makes natural the introduction of to the following

Definition 9.6. Bulk and Surface Stress Tensors

$$T(B,\Lambda):=T_{\Lambda};\qquad T(\partial B,L):=\left(\sum_{\Lambda=L+1}^{N}(-1)^{\Lambda-1-L}\,div^{\Lambda-1-L}T_{\Lambda}.n\right) \tag{66}$$

Definition 9.7. Bulk Internal Force

$$F(B,0):=\left(\sum_{\Lambda=0}^{N}(-1)^{\Lambda}\,div^{\Lambda}(T(B,\Lambda))\right)$$

By means of introduced notations the expression of internal power is meaningfully simplified as follows:

$$\mathcal{P}^{int}(B,V)=\int_B F(B,0)V+\sum_{L=0}^{N-1}\int_{\partial B}T(\partial B,L)\cdot\nabla^{L}V \tag{67}$$

We now apply the Corollary 7.8, by identifying the embedded manifold M with any of the regular parts of ∂B to transform the integrals involving surface stresses thus obtaining:(in the following formulas ν represents the unit normal to $\partial\partial B$ which is tangent to ∂B and we denoted with the symbol $\mathbb{P}_{\sigma}$ the projector operator relative to the tangent planes of every the regular parts of ∂B)

$$\int_{\partial B}T(\partial B,L)\cdot\nabla^{L}V= \tag{68}$$

$$=\sum_{J=1}^{L}(-1)^{L-J}\int_{\partial B}\left((div_{\partial B}\mathbb{P}_{\sigma})^{L-J}\,T(\partial B,L)\right)_{\perp}\cdot\left(\nabla^{J}V\right)_{\perp}+ \tag{69}$$

$$+(-1)^{L}\int_{\partial B}\left((div_{\partial B}\mathbb{P}_{\sigma})^{L}\,T(\partial B,L)\right)V+ \tag{70}$$

$$+\sum_{J=0}^{L-1}(-1)^{L-1-J}\int_{\partial\partial B}\mathbb{P}_{\sigma}((div_{\partial B}\mathbb{P}_{\sigma})^{L-1-J}\,T(\partial B,L)).\nu\cdot\nabla^{J}V \tag{71}$$

We get for the sum of all terms involving surface stresses

$$\sum_{L=0}^{N-1}\int_{\partial B} T(\partial B,L)\cdot\nabla^L V =$$

$$= \sum_{L=0}^{N-1}\sum_{J=1}^{L}(-1)^{L-J}\int_{\partial B}\Big((div_{\partial B}\mathbb{P}_\sigma)^{L-J}\,T(\partial B,L)\Big)_\perp\cdot\left(\nabla^J V\right)_\perp + \tag{72}$$

$$+\sum_{L=0}^{N-1}(-1)^L\int_{\partial B}\Big((div_{\partial B}\mathbb{P}_\sigma)^L\,T(\partial B,L)\Big)\,V$$

$$+\sum_{L=0}^{N-1}\sum_{J=0}^{L-1}(-1)^{L-1-J}\int_{\partial\partial B}\mathbb{P}_\sigma((div_{\partial B}\mathbb{P}_\sigma)^{L-1-J}\,T(\partial B,L)).\nu\cdot\nabla^J V$$

Using again associativity of summation the last equality becomes

$$\sum_{L=0}^{N-1}\int_{\partial B} T(\partial B,L)\cdot\nabla^L V = \tag{73}$$

$$=\sum_{J=1}^{N-1}\int_{\partial B}\left(\sum_{L=J}^{N-1}(-1)^{L-J}\,(div_{\partial B}\mathbb{P}_\sigma)^{L-J}\,T(\partial B,L)\right)_\perp\cdot\left(\nabla^J V\right)_\perp + \tag{74}$$

$$+\int_{\partial B}\left(\sum_{L=0}^{N-1}(-1)^L\,(div_{\partial B}\mathbb{P}_\sigma)^L\,T(\partial B,L)\right)V \tag{75}$$

$$+\sum_{J=0}^{N-2}\int_{\partial\partial B}\left(\sum_{L=J+1}^{N-1}(-1)^{L-1-J}\,\mathbb{P}_\sigma((div_{\partial B}\mathbb{P}_\sigma)^{L-1-J}\,T(\partial B,L)).\nu\right)\cdot\nabla^J V \tag{76}$$

The nomenclature introduced in the previous definitions (which are based on those introduced by Green and Rivlin) allows us a meaningful interpretation of obtained result

9.1 The terms of surface (1-)forces and surface $L+1-$ forces

Surface density $F(\partial B,0)$ of $(1-)$force appears in the addend 75

$$F(\partial B,0) := \sum_{L=0}^{N-1}(-1)^L\,(div_{\partial B}\mathbb{P}_\sigma)^L\,T(\partial B,L)$$

The introduced vector quantity $F(\partial B,0)$ generalizes Cauchy "traction" vector: it is expending power on (virtual) velocities.

Remark 9.8. In Cauchy Continua surface density of force $F(\partial B, 0)$ coincides with the only non-vanishing surface stress $T(\partial B, 0)$. In Second Gradient Continua (as established by Green and Rivlin, Mindlin and Germain) the following equality holds

$$F(\partial B, 0) = T(\partial B, 0) - (div_{\partial B}\mathbb{P}_\sigma)\, T(\partial B, 1).$$

We then consider the addend 74 which leads us to the definition

$$F(\partial B, J) := \left(\sum_{L=J}^{N-1} (-1)^{L-J} (div_{\partial B}\mathbb{P}_\sigma)^{L-J}\, T(\partial B, L) \right)_\perp \tag{77}$$

Surface $J+1-$ forces $F(\partial B, J)$ expend power on the $J-th$ transverse gradient of (virtual) velocity fields.

Finally we consider the addend 76 which leads us naturally to the following

Definition 9.9. Line Stress Tensors

$$T(\partial\partial B, L) := . \left(\sum_{L=J+1}^{N-1} (-1)^{L-1-J}\, \mathbb{P}_\sigma((div_{\partial B}\mathbb{P}_\sigma)^{L-1-J}\, T(\partial B, L)).\nu \right) \tag{78}$$

We have thus proven the following intermediate:

Lemma 9.10. *For $N-th$ gradient continua the following representation formula holds*

$$\mathcal{P}^{int}(B, V) = \int_B F(B, 0)V + \int_{\partial B} F(\partial B, 0) \cdot V \tag{79}$$

$$+\sum_{L=1}^{N-1} \int_{\partial B} F(\partial B, L) \cdot (\nabla^L V)_\perp + \sum_{L=0}^{N-2} \int_{\partial\partial B} T(\partial\partial B, L) \cdot \nabla^L V \tag{80}$$

We must now proceed by applying again the Corollary 7.8 (when this time M is any regular curve with border belonging to $\partial\partial B$) to the terms involving line stresses. Once we have introduced the projection operator $\mathbb{P}_\lambda$ relative to every regular curve composing $\partial\partial B$, and the vector $t_{\partial\partial B}$ denoting the tangent vector of the generic edge concurring in the considered wedge we get

$$\int_{\partial\partial B} T(\partial\partial B, L) \cdot \nabla^L V = \tag{81}$$

$$= \sum_{J=1}^{L} (-1)^{L-J} \int_{\partial\partial B} \left((div_{\partial\partial B}\mathbb{P}_\lambda)^{L-J} T(\partial\partial B, L) \right)_\perp \cdot (\nabla^J V)_\perp + \tag{82}$$

$$+ (-1)^L \int_{\partial\partial B} \left((div_{\partial\partial B}\mathbb{P}_\lambda)^L T(\partial\partial B, L) \right) V + \tag{83}$$

$$+ \sum_{H=0}^{L-1} (-1)^{L-1-H} \int_{\partial\partial\partial B} \mathbb{P}_\lambda((div_{\partial\partial B}\mathbb{P}_\lambda)^{L-1-H} T(\partial\partial B, L)).t_{\partial\partial B} \cdot \nabla^H V \tag{84}$$

Summing over L and using once more associativity we get

$$\sum_{L=0}^{N-2} \int_{\partial\partial B} T(\partial\partial B, L) \cdot \nabla^L V = \tag{85}$$

$$= \sum_{J=1}^{N-2} \int_{\partial\partial B} \left(\sum_{L=J}^{N-2} (-1)^{L-J} (div_{\partial\partial B}\mathbb{P}_\lambda)^{L-J} T(\partial\partial B, L) \right)_\perp \cdot (\nabla^J V)_\perp + \tag{86}$$

$$+ \int_{\partial\partial B} \left(\sum_{L=0}^{N-2} (-1)^L (div_{\partial\partial B}\mathbb{P}_\lambda)^L T(\partial\partial B, L) \right) V + \tag{87}$$

$$+ \sum_{H=0}^{N-3} \int_{\partial\partial\partial B} \left(\sum_{L=H+1}^{N-2} (-1)^{L-1-H} \mathbb{P}_\lambda((div_{\partial\partial B}\mathbb{P}_\lambda)^{L-1-H} T(\partial\partial B, L)).t_{\partial\partial B} \right) \cdot \nabla^H V \tag{88}$$

9.2 The terms of line (1-)forces and line $J+1-$ forces.

Concerning the addend in 87 we easily recognize that it must be interpreted as a line density of 1-force.

Therefore we introduce, in terms of introduced line stresses, the following

Definition 9.11. Line density of 1-force

$$F(\partial\partial B, 0) := \sum_{L=0}^{N-2} (-1)^L (div_{\partial\partial B}\mathbb{P}_\lambda)^L T(\partial\partial B, L)$$

Regarding the addend in 86, it naturally leads to the following

Definition 9.12. Line density of J + 1-force

$$F(\partial\partial B, J) := \left(\sum_{L=J}^{N-2} (-1)^{L-J} \left(div_{\partial\partial B} \mathbb{P}_{\lambda} \right)^{L-J} T(\partial\partial B, L) \right)_{\perp} \tag{89}$$

9.3 The terms of (1-)forces and $J+1-$ forces concentrated on wedges

The process of integration by parts is finally ended. We are left with forces, double forces, triple forces up to $N-3$ forces concentrated on wedges. Indeed the terms in 88, naturally lead to the following

Definition 9.13. (1−) **forces and H + 1-forces concentrated on wedges**

$$F(\partial\partial\partial B, 0) := \left(\sum_{L=1}^{N-2} (-1)^{L-1} \mathbb{P}_{\lambda}((div_{\partial\partial B} \mathbb{P}_{\lambda})^{L-1} T(\partial\partial B, L)).t_{\partial\partial B} \right) \tag{90}$$

$$F(\partial\partial\partial B, H) := \left(\sum_{L=H+1}^{N-2} (-1)^{L-1-H} \right.$$
$$\mathbb{P}_{\lambda}((div_{\partial\partial B} \mathbb{P}_{\lambda})^{L-1-H} T(\partial\partial B, L)).t_{\partial\partial B} \tag{91}$$

In conclusion in the present subsection we have finally obtained the following

Representation formula for power expended by line stresses

$$\sum_{L=0}^{N-2} \int_{\partial\partial B} T(\partial\partial B, L) \cdot \nabla^L V = \sum_{J=1}^{N-2} \int_{\partial\partial B} F(\partial\partial B, J) \cdot (\nabla^J V)_{\perp} + \tag{92}$$

$$+ \int_{\partial\partial B} F(\partial\partial B, 0) V + \sum_{J=0}^{N-3} \int_{\partial\partial\partial B} F(\partial\partial\partial B, J) \cdot \nabla^J V \tag{93}$$

9.4 The obtained expression for Internal Power in $N-th$ Gradient Continua

Gathering all results obtained in the previous subsection we get the searched representation formula for internal power

$$\mathcal{P}^{int}(B,V) = \int_B F(B,0)V + \int_{\partial B} F(\partial B,0)\cdot V + \sum_{L=1}^{N-1} \int_{\partial B} F(\partial B,L)\cdot \left(\nabla^L V\right)_\perp + \tag{94}$$

$$+ \int_{\partial\partial B} F(\partial\partial B,0)V + \sum_{J=1}^{N-2} \int_{\partial\partial B} F(\partial\partial B,J)\cdot \left(\nabla^J V\right)_\perp + \sum_{J=0}^{N-3} \int_{\partial\partial\partial B} F(\partial\partial\partial B,J)\cdot \nabla^J V \tag{95}$$

where all types of forces are represented in terms of bulk, surface and line stresses by means of the relationships which we recall for convenience in the following summary.

Remark 9.14. Surface stresses are defined in terms of bulk stresses and line stresses in terms of surface stresses by means of equations 78 and 66. Therefore All kind of stresses are defined in terms of bulk stresses. This implies that all type of forces are represented in terms of bulk stresses.

We list now all introduced definitions for forces and stresses

$$T(\partial B,L) := \left(\sum_{\Lambda=L+1}^{N} (-1)^{\Lambda-1-L}\, div^{\Lambda-1-L} T_\Lambda . n\right) \tag{96}$$

$$T(\partial\partial B,J) := .\left(\sum_{L=J+1}^{N-1} (-1)^{L-1-J}\, \mathbb{P}_\sigma((div_{\partial B}\mathbb{P}_\sigma)^{L-1-J}\, T(\partial B,L)).\nu\right) \tag{97}$$

$$F(\partial B,0) := \sum_{L=0}^{N-1} (-1)^L\, (div_{\partial B}\mathbb{P}_\sigma)^L\, T(\partial B,L)$$
$$F(\partial B,J) := \left(\sum_{L=J}^{N-1} (-1)^{L-J}\, (div_{\partial B}\mathbb{P}_\sigma)^{L-J}\, T(\partial B,L)\right)_\perp \tag{98}$$

$$
\begin{aligned}
F(\partial\partial B, 0) &:= \sum_{L=0}^{N-2} (-1)^{L} \left(div_{\partial\partial B}\mathbb{P}_{\lambda}\right)^{L} T(\partial\partial B, L) \\
F(\partial\partial B, J) &:= \left(\sum_{L=J}^{N-2} (-1)^{L-J} \left(div_{\partial\partial B}\mathbb{P}_{\lambda}\right)^{L-J} T(\partial\partial B, L)\right)_{\perp}
\end{aligned}
\tag{99}
$$

$$
F(\partial\partial\partial B, H) := \left(\sum_{L=H+1}^{N-2} (-1)^{L-1-H} \mathbb{P}_{\lambda}((div_{\partial\partial B}\mathbb{P}_{\lambda})^{L-1-H} T(\partial\partial B, L)).t_{\partial\partial B}\right.
\tag{100}
$$

Finally it is easy to prove the following corollaries:

Corollary 9.15. *The tensor T_N determines the surface $N-1$ forces, the edge $N-2$ forces and the wedge $N-3$ forces as established by means of the following formulas:*

$$F\left(\partial B, N-1\right) = T(\partial B, N-1) = (T_N.n)_{\perp} \tag{101}$$

$$F(\partial\partial B, N-2) := T(\partial\partial B, n-2) = (\mathbb{P}_{\sigma}(T_N.n).\nu)_{\perp} \tag{102}$$

$$F(\partial\partial\partial B, N-3) = (\mathbb{P}_{\lambda}\left(\mathbb{P}_{\sigma}(T_N.n)).\nu)).t_{\partial\partial B}\right) \tag{103}$$

Clearly

Corollary 9.16. *The set of equations 96, 97, 98, 99 and 100 allow us to define three functionals mapping the contact surface ∂B and the $N-tuple$ of tensor fields $\{T_{\Lambda}\}$ on to the set of corresponding contact actions*

$$
\left\{
\begin{array}{c}
\left(F\left(\partial B, J\right), J = 0, ...N-1\right) \\
F(\partial\partial B, J), J = 0, ...N-1 \\
F(\partial\partial\partial B, J), J = 0, ..., N-2
\end{array}
\right\}
\tag{104}
$$

This map, for a fixed contact surface ∂B, is linear. We will denote the introduced linear functionals with the symbols introduced by means of the following definitions$\mathfrak{F}(\partial B)$, $\mathfrak{F}(\partial\partial B)$ and $\mathfrak{F}(\partial\partial\partial B)$

$$
\begin{aligned}
\mathfrak{F}(\partial B, \{T_{\Lambda}\}, J) &:= F\left(\partial B, J\right) \\
\mathfrak{F}(\partial\partial B, \{T_{\Lambda}\}, J) &:= F(\partial\partial B, J) \\
\mathfrak{F}(\partial\partial\partial B, \{T_{\Lambda}\}, J) &:= F(\partial\partial\partial B, J)
\end{aligned}
\tag{105}
$$

Proof. Simply consider the previous formulas 100, 96 and 97 and evaluate them when $H = N-3$, $L = N-2$, $L = N-1$. □

We will use the previous corollaries later, when we will consider the consequence of the Cauchy tetrahedron argument on the structure of contact actions in $N-th$ gradient continua.

9.5 Some final considerations about obtained results:

Using the Postulate 8 and assuming that the bulk, surface and line stresses are sufficiently regular so that all needed divergences can be estimated, then from the representation formulas 94 and 57 and arbitrariness of test functions V we can get the following "strong' version of the Principle of Virtual Powers

$$F(B,0) = E \tag{106}$$

$$\forall L = 0, ...N-1 \quad F(\partial B, L) = F_L \tag{107}$$

$$\forall J = 0, ...N-2 \quad F(\partial\partial B, J) = G_J \tag{108}$$

$$\forall J = 0, ...N-3 \quad F(\partial\partial\partial B, J) = H_J \tag{109}$$

The first of the listed equations is a system of PDE defined inside the reference configuration of the body B,while the others are appropriate boundary conditions to be verified on regular surfaces, edges and wedges of the topological boundary of B.

We explicitly remark that

- From equations 98 and 96 it is evident the very particular nature of so called "Cauchy Postulate". Indeed the contact force per unit surface at any regular part of a surface regarded as a Cauchy cut in general do not depend only on the normal n of such surface. In $N-th$ gradient continua such contact force in general depends also on ∇n up to $\nabla^{N-1}n$.
- From equations 99, 96 and 97 it is evident that line contact force depends in general on the vectors and tensors ν and $\nabla^L n$ relative to all regular surfaces concurring on considered edge, and on their derivatives along the edges.
- Concentrated $J-$forces are present at wedges depending on the geometry of concurring edges and regular surfaces, and more precisely on edges tangent vectors, on surface normals and their gradients, on edges exterior normals and their gradients.

In the present section we have completely characterized the structure of contact actions in $N-th$ gradient continua by using the first method delineated in the section 2.

10 Bodies, Contact Surfaces and Shapes ??

In this section we start attacking the problem of characterizing the structure of contact actions in $N-th$ gradient continua by using the second method delineated in the section 2: in other words we use the approach *à la Cauchy.*

Following Truesdell (1977) we call domains the closures of open Kellogg (1929) regular regions. We want to describe the contact actions exerted on a body, identified with the domain $\mathcal{C}$ it occupies in Euclidean space, through its topological boundary $\mathcal{S} := \partial\mathcal{C}$ (which we will call *the contact surface*).

Only when we will need to represent with integral formulas the power expended by contact actions we will choose to limit our attention to a particular class of contact surfaces: those which are *piecewise regular surfaces with edges and wedges*, as defined in the previous section 4.

Remark 10.1. The definition of piecewise regular surface with edges and wedges will need to be generalized. In a forthcoming paper the present analysis will be generalized and we will introduce edges and wedges of order $k+1$, i.e. curves (edges) on piecewise regular surfaces on which the $k-th$ order gradient of the normal suffers a jump (discontinuity of the first kind, or jump discontinuity), and points (wedges) at which edges of order k are concurring.

Remark 10.2. The reasonings developed in the previous section 4.prove that on edges and wedges of order $k+1$ there may be concentrated $h-$forces with $h \geq k+1$.

We are interested in the dependence of contact actions on the shape of $\mathcal{S}$.

Then we have to define precisely what we call *shape* of $\mathcal{S}$: it will be a local concept.

Definition 10.3. Surfaces locally having the same shape We say that the shape of the contact surface $\mathcal{S}$ at the point $x \in \mathcal{S}$ is the same as the shape of the contact surface $\mathcal{S}'$ at the point $x' \in \mathcal{S}'$ if and only if there exist two open sets $I(x)$ and $I(x')$ such that

$$i) \quad x \in I(x), x' \in I(x') \quad \text{and } ii) \quad t_{x'-x}(I(x) \cap \mathcal{S}) = I(x') \cap \mathcal{S}' \tag{110}$$

where t_u denotes the translation of points, vectors and sets induced by the vector u. To say that the shape of the contact surface $\mathcal{S}$ at the point $x \in \mathcal{S}$ is the same as the shape of the contact surface $\mathcal{S}'$ at the point $x' \in \mathcal{S}'$ we will occasionally use the notation

$$(x, \mathcal{S})\, \mathfrak{R}\, (x', \mathcal{S}') \, . \tag{111}$$

Definition 10.4. Shapes We call shape of $\mathcal{S}$ at the point $x \in \mathcal{S}$ the equivalence class with respect to the above defined equivalence relation $\mathfrak{R}$ to which belongs the ordered pair $(x, \mathcal{S})$.We denote this equivalence class by means of the symbol $[(x, \mathcal{S})]$.

Remark 10.5. Note that, according to this definition, when a surface is rotated it do change its shape.

Definition 10.6. Sets of shapes: Let $\mathcal{S}$ be a contact admissible surface. We introduce the set of the shapes of $\mathcal{S}$

$$\Phi(\mathcal{S}) := \{[(x, \mathcal{S})] / \quad x \in \mathcal{S}\} \tag{112}$$

We will also need to consider the set Φ of all admissible shapes:

$$f \in \Phi \Leftrightarrow (\exists \mathcal{S} \text{ admissible}) \, (f \in \Phi(\mathcal{S})) \tag{113}$$

Definition 10.7. Plane shapes Obviously the shape of a plane π at any of its points depends only on the normal n to the plane. When there is no ambiguity, we denote the shape of a plane simply by n.

Definition 10.8. Dihedral shapes Let us consider a non-degenerate dihedron. We denote by n_1 and n_2 the external normals to the half-planes forming it, and by τ the unit vector tangent to the edge of the dihedron, i.e. the line which is the intersection of the two half-planes. We assume that the ordered triple (n_1, n_2, τ) verifies the condition

$$(n_2 \times n_1) \cdot \tau > 0.$$

Each of considered two half-planes is a two-dimensional manifold with border: the outer normal to this border is given respectively by the vectors $\nu_1 = n_1 \times \tau$ and $\nu_2 = -n_2 \times \tau$.On the edge of such a dihedron, the shape is constant and is determined by n_1, n_2 and τ. This shape will be denoted by (n_1, n_2, τ). Note that $(n_1, n_2, \tau) = (n_2, n_1, -\tau)$.The angle $(-n_1, n_2)$ in the plane oriented by τ will be called the dihedral angle of (n_1, n_2, τ); it is different from $0, \pi$ and 2π.

Definition 10.9. Nondegenerate $k-$*tuple* **of vectors.** Let us consider a point x and a $k-$ *tuple* of vectors $(n_1, ..., n_k)$ applied at x. The planes π_i are defined by the conditions: i) $x \in \pi_i$ and ii) $n_i \perp \pi_i$. The unit tangent vector τ_i to the the edge $\gamma_i := \pi_i \cap \pi_{i+1}$ is given by $\tau_i = n_i \times n_{i+1}$ (we define $n_{k+1} := n_1$).Let us denote P_π the projection operator in the plane π. The $k-$ *tuple* $(n_1, ..., n_k)$ of vectors is said to be nondegenerate when there exists a plane π such that i) $x \in \pi$, ii) for a unit vector e in π the following property holds

$$(\forall i \leq k)\,(\forall j \leq k)\left((i<j) \Rightarrow \left(0 < \frac{P_\pi \tau_i}{\|P_\pi \tau_i\|}\cdot e < \frac{P_\pi \tau_j}{\|P_\pi \tau_j\|}\cdot e\right)\right). \tag{114}$$

Definition 10.10. (Nondegenerate) **Polihedral Wedge Shapes** or Multiple planar shapes with edges concurring in a wedge. Let us consider a point x and a nondegenerate $k-tuple$ of vectors $(n_1, ..., n_k)$ applied at x.We call A_i the angle having vertex in x and as sides the half-lines originating from x in the direction of τ_i. Let us consider the piecewise regular surface $\bigcup_i A_i$. We call polihedral wedge shape or multiple planar shape with edges concurring in a wedge the equivalence class $[(x, \bigcup_i A_i)]$. When this will not lead to confusion we will denote a polihedral wedge shape with the symbol $(\widehat{n_1, ..., n_k})$.

Definition 10.11. Cuts. Let $D_1, ...D_p$ be p compact domains having as (topological) boundaries the contact surfaces denoted $\mathcal{S}_1, ..., \mathcal{S}_p$. Let $\mathcal{S}^\cap$ be the topological boundary of $\bigcap_{i=1}^{p} D_i$. At each point x in $\bigcap_{i=1}^{p} \mathcal{S}_i \cap \mathcal{S}^\cap$ the shape of $\mathcal{S}^\cap$ depends only on the shape of all $\mathcal{S}_i$. For every $x \in \bigcap_{i=1}^{p} \mathcal{S}_i \cap \mathcal{S}^\cap$ we denote the shape of $\bigcap_{i=1}^{p} \mathcal{S}_i \cap \mathcal{S}^\cap$ at x by the symbols

$$Cut\left([(x, \mathcal{S}_i)], i = 1, .., p\right); \quad \text{or} \quad Cut\left([(x, \mathcal{S}_1)], [(x, \mathcal{S}_2)], ..., [(x, \mathcal{S}_p)]\right). \tag{115}$$

Example 10.12. Plane Cuts. Let D be a domain the boundary of which is the contact surface $\mathcal{S}$. Let π be a plane and H the half-space with boundary π and outer unit normal n. Let $x \in \pi \cap \mathcal{S}$. We will call the $Cut\left([(x, \mathcal{S})], [(x, \pi)]\right)$ a plane cut and we will shortly denote it by $Cut\left([(x, \mathcal{S})], n\right)$.

Example 10.13. Multiple Plane Cuts. Let D be a domain the boundary of which is the contact surface $\mathcal{S}$. Let $\pi_i\ ; i = 1, ...p$ be planes having as outer unit normal the vectors n_i respectively. Let $x \in \bigcap_{i=1}^{p} \pi_i \cap \mathcal{S}$. We will call the $Cut\left([(x, \mathcal{S})], [(x, \pi_1)], .., [(x, \pi_p)]\right)$ a plane cut and we will shortly denote it by $Cut\left([(x, \mathcal{S})], n_1, n_2, ..., n_p\right)$.

Definition 10.14. Admissible domains and contact surfaces. We only consider domains the topological boundary of which $\mathcal{S}$ (**contact admissible surface**) is piecewise regular. In other words we assume that i) $\mathcal{S}$

is a finite union of two-dimensional suitably regular compact manifolds with border (called the **faces** of $\mathcal{S}$) ii) the union of the borders of these faces is a finite union of one-dimensional suitably regular compact manifolds with boundary (called the **edges** of $\mathcal{S}$), iii) the edges are concurring in wedges and iv) the set of wedges is finite.

Definition 10.15. Regular points of contact surfaces All internal points of the faces (i.e. those points which do not belong to the border of the faces) are called regular points of the face, or points where $\mathcal{S}$ has a regular shape. The set of regular points is denoted by $\mathcal{S}_r$; the set of all internal points of the edges (i.e. those points which do not belong to the border of the faces) is denoted by $\mathcal{L}_r$. An internal point of an edge is also called regular point of the edge, or point where $\mathcal{S}$ has an edge shape. The set of all wedges will be denoted by $\mathcal{W}_S$. Moreover, we assume that, i) everywhere in $\mathcal{L}_r$, $\mathcal{S}$ is tangent to a non degenerate dihedron and ii) at every wedge in $\mathcal{S}$ there exists a tangent polihedral plane surface (i.e. a surface having a polihedral wedge shape or, which is the same, a multiple planar shape with edges concurring in it).

Remark 10.16. When we say that at a edge regular point the contact surface is tangent to a nondegenerate dihedron we mean that the two regular surfaces concurrent at the edge have both a tangent plane and that these two plane form a nondegenerate dihedron. Similarly when we say that at every wedge in $\mathcal{S}$ there exists a tangent polihedral plane surface we mean that every edge concurring in the wedge has a tangent line and that every regular surface concurring in the wedge has a tangent plane and the set of these tangent lines and planes form a surface having a polihedral wedge shape.

Definition 10.17. We denote by $\Phi^F(\mathcal{S})$ the set of all face shapes of $\mathcal{S}$, by $\Phi^E(\mathcal{S})$ the set of all edge shapes of $\mathcal{S}$ and by $\Phi^W(\mathcal{S})$ the set of all wedge shapes of $\mathcal{S}$: in formulas

$$\Phi^F(\mathcal{S}) := \{[(x,\mathcal{S})]\,, x \in \mathcal{S}_r\}\,; \quad \Phi^E(\mathcal{S}) := \{[(x,\mathcal{S})]\,, x \in \mathcal{L}_r\}\,; \tag{116}$$

$$\Phi^W(\mathcal{S}) := \{[(x,\mathcal{S})]\,, x \in \mathcal{W}_S\}\,. \tag{117}$$

We remark that in this way we have found a partition of the set of shapes of a contact admissible surface:

$$\Phi(\mathcal{S}) = \Phi^F(\mathcal{S}) \cup \Phi^E(\mathcal{S}) \cup \Phi^W(\mathcal{S}) \tag{118}$$

Definition 10.18. Prescribed shapes: A set of shapes E is called a set of prescribed shapes if there exists a finite set $\{\mathcal{S}_1, \mathcal{S}_2, ...\mathcal{S}_m\}$ of compact admissible contact surfaces such that $E \subseteq \bigcup_{i=1}^{m} \Phi(\mathcal{S}_i)$.

The following examples should make clearer the meaning of the just introduced definition.

Example 10.19. The set of images C_t of a cube C_1 under a family of homothetic transformations of ratio $t \in]0,1]$ is a set of prescribed shapes;

Example 10.20. The set of images C'_t of a cube C_1 under a family of rotations of angle $t \in [0,\pi]$ around a given axis is not a set of prescribed shapes;

Example 10.21. The family of spheres S_t of centre x and radius $t \in]0,1]$ has not prescribed shapes.

Definition 10.22. Let us consider a finite set $\{\mathcal{S}_1, \mathcal{S}_2, ...\mathcal{S}_m\}$ of compact admissible surfaces and a finite set of unit vectors $U := \{u_1, ..., u_M\}$. A set of shapes E is called a set of **prescribed (multiple) plane cuts**

$$E \subseteq \bigcup_{l=1}^{M} \{Cut\,(f, u_1, ..., u_l)\,; f \in \Phi\,(\mathcal{S}_i)\,; u_i \in U, i = 1, ..., l\} \tag{119}$$

In what follows we will need to consider functions defined in some set of shapes. It would be desirable to introduce in the set of shapes a topological structure: in the present context we refrain from such a task which is really challenging. Instead, by introducing the concept of "set of prescribed shapes" we have considered a kind of "compactness" in the set of all admissible shapes. This definition will reveal itself to be sufficient to our aims.

10.1 Contact Actions

In this paper we consider the class of continua which were introduced by Green and Rivlin (1964) or Mindlin (1965). In these continua there are contact actions which are not forces. An example of such generalized continua is given by those studied by Germain, which called them second gradient continua, as their deformation energy density depends not only on the first gradient of displacement but also on the second gradient of displacement. However in generalized continua forces represent an important class of contact actions the properties of which are somehow different from those of the other contact actions.

As first step we will start by accepting the following:

Generalized Cauchy Postulate: Part I

Let us consider a continuous body B occupying, in a given configuration, a domain D having as topological boundary the admissible contact surface S . Let D the set of admissible virtual displacement of B starting from the considered configuration. Each of these displacements is a field defined in D and having trace on S and is a vector valued test function. We assume that the power expended by contact actions exerted on B is a distribution in the sense of Schwartz.

This postulate has been already introduced in continuum mechanics (in a slightly less general form) at least already by Cosserat brothers (1908)-(1909). A detailed historic review of the relevant literature can be found e.g. in Maugin (1980). In the textbook of Salençon (1988-2005) this postulate is presented in a attractive didactical form.

Because of the just introduced postulate we can be guided in the formulation of the needed definitions by the already cited (see previous sections) representation theorem for distributions due to Laurent Schwartz (quoted in the previous section 4).

Therefore we start by considering contact actions which indeed are $(1-)$ forces and then proceed by considering $k-$forces.

Definition 10.23. Surface and Line density of contact $(1-)$forces. Contact $(1-)$forces concentrated on wedges. Let us consider any admissible domain D.Its boundary $\mathcal{S}$ is a contact piecewise regular surface, including edges and wedges. We call contact $(1-)$force a contact action for which the power expended on a virtual velocity field V is given by the following expression

$$\mathcal{P}_0^{con}(V) = \int_{\mathcal{S}_r} \mathsf{F}_0^s V + \int_{\mathcal{L}_r} \mathsf{F}_0^l V + \int_{\mathcal{W}_S} \mathsf{F}_0^c V. \tag{120}$$

the vector functions, F_0^s,F_0^l and F_0^c defined respectively on the faces, the edges and the wedges of $\mathcal{S}$ are called i) surface density of contact $(1-)$forces, ii) line density of contact $(1-)$forces, iii) contact $(1-)$forces concentrated on wedges.

Remark 10.24. In the following we will be willing to consider surface contact forces depending continuously on curvature. We will assume that such forces densities are bounded on a set of prescribed shapes but not on the set of all admissible shapes.

Definition 10.25. Surface and Line density of contact $k-$forces. Contact $k-$forces concentrated on wedges. Let us consider any admissible domain D.Its boundary $\mathcal{S}$ is a contact piecewise regular surface, including edges and wedges. We call contact ($k-$)force a contact action for which the power expended on a virtual velocity field V is given by the following expression (where we used the notations introduced in the previous section 4)

$$\mathcal{P}_k^{con}(V) = \int_{\mathcal{S}_r} \mathsf{F}_k^s \cdot \left(\nabla^k V\right)_\perp + \int_{\mathcal{L}_r} \mathsf{F}_k^l \cdot \left(\nabla^k V\right)_\perp + \int_{\mathcal{W}_S} \mathsf{F}_k^c \cdot \nabla^k V. \quad (121)$$

the tensor functions, F_k^s,F_k^l and F_k^c defined respectively on the faces, the edges and the wedges of $\mathcal{S}$ are called i) surface density of contact $k-$forces, ii) line density of contact $k-$forces, iii) contact $k-$forces concentrated on wedges.

On occasion we will denote a generic vector or tensor field in the set $\left\{\mathsf{F}_k^s, \mathsf{F}_k^l, \mathsf{F}_k^c; k = 0, ..., L\right\}$ simply with the symbol F,adding some super or sub script when needed.

All the considerations which we will develop in this work are based on the following further hypothesis:

Generalized Cauchy Postulate: Part II

Let us consider a body, occupying a region $\mathcal{C}$ included in the Euclidean three-dimensional space and denote by $\mathcal{V}$ its space of translations. We assume that all tensor functions F previously introduced are univocally determined in terms of the position x and on the shape of $\mathcal{S}$ at x.

In other words:

we assume that there exists a function $\hat{\mathsf{F}}_k$ (which we will call $k-th$ order stress function) of the two independent variables x, and f, where f is a shape

$$\hat{\mathsf{F}}_k\colon \quad (x,f) \in \mathcal{C} \times \oplus \mapsto \quad \hat{\mathsf{F}}_k\,(x,f) \in \quad \underbrace{\mathcal{V} \otimes ... \otimes \mathcal{V}}_{k \text{ times}}$$

such that

$$\mathsf{F}_k(x) = \hat{\mathsf{F}}_k(x, [(x, \mathcal{S})]).$$

Definition 10.26. We call **generalized stress state** of the considered body the set of the functions $\hat{\mathsf{F}}$ the existence of which we have just postulated.

Remark 10.27. We do not assume uniform boundedness of all tensor partial functions $\hat{\mathsf{F}}(x, \cdot)$ in the set of all shapes Φ, i.e. -once x is fixed- with respect to the variable f.

Remark 10.28. Because of the previous definition the fields F_k^l may depend not only on the geometry of the edge itself, but also on the "limit properties" of the faces common border of which is the considered edge.

Remark 10.29. Similarly the $k-$forces F_k^c depend on all the "limit properties" of the faces and edges concurring in the considered wedge.

In the same way as the Cauchy's construction of the stress tensor presumes the continuity of contact forces, our construction will require a list of

Regularity assumptions.

The functions $\hat{\mathsf{F}}$ verifies all the conditions which follows:

1. Let $\mathcal{S}_F$ be a face and $\mathcal{L}$ an edge of an admissible contact surface $\mathcal{S}$. The functions

$$x \in \mathcal{S}_F \mapsto \hat{\mathsf{F}}^s(x, [(x, \mathcal{S}_F)]); \quad x \in \mathcal{L} \mapsto \hat{\mathsf{F}}^l(x, [(x, \mathcal{L})])$$

are continuous.

2. Let E be a set of prescribed shapes or prescribed plane cuts. We partition E into three disjoint subsets, E^l, E^s, E^c which contain respectively the edge shapes, the regular shapes and the wedge shapes in E. For a suitably fixed shape f (i.e. for a regular, edge or wedge shape depending on the type of $\hat{\mathsf{F}}$ we are considering) we can introduce the partial function

$$\hat{\mathsf{F}}(\cdot, f) : x \mapsto \hat{\mathsf{F}}(x, f).$$

We assume the equi-continuity of the families of functions

$$\left\{\hat{\mathsf{F}}^s(\cdot, f), f \in E^s\right\}, \quad \left\{\hat{\mathsf{F}}^l(\cdot, f), f \in E^l\right\}, \quad \left\{\hat{\mathsf{F}}^c(\cdot, f), f \in E^c\right\}$$

that is, we assume that (the superscript α can assume the values l, s, c)

$$(\forall \varepsilon > 0)\, (\exists \eta > 0)\, (\forall x_0)\, (\forall x)\, (\forall f \in E^\alpha)$$
$$\left(\|x - x_0\| < \eta \Rightarrow \left\|\hat{\mathsf{F}}^\alpha(x_0, f) - \hat{\mathsf{F}}^\alpha(x, f)\right\| < \varepsilon\right). \quad (122)$$

3. Let $\mathcal{S}_F$ be a given face of an admissible surface $\mathcal{S}$, let u be a unit vector nowhere normal to $\mathcal{S}_F$. We assume that the function which maps the variables x into the vector

$$\hat{\mathsf{F}}^l(x, Cut\left(\left[(x, \mathcal{S}_F)\right], u\right) \quad (123)$$

is a continuous function. In formulas

$$\left(\begin{array}{c}(\forall\varepsilon>0)\,(\forall x_0\in\mathcal{S}_F)\,(\exists\delta>0)\,(\forall x\in\mathcal{S}_F)\\ \|x-x_0\|<\delta\Rightarrow\\ \Rightarrow\left\|\hat{\mathsf{F}}^l(x,Cut\left(\left[(x,\mathcal{S}_F)\right],u\right)-\hat{\mathsf{F}}^l(x_0,Cut\left(\left[(x_0,\mathcal{S}_F)\right],u\right)\right\|<\varepsilon\end{array}\right) \tag{124}$$

4. Let $\mathcal{S}_F$ be a given face of an admissible surface $\mathcal{S}$, let $u_1,...,u_p$ be unit vectors nowhere normal to $\mathcal{S}_F$. Let $\mathcal{L}$ be a given edge of an admissible surface $\mathcal{S}$, let $u_1,...,u_q$ be unit vectors nowhere normal to $\mathcal{L}$. Let us assume that for $x\in\mathcal{S}_F$ and $y\in\mathcal{L}$, $Cut\left(\left[(x,\mathcal{S}_F)\right],u_1,...,u_p\right)$ and $Cut\left(\left[(y,\mathcal{S})\right],u_1,...,u_q\right)$ are wedge shapes. We assume that the functions

$$F_{\mathcal{S}_F}:(x,u_1,...,u_p)\mapsto\hat{\mathsf{F}}^c(x,Cut\left(\left[(x,\mathcal{S}_F)\right],u_1,...,u_p\right) \tag{125}$$

$$F_{\mathcal{L}}:(y,u_1,...,u_q)\mapsto\hat{\mathsf{F}}^c(y,Cut\left(\left[(y,\mathcal{S})\right],u_1,...,u_q\right) \tag{126}$$

are continuous functions respectively with respect to the variables x or y.

Proposition 10.30. *Let $\mathcal{S}_F$ and $\mathcal{L}$ be the intersection with open subsets of a face and an edge included in an admissible surface $\mathcal{S}$, let u be a unit vector nowhere normal to $\mathcal{S}_F$ and let B be a compact set. Let us consider the following functions, determined by the generalized stress state $\hat{\mathsf{F}}$:*

$$F^s:(x,y)\in B\times(B\cap\mathcal{S}_F)\mapsto\hat{\mathsf{F}}^s\left(x,\left[(y,\mathcal{S})\right]\right) \tag{127}$$

$$F^l:(x,y)\in B\times(B\cap\mathcal{L})\mapsto\hat{\mathsf{F}}^l\left(x,\left[(y,\mathcal{S})\right]\right) \tag{128}$$

$$F^l_{Cut}:(x,y)\in B\times(B\cap\mathcal{S}_F)\mapsto\hat{\mathsf{F}}^l\left(x,Cut\left(\left[(y,\mathcal{S})\right],u\right)\right). \tag{129}$$

$$F^c_{Cut,\mathcal{S}}:(x,y)\in B\times(B\cap\mathcal{S}_F)\mapsto\hat{\mathsf{F}}^c(x,Cut\left(\left[(x,\mathcal{S}_F)\right],u_1,...,u_p\right) \tag{130}$$

$$F^c_{Cut,\mathcal{L}}:(x,y)\in B\times(B\cap\mathcal{L})\mapsto\hat{\mathsf{F}}^c(y,Cut\left(\left[(y,\mathcal{S})\right],u_1,...,u_q\right) \tag{131}$$

The regularity assumptions which we have accepted imply that F^s, F^l and F^l_{Cut} are uniformly continuous.

Proof. To prove this proposition for the function 127 let us consider $(x,y)\in B\times(B\cap\mathcal{S}_F)$ and $(\bar{x},\bar{y})\in B\times(B\cap\mathcal{S}_F)$, the vector $u=x-y$,the point $\bar{y}'=\bar{y}+u$ and the translated contact surface $\mathcal{S}':=t_u(\mathcal{S})$. Obviously as $(y,\bar{y})\in\mathcal{S}_F\times\mathcal{S}_F$ then $x=y+u\in\mathcal{S}'$, $\bar{y}'=\bar{y}+u\in\mathcal{S}'$, $[(\bar{y},\mathcal{S})]=[(\bar{y}',\mathcal{S}')]$ and

$[(y, \mathcal{S})] = [(x, \mathcal{S}')]$. It is easy to verify that the following chain of equalities and inequalities hold

$$\left\| \hat{\mathsf{F}}^s\left(x, [(y, \mathcal{S})]\right) - \hat{\mathsf{F}}^s\left(\bar{x}, [(\bar{y}, \mathcal{S})]\right) \right\| = \tag{132}$$

$$= \left\| \hat{\mathsf{F}}^s\left(x, [(x, \mathcal{S}')]\right) - \hat{\mathsf{F}}^s\left(\bar{x}, [(\bar{y}', \mathcal{S}')]\right) \right\| \leq \tag{133}$$

$$\leq \left\| \hat{\mathsf{F}}^s\left(x, [(x, \mathcal{S}')]\right) - \hat{\mathsf{F}}^s\left(\bar{y}', [(\bar{y}', \mathcal{S}')]\right) \right\| + \tag{134}$$

$$+ \left\| \hat{\mathsf{F}}^s\left(\bar{y}', [(\bar{y}', \mathcal{S}')]\right) - \hat{\mathsf{F}}^s\left(\bar{x}, [(\bar{y}', \mathcal{S}')]\right) \right\| . \tag{135}$$

Finally one gets the required uniform continuity by finding separately the upper bounds for the addends 134 and 135. The first of these upper bounds is easily found by recalling the assumed continuity (numbered item 1. in the regularity assumptions) which becomes uniform continuity as the set B is assumed to be compact. The second upper bound is estimated by using the equi-uniform continuity (numbered item 2.) when recalling that

$$\|\bar{y}' - \bar{x}\| = \|\bar{y} + u - \bar{x}\| \leq \|y - x\| + \|\bar{y} - \bar{x}\| .$$

With similar arguments we can prove the statement for the functions 128, 129, 130 and.131. □

Proposition 10.31. . *Let us consider a family of admissible contact surfaces* $\{\mathcal{S}_\lambda, \lambda \in I\}$ *where* I *is a suitable set of indices and the set of shapes*

$$\Phi_I := \{f \in \Phi / \left(\exists \lambda \in I\right) \left(f \in \Phi(\mathcal{S}_\lambda)\right)\} . \tag{136}$$

Let us assume that Φ_I *is a set of prescribed shapes or prescribed plane cuts. Then the regularity hypotheses 1., 2. and 3. imply the uniform boundedness of the families of functions (where* k *is the needed tensoriality order)*

$$\left\{ \hat{\mathsf{F}}(\cdot, f) : x \in \mathcal{C} \mapsto \hat{\mathsf{F}}(x, f) \in \underbrace{\mathcal{V} \otimes ... \otimes \mathcal{V}}_{k\ times}, \quad f \in \Phi_I \right\} . \tag{137}$$

Proof. Indeed, Proposition 10.30 implies that, for every admissible surface $\mathcal{S}_\lambda$, the functions 127,128,129,130 and 131 (in the definitions replace the faces, edges and wedges of $\mathcal{S}$ with those of $\mathcal{S}_\lambda$) are bounded in their domains of definitions. The proposition is then proven simply recalling the definitions of prescribed shapes and prescribed plane cuts. □

Remark 10.32. Proposition 10.31 states a kind of relative compactness of the families 137 (see the Corollary to the Arzelà-Ascoli theorem in Abraham, Marsden and Ratiu (1988)).

Remark 10.33. Our hypotheses allow for any continuous dependence of contact $k-$forces densities F on the curvature tensor, edge curvature or torsion or on any other higher order shape operator of the contact surface, at any surface, edge or wedge point.

11 A Seeming Impossibility For Edge Forces

This section is very close to section 3. of the paper dell'Isola and Seppecher (1997). It is aimed to persuade the reader of the validity of an important statement:

Not all types of Contact Actions are physically admissible.

Indeed we will start by proving that

If a stress state is constituted only by contact surface and line forces then it is physically admissible only when contact line forces are always vanishing.

We must, obviously, specify what we mean with the expression: **physically admissible.**

We will say that a stress state is physically admissible when it verifies the regularity assumptions **10.1** and quasi-balance of contact powers.

Quasi-balance of powers expended by contact actions is a condition which has solid physical grounds: it implies stringent restrictions upon the dependence of contact actions on shape It is a generalization of the hypothesis of 'quasi-balance of contact forces' formulated by Noll and Virga (1990). This last states that

Hypothesis of quasi-balance of contact forces

There exists a positive scalar K such that, for any admissible domain V, the following inequality holds

$$\|F^c(V)\| < K\,|V|\,. \tag{138}$$

In (**?**) it is also assumed the following assumptions

Contact actions reduce to surface and edge forces

In formulas:

$$F^c(V) = \int_{\mathcal{S}^r} \hat{\mathsf{F}}_0^s(x, [(x, \mathcal{S})])\, ds + \int_{\mathcal{L}^r} \hat{\mathsf{F}}_0^l(x, [(x, \mathcal{S})])\, dl. \tag{139}$$

Using the language which we have introduced in this work: in the paper Noll and Virga (1990) it is assumed that

i) the stress state is e completely specified by the two functions

$$\hat{\mathsf{F}}_0^s, \quad \hat{\mathsf{F}}_0^l \tag{140}$$

ii) the only physically allowed stress states are those for which condition 138 holds.

In this section we prove that by means of its theoretical framework actually the aforementioned paper does not manage to abandon the Cauchy format of continuum mechanics.

Instead: to give a framework where the searched generalization can be obtained we must withdraw from the format of mechanics in which the most fundamental concept is that of "force" (so called Newtonian Axiomatics) and we must embrace the point of view of D'Alembert whose Axiomatics is based on the fundamental concept of power and who considers the concept of force as a derived one.

By following the beautiful presentation of continuum mechanics due to Salençon (1988-2005) we improve the preceding hypothesis 138, assuming that

Hypothesis of quasi-balance of power expended by contact actions

The power $\mathcal{P}_U^c$ of contact actions distribution expended on any C^∞ velocity field U is quasi-balanced. This means that, for any C^∞ field U, we assume the existence of a positive K_U such that, for any admissible domain V , the following inequality holds

$$|\mathcal{P}_U^c| < K_U\, |V| \tag{141}$$

Remark 11.1. Everybody who wants to develop an axiomatization of mechanics must be willing to use power at least as a derived concept. Indeed nobody is able to refrain from the use of the concept of power or may doubt about the validity of equation 141. The most faithful followers of so called Newtonian approach may be willing to deduce (!) 141 from a most fundamental (!) assumption but nobody can dare to state that it should not be accepted as valid.

Remark 11.2. The condition expressed by (138) is implied by the one expressed by (142). This can be simply verified by considering that, when considering in it three linearly independent constant fields, formula (142) reduces to (138).

Once we accept the hypothesis 140 about contact actions then the hypothesis 141 reads:

Quasi balance of Powers for Contact Actions reducing to Surface and Line Forces

$$\left|\mathcal{P}_U^c(V)\right| = \left| \begin{array}{c} \int_{\mathcal{S}^r} \hat{\mathsf{F}}_0^s(x,[(x,\mathcal{S})]).U(x)\,ds \\ +\int_{L^r} \hat{\mathsf{F}}_0^l(x,[(x,\mathcal{S})]).U(x)\,dl \end{array} \right| < K_U\,|V|\,. \tag{142}$$

for every field U, ($v.w$ denotes the inner product between the vectors v and w).

Remark 11.3. The dependence of K_U on U will be immaterial in what follows, as in our arguments the field U is always kept fixed.

In this section we want to show that 142 implies that on every edge $\mathcal{L}^r$ of a contact surface $\mathcal{S}$

$$\hat{\mathsf{F}}_0^l(x,[(x,\mathcal{S})]) = 0.$$

This means that actually by assuming 142 one does not obtain any generalization of Cauchy format of continuum mechanics.

In the language which we have introduced in this work one can state that: Quasi-balance of contact power and the assumption 140 about stress state implies that no contact edge forces are possible.

Remark 11.4. The quasi-balance of moment of forces can be obtained from condition 142 by considering three independent spins.

As hypothesis (142) is stronger than (138) it will imply more stringent restrictions upon the dependence of contact actions on shape.

Our goal now is to study its consequences on the functions $\hat{\mathsf{F}}$. We begin by considering edges whose shape is dihedral; we will then extend our results to general edges.

11.1 Condition 142 implies that Forces on Dihedral Edges must vanish.

From the point of view of the logical flow of the presentation, *stricto sensu,* the reader can simply jump this section. However the rigorous argument which is presented here gives a more intuitive ground to the following Theorem 11.7.

In this subsection we see how quasi balance of power expended by contact surface and line forces implies that line forces must vanish.

Theorem 11.5. *For stess states of the form 140, and verifying the regularity assumption 10.1 inequality (142) is incompatible with nonzero contact line forces on dihedral edges: indeed (142) implies that for any dihedral shape d, the following equality holds*

$$\hat{\mathsf{F}}_0^l(\cdot, d) = 0. \tag{143}$$

Proof. Let $d = (n_1,n_2,\tau)$.

We use the orthogonal coordinate system $(x_0;e_1,e_2,e_3)$, with

$$e_2 = \tau, \quad e_3 = \frac{n_1 + n_2}{\|n_1 + n_2\|} \tag{144}$$

We consider a family of domains, parametrized by the set of integers greater than a positive N. The general element V_N (the boundary of which we denote by S_N) of this family is a thin slab with a grooved surface (see Figure 1 in dell'Isola and Seppecher (1997)). This domain is a polyhedron conceived in such a way that the set of shapes of its boundary is finite and is independent of N (This set contains exactly 7 different plane shapes and 16 different dihedral shapes.) Its volume $|V_N|$ is of the same order as N^{-4} the total area of its boundary $|S_N|$ is of the same order as N^{-2}, when N tends to infinity. Let us define the following unions of edges:

$$L_N^1 = \{x \in S_N : [(x, S)] = (n_1, n_2, \tau)\} \tag{145}$$

$$L_N^2 = \{x \in S_N : [(x, S)] = (n_2, n_1, \tau)\} \tag{146}$$

and

$$L_N^3 = L_N \diagdown \left(L_N^1 \cup L_N^2\right). \tag{147}$$

The total length of L_N^3 is of the same order as N^{-1} and the total lengths of L_N^1 and L_N^2 tend to 1. Then the forces on the double array of edges $L_N^1 \cup L_N^2$ are dominant. As the shapes of V_N are prescribed, contact force densities are bounded independently of N. Inequality (138) applied to V_N implies

$$\lim_{N\longrightarrow\infty}\left\{\int_{L_N^1}\hat{\mathsf{F}}_0^l\left(x,(n_1,n_2,\tau)\right)\,dl+\int_{L_N^2}\hat{\mathsf{F}}_0^l\left(x,(n_2,n_1,\tau)\right)\,dl\right\}=0.$$

Using the mean value theorem for each component of the last equality and again the continuity of $\mathcal{F}$ with respect to x we get an action-reaction principle

$$\hat{\mathsf{F}}_0^l\left(x_0,(n_1,n_2,\tau)\right)=-\hat{\mathsf{F}}_0^l\left(x_0,(n_2,n_1,\tau)\right). \tag{148}$$

Consider the field

$$U:x\mapsto(x.e_3)U_0,$$

U_0 being a fixed vector. On V_N, N^2U is bounded independently of N. The same reasoning as before shows that inequality (142) implies

$$\lim_{N\longrightarrow\infty}N^2\left\{\int_{L_N^1}\hat{\mathsf{F}}_0^l\left(x,(n_1,n_2,\tau)\right).U\left(x\right)\,dl\right.$$
$$\left.+\int_{L_N^2}\hat{\mathsf{F}}_0^l\left(x,(n_2,n_1,\tau)\right).U\left(x\right)\,dl=0.\right. \tag{149}$$

On L_N^1 and L_N^2, N^2U does not depend either on N or on x as it is equal respectively to U_0 and $2U_0$. Then we obtain, because of arbitrariness of U_0,

$$\lim_{N\longrightarrow\infty}\left\{\int_{L_N^1}\hat{\mathsf{F}}_0^l\left(x,(n_1,n_2,\tau)\right)\,dl+2\int_{L_N^2}\hat{\mathsf{F}}_0^l\left(x,(n_2,n_1,\tau)\right)\,dl\right\}=0.$$

Using the continuity of $\hat{\mathsf{F}}_0^l$ with respect to x, the mean value theorem for each component of the previous equality and Equation (148) we get

$$\hat{\mathsf{F}}_0^l\left(x_0,(n_1,n_2,\tau)\right)=0. \tag{150}$$

□

Remark 11.6. This proof is not the simplest one can conceive (see the proof of the following Theorem 11.7). However, we present it here because it is suggestive: our construction shows that a limit of pairs of opposite edge forces cannot be quasi-balanced. This means that in generalized continua contact actions can include nonvanishing edge forces only if they also include nonvanishing surface 2−forces. More precisely: the theorem we have just proven shows how a sequence of systems of edge forces can converge to a surface double force: we have thus rigorously illustrated the terms 'double forces' and 'double normal traction' introduced by Germain (1973).

11.2 Condition 142 implies that Forces on General Edges must vanish.

The Theorem proven in the previous subsection can be actually generalized to any edge of a contact surface.

Theorem 11.7. *Let $\mathcal{S}$ be an admissible surface. For stess states of the form 140, and verifying the regularity assumptions 10.1, inequality (142) imply that, at every regular point x_0 of an edge, we have*

$$\hat{\mathsf{F}}_0^l\left(x_0,[(x_0,\mathcal{S})]\right)=0. \tag{151}$$

In other words: inequality (142) implies that Cauchy stress states including only surface contact forces cannot be generalized simply by adding contact edge forces.

Proof. Let V be an admissible domain. Let its boundary $\mathcal{S}$ contain an edge $\mathcal{L}$, let x_0 be a regular point of this edge. $\mathcal{S}$ is tangent at the point x_0 to the dihedral shape (n_1,n_2,τ). In this proof we consider the case when the dihedral angle belongs to $]0,\pi[$ (The proof has to be slightly modified if the angle is greater than π). We use the coordinate system $(x_0;e_1,e_2,e_3)$ with

$$e_2=\tau,\quad e_3=n_1+n_2/\left\|n_1+n_2\right\|.$$

For any $\varepsilon>0$, let us translate V and its contact surface relatively to the vector $\varepsilon^2 e_3$:

$$V':=t_{\varepsilon^2 e_3}(V),\quad \mathcal{S}':=t_{\varepsilon^2 e_3}(\mathcal{S}),\quad \mathcal{L}'=t_{\varepsilon^2 e_3}(\mathcal{L}) \tag{152}$$

and let us define V_ε as the intersection of the domain V' with the parallelepiped (represented in the introduced coordinate system)

$$P_\varepsilon=\left[-c\varepsilon^2,c\varepsilon^2\right]\times[0,\ell\varepsilon]\times\left[0,2\varepsilon^2\right] \tag{153}$$

(see Figure 2 in dell'Isola and Seppecher (1997)). The dihedral angle belongs to $]0,\pi[$ and the curvatures of the faces of $\mathcal{S}$ and of the edge $\mathcal{L}$ are bounded. Then c and ℓ may be chosen in such a way that, for ε small enough: i) $\mathcal{L}'$ meets ∂P_ε on the surfaces $\{x.e_2=0\}$ and $\{x.e_2=\ell\,\varepsilon\}$, so that at every point x on $\mathcal{L}'\cap P_\varepsilon$, we have $x.e_3>0$ ii) $\mathcal{S}'$ meets ∂P_ε on surfaces $\{x.e_2=0\}$,$\{x.e_2=\ell\,\varepsilon\}$ and on the surface $\{x.e_3=0\}$. The geometry of considered construction is illustrated by Figure 2 in dell'Isola and Seppecher (1997). We denote by $\mathcal{S}_\varepsilon$ the boundary of V_ε and by $\mathcal{L}_\varepsilon$ the upper edge of V_ε : in formulas

$$\mathcal{L}_\varepsilon=\mathcal{L}'\cap P_\varepsilon. \tag{154}$$

The family of shapes $\bigcup\limits_{\varepsilon>0} \Phi(\mathcal{S}_\varepsilon)$ is a set of prescribed shapes or prescribed plane cuts. Then the surface force and the line force densities

$$\hat{\mathsf{F}}_0^l\left(x_0, [(x_0, \mathcal{S}_\varepsilon)]\right) \quad \hat{\mathsf{F}}_0^s\left(x_0, [(x_0, \mathcal{S}_\varepsilon)]\right) \tag{155}$$

are uniformly bounded with respect to ε. Let us consider the vector field (this vector field will be kept fixed in our limit process)

$$U : x \mapsto (x.e_3)U_0 \tag{156}$$

U_0 being a generic vector. The geometry of V_ε assures that $\varepsilon^{-2}U$ is bounded independently of ε. On the other hand, as U vanishes on the plane (x_0, e_1, e_2), we do not have to consider the forces exerted on the edges which are included in this plane. Considering the measure of each face and edge, we get from inequality (142)

$$\lim_{\varepsilon\longrightarrow 0} \varepsilon^{-3} \int_{\mathcal{L}_\varepsilon} \hat{\mathsf{F}}_0^l\left(x, [(x, \mathcal{S}_\varepsilon)]\right) .U\left(x\right) dl = 0.$$

The length of $\mathcal{L}_\varepsilon$ is equal to $\varepsilon\ell$ within higher order terms. On the other hand, there exists a positive scalar k (depending on the curvature of the edge at x_0 and on ℓ but independent of ε) such that

$$\lim_{\varepsilon\longrightarrow 0} \varepsilon^{-3} \int_{\mathcal{L}_\varepsilon} (x.e_3)\, dl = k.$$

Let $\delta > 0$, the geometry of the domain and Proposition 10.30 – which is a consequence of our regularity assumptions – imply that, for ε small enough,

$$\forall x \in \mathcal{L}_\varepsilon, \left\| \hat{\mathsf{F}}_0^l\left(x, [(x, \mathcal{S}_\varepsilon)]\right) - \hat{\mathsf{F}}_0^l\left(x_0, [(x_0, \mathcal{S})]\right) \right\| < \delta. \tag{157}$$

Then

$$\left| k\hat{\mathsf{F}}_0^l\left(x_0, [(x_0, \mathcal{S})]\right) .U_0 - \lim_{\varepsilon\longrightarrow 0} \varepsilon^{-3} \int_{L_\varepsilon} \hat{\mathsf{F}}_0^l\left(x, [(x, \mathcal{S}_\varepsilon)]\right) .U\left(x\right) dl \right| < k\delta \left\|U_0\right\|. \tag{158}$$

This result holds for any δ and for any U_0, so that we can conclude with the searched equality

$$\hat{\mathsf{F}}_0^l\left(x_0, [(x_0, \mathcal{S})]\right) = 0.$$

□

12 Generalized Noll Theorem

Theorem 11.7 states that when stress state is of the form specified as in 140 and if one accepts the quasi-balance of contact powers 142 then there are no contact edge forces.

In this section we prove a generalization of Noll theorem for such stress states.

Remark 12.1. In Noll (1959) it is proven (we use the language introduced in the present work) that given a body i) occupying in a configuration the compact region D, ii) for which the stress state is characterized by the only function $\hat{\mathsf{F}}_0^s$, and iii) such that $\left\|\hat{\mathsf{F}}_0^s\right\|$ is bounded in the set $D \times \Phi^r$ (i.e. if contact surface forces are uniformly bounded in the set Φ^r of all possible regular shapes of contact surfaces!!) then for every regular shape f tangent to the plane π in the point x the following equality holds

$$\hat{\mathsf{F}}_0^s(x, f) = \hat{\mathsf{F}}_0^s(x, [(x, \pi)]) . \tag{159}$$

Remark 12.2. The cited result by Noll cannot be applied in those circumstances in which one wants, e.g., to allow for a linear dependence of contact surface forces on local curvature of contact surface. In Noll and Virga (1990) it was announced for a future paper a generalization of Noll's result by relaxing the previously considered condition iii). We could not find such a paper in the literature.

Remark 12.3. The regularity assumptions for stress state proposed in dell'Isola and Seppecher (1997) and generalized in the present work actually relaxed the aforementioned condition iii) and allow for a more general dependence of stress state functions on shapes. In particular our assumptions allow for a linear dependence of surface contact forces on curvature tensor of contact surface.

Remark 12.4. In dell'Isola and Seppecher (1997) it is proved a generalization of Noll's result under the weaker regularity assumptions **10.1** on which also the present work is based.

We give now a proof of property 159 when stress state is characterized only by a surface contact force function $\hat{\mathsf{F}}_0^s$ which, however, is not bounded in the set of all regular shapes.

Theorem 12.5. *When no edge forces are present, and under the regularity assumptions* ***10.1****, for all regular shapes f tangent to the plane shape n the validity of quasi-balance of power 142 implies*

$$\hat{\mathsf{F}}_0^s(x, f) = \hat{\mathsf{F}}_0^s(x, n) . \tag{160}$$

Proof. Let S be the boundary of an admissible domain V and let x_0 be a regular point of S^r. We call n_0 the normal to S at x_0. We use the coordinate system (x_0, e_1, e_2, e_3) (with $e_3 = n_0$). Let us consider the family of parallelepipeds

$$C_\varepsilon = [0,\varepsilon] \times [0,\varepsilon] \times [-c\varepsilon^2, c\varepsilon^2].$$

Let us define the sets

$$V_\varepsilon := V \cap C_\varepsilon, \quad S_\varepsilon = S \cap C_\varepsilon.$$

As the curvature of S is bounded in a neighborhood of x_0, a positive scalar c can be found such that, for ε sufficiently small, S_ε does not intersect one of the faces

$$S^+ := \left\{x_3 = c\varepsilon^2\right\}, \quad S^- = \left\{x_3 = -c\varepsilon^2\right\} \tag{161}$$

of C_ε (see Figure 3 in dell'Isola and Seppecher (1997)). The shapes of the boundary of V_ε are either prescribed shapes or prescribed plane cuts. Inequality (138) when applied to C_ε implies

$$\lim_{\varepsilon \longrightarrow 0} \varepsilon^{-2} \int_{S^+} \hat{\mathsf{F}}_0^s\left(x, n_0\right)\, ds + \lim_{\varepsilon \longrightarrow 0} \varepsilon^{-2} \int_{S^-} \hat{\mathsf{F}}_0^s\left(x, -n_0\right)\, ds = 0,$$

which leads to

$$\hat{\mathsf{F}}_0^s\left(x, n_0\right) = -\hat{\mathsf{F}}_0^s\left(x, -n_0\right). \tag{162}$$

Inequality (138) when applied to V_ε implies

$$\lim_{\varepsilon \longrightarrow 0} \varepsilon^{-2} \int_{S_\varepsilon} \hat{\mathsf{F}}_0^s\left(x, [(x,S)]\right)\, ds + \lim_{\varepsilon \longrightarrow 0} \varepsilon^{-2} \int_{S^-} \hat{\mathsf{F}}_0^s\left(x, -n_0\right)\, ds = 0,$$

which leads to

$$\hat{\mathsf{F}}_0^s\left(x, [(x,S)]\right) + \hat{\mathsf{F}}_0^s\left(x, -n_0\right) = 0. \tag{163}$$

Equation (160) is then obtained when recalling (162). □

Remark 12.6. In the previous proof we just had to modify the argument used in Truesdell (1977) by using a cylinder whose basis is a square instead of a circle. The important difference is that -in this way- we use only prescribed shapes (in the sense defined in ??).

12.1 CAUCHY THEOREM

Stress states verifying the hypotheses of Theorem 11.7 cannot include nonvanishing contact edge forces. Then the Cauchy's construction of stress tensor, which is strongly depending on the absence of edge contact forces, (refer to Truesdell (1977)) is valid.

Theorem 12.7. *When stress state is of the form specified in 140 and if one accepts regularity assumptions* ***10.1 and*** *the quasi-balance of contact powers 142 there exists a continuous tensor field T of order two and symmetric such that, for any plane shape n,*

$$\hat{\mathsf{F}}_0^s(x,n) = T(x).n. \tag{164}$$

13 The structure imposed on stress states by the regularity assumptions 10.1 and quasi-balance of contact power

In this section we show how to generalize the results found in dell'Isola and Seppecher (1995)-(1997) to $N-th$ gradient continua. The complete proof strategy is simply drafted: all the details of the mathematical concepts and proofs will be presented in a forthcoming separate paper.

We start with a "rough" statement which will be followed immediately by a more formal one.

Balance of power implies that the stress state in a $n-th$ order generalized continuum B must verify the following conditions

1. The existence of $k-$forces concentrated on wedges implies the existence of line $k+1-$forces distributed on edges
2. The existence of lines $k+1-$forces distributed on edges implies the existence of surfaces $k+2-$forces distributed on faces
3. The existence of surfaces $k+2-$forces distributed on faces implies the existence of $n=k+3-$stresses in the volume occupied by the body.

More formally after the following

Remark 13.1. When considering a stress state, instead of using the previously introduced notation

$$\left\{\mathsf{F}_k^s, \mathsf{F}_k^l, \mathsf{F}_k^c; k=0,...,L\right\}, \tag{165}$$

we will use another one slightly different. Indeed: instead of the superscripts ${}^{s},{}^{l},{}^{c}$ we want to use the superscripts ${}^{2},{}^{1},{}^{0}$.

This change of notation will be useful in the formulation of the following definitions and theorem.

Definition 13.2. Let B a body occupying in one of its configurations the domain D. Let us consider a stress state $\mathfrak{S}$ of the form

$$\mathfrak{S} = \left\{ \mathsf{F}_k^2, \mathsf{F}_k^1, \mathsf{F}_k^0; k = 0, ..., L \right\}. \tag{166}$$

where all F are defined in D. Let $\mathcal{S}$ be the contact surface of D, and let $\mathcal{S}_r$, $\mathcal{L}_r$ and $\mathcal{W}_S$ denote the subsets of $\mathcal{S}$ constituted by regular surface and edge points and wedges respectively. The power expended on the velocity field U by contact actions exerted on D through $\mathcal{S}$ (when B is in the stress state $\mathfrak{S}$) is given by the following formula:

$$\mathcal{P}_U(\mathfrak{S}, D) := \left(\int_{\mathcal{S}_r} \mathsf{F}_0^2 \cdot U + \int_{\mathcal{L}_r} \mathsf{F}_0^1 \cdot U + \int_{\mathcal{W}_S} \mathsf{F}_0^0 \cdot U \right) + \tag{167}$$

$$+ \sum_{k=1}^{L} \left(\int_{\mathcal{S}_r} \mathsf{F}_k^2 \cdot \left(\nabla^k U\right)_\perp + \int_{\mathcal{L}_r} \mathsf{F}_k^1 \cdot \left(\nabla^k U\right)_\perp + \int_{\mathcal{W}_S} \mathsf{F}_k^0 \cdot \nabla^k U \right) \tag{168}$$

When this will not lead to confusion we will skip the argument $\mathfrak{S}$ in the previous expression.

Definition 13.3. Let us consider a stress state $\mathfrak{S}$.We say that it is **physically admissible** when it verifies i) the regularity assumptions 10.1 and ii) the quasi-balance of power

$$\left| \mathcal{P}_U \left(\mathfrak{S}, D \right) \right| \leq K_U \left| D \right|. \tag{169}$$

Theorem 13.4. *Let us consider a stress state $\mathfrak{S}$ of the form*

$$\mathfrak{S} = \left\{ \mathsf{F}_k^2, \mathsf{F}_k^1, \mathsf{F}_k^0; k = 0, ..., L \right\}. \tag{170}$$

If $\mathfrak{S}$ is physically admissible then there exists a (unique) natural number $N - 1 \leq L$ such that

$$\mathsf{F}_{N-1}^2 \neq 0 \tag{171}$$

$$(\forall k \in \{0, ..., L\}) \, (\forall a \in \{0, 1, 2\}) \, ((k + 2 \geq N + a) \Rightarrow (\mathsf{F}_k^a = 0)) \tag{172}$$

Proof. It can be sketched by paralleling the previous proof of Theorem 11.7. □

Definition 13.5. Let $\mathfrak{S}$ be a physically admissible stress state. We call **grade** of $\mathfrak{S}$ the unique natural number $N_\mathfrak{S}$ the existence of which has been proven in the previous theorem.

It can also be proven the following:

Theorem 13.6. *Let $\mathfrak{S}$ be a physically admissible stress state of grade N. If $\mathsf{F}^1_{N-2} = 0$ then $\mathsf{F}^0_{N-3} = 0$.*

Proof. The searched implication is obtained by means of a construction generalizing the one used for Theorem 11.7. □

Definition 13.7. Let B be a body. We say that B **is a** $N - th$ **gradient continuum** if i) the set of the stress states admissible for B is included in the set of physically admissible stresses of grade N, and ii) there exists at least one nonvanishing admissible stress state for B the grade of which is exactly N.

14 The tetrahedron argument applied to N-th gradient continua

In this final section we draft a line of thought which proves how the tetrahedron argument valid for second gradient materials which was presented in dell'Isola and Seppecher (1995),(1997) actually can be extended to encompass all those continua imagined by Green and Rivlin (1964) and described in the previous sections.

14.1 On contact actions including forces of order greater than one.

Theorems 13.4 and 13.6 have shown that not all stress states are physically admissible. In the present subsection we want to establish a characterization theorem, which states how to construct physically admissible stress states by means of suitably introduced (hyper-)stress tensor fields. We start with second gradient continua.

Surface Double Forces

In dell'Isola and Seppecher (1995)-(1997) it was proven the impossibility -under the assumption (142)- of considering stress states including only nonvanishing surface and contact edge forces. The idea developed there was the following: the power of contact actions is actually quasi-balanced but the expression for this power used in inequality (142) is too naive. Indeed the contact actions must be endowed with a more complex structure: to be precise (using the presently used notation) the simplest stress states which can be considered -after those studied by Cauchy- have the following structure (typical of Germain's second gradient continua)

$$\mathfrak{S} = \{\mathsf{F}^2_0, \mathsf{F}^2_1, \mathsf{F}^1_0\}\,. \tag{173}$$

Remark 14.1. From a physical point of view, the need of introducing a contact double force or contact '1-normal force' can be justified as follows: in the balance of energy an additional term must be considered which does not appear in the balance of forces. An alternative approach trying to meet this need is due to Dunn and Serrin (1985), who introduced directly a supplementary flux of energy ('interstitial working'). Our approach, based on the concept of Schwartz distributions, has the following advantages when compared with that found there: (i) it does not assume *a priori* that the extra energetic term is a flux, (ii) it shows the mechanical nature of this term, its linear dependence on the velocity field being a basic assumption, (iii) it naturally yields general and physically meaningful boundary conditions (see e.g. Seppecher (1989)).

Most mechanicians will not be surprised by the introduction of contact double forces distributions, as contact couples are needed already in the standard theories of beams and shells. Another example of '1-normal' contact forces distribution found in the literature (this time for 3-D continua) is given by couple stresses introduced by Cosserat (1908)-(1909). The microscopic meaning of contact forces of order greater than one can be understood by considering the asymptotic limit of non-local short range interactions.

Example 14.2. A system of non-local short range forces converging to a surface contact double force (1-normal distribution). Using the Cartesian coordinates $(x_1; x_2; x_3)$, the domain

$$V = \{x : x_1 < 0\}$$

is in contact with the external world through the plane

$$S = \{x : x_1 = 0\}\,.$$

Assume that the external forces exerted on V have short range $\varepsilon \ll 1$ (compared with some other characteristic length) and that these forces are represented by the volume density

$$f_\varepsilon(x) = f_0^{-\gamma} \varphi\left(\varepsilon^{-1} x_1\right)$$

where φ is a function whose support is a compact set included in $]-\infty, 0[$ and f_0 is a given vector. If $\gamma = 1$ and φ is a non negative function whose integral is equal to 1, the distribution tends, as ε tends to 0, to the vector measure on S having a surface density equal to f_0: this is the classical case of surface force density. However, if $\gamma = 2$ and φ is the derivative of a non negative function whose integral is equal to 1, the distribution tends to a '1-normal' distribution D such that

$$D(U) = \int_S f_0 \cdot \partial U / \partial x_1 ds.$$

This force distribution i) is localized on S, ii) has no influence upon balance of forces, iii) supplies energy in presence of velocity fields in V , even when these fields vanish on S .

The assumption of quasi-balance of contact actions for stress states of the form 173 takes the following particular form:

for all C^∞ field U , there exists a scalar K_U such that, for any admissible domain V having contact surface S with surface and edge regular point sets denoted respectively by S_r and L_r , we have

$$\left| \begin{array}{c} \int_{S^r} \mathsf{F}_1^2(x,[(x,S)]).\frac{\partial U}{\partial n}(x)\,ds + \int_{S^r} \mathsf{F}_0^2(x,[(x,S)]).U(x)\,ds \\ + \int_{L^r} \mathsf{F}_0^1(x,[(x,S)]).U(x)\,dl \end{array} \right| < K_U\,|V|. \tag{174}$$

Remark 14.3. This assumption is less stringent on F_0^1 than the corresponding hypothesis (142). It will imply less stringent restrictions upon edge contact forces. Indeed condition (174) does not imply that edge contact forces must be vanishing.

Remark 14.4. The fact that (174) implies the quasi-balance of forces (138) is still true (it can be again verified by considering three linearly independent constant fields U). Again, we do not need any assumption on the behavior of K_U with respect to U.

Remark 14.5. Note that, in each proof we present, a limit in inequality (174) is calculated with a fixed field U . For this reason we do not need any assumption on the behavior of K_U with respect to U .

Dependence of second order stress functions on the shape of the contact surface. A theorem analogous to Noll Theorem.

We first prove a version of action-reaction principle valid for double forces i.e. for second order stress functions.

The theorem, at this stage, concerns only plane shapes.

Theorem 14.6. *Under regularity assumptions 10.1 and having postulated the quasi-balance of contact actions, at every point* x*, for all plane shape* n *and for all dihedral shape* (n, e_1, e_2) *we have*

$$\mathsf{F}_1^2(x,n) - \mathsf{F}_1^2(x,-n) = 0. \tag{175}$$

$$\begin{aligned} \mathsf{F}_0^1(x_0,(n,-e_1,e_2)) + \mathsf{F}_0^1(x_0,(n,e_1,e_2)) = 0; \\ \mathsf{F}_0^1(x_0,(-n,e_1,e_2)) + \mathsf{F}_0^1(x_0,(n,e_1,e_2)) = 0; \\ \mathsf{F}_0^1(x_0,(n,e_1,-e_2)) + \mathsf{F}_0^1(x_0,(n,e_1,e_2)) = 0 \end{aligned} \tag{176}$$

Proof. Using the coordinate system (x_0, e_1, e_2, e_3) (with $e_3 = n$: the vectors e_i are unitary, but not necessarily orthogonal), let introduce for every point its coordinates (x_1, x_2, x_3) in the introduced system and let us consider the domain

$$C_\varepsilon = [0, \varepsilon^\alpha] \times \left[0, \varepsilon^\beta\right] \times [0, \varepsilon^\gamma]$$

(where α, β and γ are positive) and the vector field

$$\bar{U} : x \mapsto x_3 U_0 \tag{177}$$

where U_0 is a given vector. We consider each face or edge of ∂C_ε on which some addends of LHS in inequality (174) when calculated on the velocity field 177 are nonvanishing i.e.

$$S_\varepsilon^+ := \{x : x_3 = +\varepsilon^\gamma\}, \quad S_\varepsilon^- := \{x : x_3 = 0\} \tag{178}$$

$$\begin{aligned} L_{\alpha 0}^+ &:= [0, \varepsilon^\alpha] \times \{0\} \times \{\varepsilon^\gamma\} & L_{\alpha\beta}^+ &:= [0, \varepsilon^\alpha] \times \left\{\varepsilon^\beta\right\} \times \{\varepsilon^\gamma\} \\ L_{\beta 0}^+ &:= \{0\} \times \left[0, \varepsilon^\beta\right] \times \{\varepsilon^\gamma\} & L_{\beta\alpha}^+ &:= \{\varepsilon^\alpha\} \times \left[0, \varepsilon^\beta\right] \times \{\varepsilon^\gamma\} \end{aligned} \tag{179}$$

$$\begin{aligned} L_{00}^{\rightarrow} &:= \{(0,0)\} \times [0, \varepsilon^\gamma] & L_{0\beta}^{\rightarrow} &:= \left\{\left(0, \varepsilon^\beta\right)\right\} \times [0, \varepsilon^\gamma] \\ L_{\alpha 0}^{\rightarrow} &:= \{(\varepsilon^\alpha, 0)\} \times [0, \varepsilon^\gamma] & L_{\alpha\beta}^{\rightarrow} &:= \left\{\left(\varepsilon^\alpha, \varepsilon^\beta\right)\right\} \times [0, \varepsilon^\gamma] \end{aligned} \tag{180}$$

the area or length of which are respectively given by

$$\mu_2(S_\varepsilon^+) = \mu_2(S_\varepsilon^-) = \varepsilon^{\alpha\beta}; \quad \mu_1(L_{\alpha 0}^+) = \mu_1(L_{\alpha\beta}^+) = \varepsilon^\alpha; \tag{181}$$

$$\mu_1(L_{\beta 0}^+) = \mu_1(L_{\beta\alpha}^+) = \varepsilon^\beta; \tag{182}$$

$$\mu_1(L_{00}^{\rightarrow}) = \mu_1(L_{0\beta}^{\rightarrow}) = \mu_1(L_{\alpha 0}^{\rightarrow}) = \mu_1(L_{\alpha\beta}^{\rightarrow}) = \varepsilon^\gamma. \tag{183}$$

As

$$\left(\forall x \in S_\varepsilon^+\right) (x_3 = \varepsilon^\gamma); \tag{184}$$

$$\left(\forall x \in S_\varepsilon^-\right) (x_3 = 0); \tag{185}$$

$$\left(\forall x \in S_\varepsilon^\pm\right) \left(\partial \bar{U} / \partial n = \pm U_0\right) \tag{186}$$

and $\mu_3(C_\varepsilon) = \varepsilon^\alpha \varepsilon^\beta \varepsilon^\gamma$; the inequality (174) applied to C_ε and the field 177 becomes:

$$\begin{gathered} \left| \int_{S_\varepsilon^+} \mathsf{F}_1^2(x,n).U_0 ds + \int_{S_\varepsilon^-} \mathsf{F}_1^2(x,-n).\left(-U_0\right) ds + \right. \\ + \int_{S_\varepsilon^+} \mathsf{F}_0^2(x,n).\varepsilon^\gamma U_0 ds + \int_{L_\alpha^+} \mathsf{F}_0^1(x,[(x,\partial C_\varepsilon)]).\varepsilon^\gamma U_0 dl + \\ + \int_{L_\beta^+} \mathsf{F}_0^1(x,[(x,\partial C_\varepsilon)]).\varepsilon^\gamma U_0 dl + \\ \left. + \int_{L_\varepsilon^\rightarrow} \mathsf{F}_0^1(x,[(x,\partial C_\varepsilon)]).\left((x-x_0)\cdot n\right) U_0 dl \right| < K_{\bar{U}} \left(\varepsilon^\alpha \varepsilon^\beta \varepsilon^\gamma\right). \end{gathered} \tag{187}$$

where

$$L_\alpha^+ = L_{\alpha 0}^+ \cup L_{\alpha\beta}^+; \quad L_\beta^+ = L_{\beta 0}^+ \cup L_{\beta\alpha}^+; \quad L_\varepsilon^\rightarrow = L_{00}^\rightarrow \cup L_{0\beta}^\rightarrow \cup L_{\alpha 0}^\rightarrow \cup L_{\alpha\beta}^\rightarrow. \tag{188}$$

Remarking that

$$\begin{gathered} x \in L^+ \Rightarrow [(x,S)] = (e_1, n, n\times e_1) \\ \text{or } (e_2, n, n\times e_2) \text{ or } (-e_1, n, -n\times e_1) \text{ or } (-e_2, n, -n\times e_2) \end{gathered}$$

$$\begin{gathered} x \in L_\varepsilon^\rightarrow \Rightarrow [(x,S)] = (-e_2, -e_1, n) \\ \text{or } (e_2, -e_1, n) \text{ or } (e_2, e_1, n) \text{ or } (e_1, -e_2, n) \end{gathered}$$

a simple inspection of previous formulas allows us to recognize that $\Phi(\partial C_\varepsilon)$ is a set of prescribed shapes and therefore that all F are uniformly bounded on the set

$$\bigcup_{\varepsilon>0} \left(C_\varepsilon \times \Phi(\partial C_\varepsilon)\right). \tag{189}$$

As a consequence we can state that there exist suitable constants K such that

$$\left| \int_{S_\varepsilon^\pm} \mathsf{F}_1^2(x,n).U_0 ds \right| < K_1^2 \left\| U_0 \right\| \varepsilon^\alpha \varepsilon^\beta \tag{190}$$

$$\left| \int_{S_\varepsilon^+} \mathsf{F}_0^2(x,n).\varepsilon^\gamma U_0 ds \right| < K_0^2 \left\| U_0 \right\| \varepsilon^\alpha \varepsilon^\beta \varepsilon^\gamma \tag{191}$$

$$\left| \int_{L_\alpha^+} \mathsf{F}_0^1(x,[(x,\partial C_\varepsilon)]).\varepsilon^\gamma U_0 dl \right| < K_0^1 \left\| U_0 \right\| \varepsilon^\alpha \varepsilon^\gamma \tag{192}$$

$$\left| \int_{L_\beta^+} \mathsf{F}_0^1(x,[(x,\partial C_\varepsilon)]).\varepsilon^\gamma U_0 dl \right| < K_0^1 \left\| U_0 \right\| \varepsilon^\beta \varepsilon^\gamma \tag{193}$$

$$\left| \int_{L_\varepsilon^\to} \mathsf{F}_0^1(x,[(x,\partial C_\varepsilon)]).\left((x-x_0)\cdot n\right) U_0 dl \right| < K_0^1 \left\| U_0 \right\| \varepsilon^{2\gamma}. \tag{194}$$

If one chooses $\gamma > \beta$ and $\gamma > \alpha$ the inequality 187 (multiplied times $\varepsilon^{-\alpha}\varepsilon^{-\beta}$) implies that

$$\lim_{\varepsilon\to 0} \varepsilon^{-\alpha}\varepsilon^{-\beta} \left(\int_{S_\varepsilon^+} \mathsf{F}_1^2(x,n).U_0 ds + \int_{S_\varepsilon^-} \mathsf{F}_1^2(x,-n).\left(-U_0\right) ds \right) = 0$$

The continuity properties of the partial function $\mathsf{F}_1^2(\cdot,n)$ and the arbitrariness of U_0 finally imply

$$\mathsf{F}_1^2\left(x_0,n\right) = \mathsf{F}_1^2\left(x_0,-n\right). \tag{195}$$

On the other hand let us multiply 187 times $\varepsilon^{-\beta}\varepsilon^{-\gamma}$, let us choose arbitrarily $\beta > 0$ and consequently α and γ verifying the conditions:

$$\alpha > \gamma \quad \alpha > \beta > 0 \quad (\beta > 0) \wedge \left(\gamma^2 - \beta - \gamma\right) > 0 \Leftrightarrow \gamma > \tfrac{1}{2}\sqrt{4\beta+1} + \tfrac{1}{2}$$

and finally calculate the limit for $\varepsilon \to 0$, of the LHS of obtained inequality. Recalling 188 we get

$$\lim_{\varepsilon\to 0} \varepsilon^{-\beta}\varepsilon^{-\gamma} \left(\int_{L_\beta^+ = L_{\beta 0}^+ \cup L_{\beta\alpha}^+} \mathsf{F}_0^1(x,[(x,\partial C_\varepsilon)]).\varepsilon^\gamma U_0 dl \right) = \tag{196}$$

$$= \lim_{\varepsilon\to 0} \varepsilon^{-\beta} \left(\int_{L_{\beta 0}^+} \mathsf{F}_0^1(x,(n,-e_1,e_2)).U_0 dl + \int_{L_{\beta\alpha}^+} \mathsf{F}_0^1(x,(n,e_1,e_2)).U_0 dl \right) = 0 \tag{197}$$

The continuity properties of the partial function $\mathsf{F}_0^1(\cdot,(u,v,w))$ (where (u,v,w) is a given dihedral shape) and the arbitrariness of U_0 finally imply

$$\mathsf{F}_0^1(x_0,(n,-e_1,e_2)) + \mathsf{F}_0^1(x_0,(n,e_1,e_2)) = 0 \tag{198}$$

The other two relations in 176 are obtained with similar constructions. □

We are now able to prove a theorem for '1-normal' distributions analogous to the theorem of Noll (?), (?).

Theorem 14.7. *Under regularity assumptions 10.1 and having postulated the quasi-balance of contact actions, at every point* x *and for every regular shape* f *tangent to the plane shape* n *we have*

$$\mathsf{F}_1^2\left(x, f\right) = \mathsf{F}_1^2\left(x, n\right) \tag{199}$$

that is F_1^2*, depends on the shape of the contact surface only through its normal.*

Proof. The proof is close to that we have given for the Theorem 12.5. At a regular point x_0 of the boundary S of an admissible domain, we consider the family of domains V_ε described in the proof of the Theorem 12.5 (see also Figure 3 in dell'Isola and Seppecher (1997)). We consider the vector field

$$U : x \longrightarrow (x \cdot n_0)\, U_0$$

where n_0 denotes the normal to S at x_0 and U_0 is any vector. The shapes of ∂V_ε are either prescribed shapes or prescribed plane cuts. Moreover $\varepsilon^{-2}U$ is bounded in V_ε independently of ε. The inequality (174) applied to V_ε leads to

$$\lim_{\varepsilon \longrightarrow 0} \varepsilon^{-2} \int_{S \cap C_\varepsilon} \mathsf{F}_1^2(x, [(x,S)] \cdot U_0 \, (n \cdot n_0) \; ds +$$
$$+ \lim_{\varepsilon \longrightarrow 0} \varepsilon^{-2} \int_{S^-} \mathsf{F}_1^2\left(x, -n_0\right) \cdot \left(-U_0\right) \; ds = 0.$$

As $n \cdot n_0$ is a continuous function with respect to x on S, the regularity properties of F_1^2 and the arbitrariness of U_0 imply

$$\mathsf{F}_1^2(x_0, [(x_0,S)] = \mathsf{F}_1^2\left(x_0, -n_0\right) = \mathsf{F}_1^2\left(x_0, n_0\right). \tag{200}$$

□

14.2 A Representation Theorem for surface 1−forces generalizing the Cauchy tetrahedron Theorem.

The celebrated Cauchy tetrahedron argument can be used also to get a representation Theorem which is valid for $N - th$ gradient continua. Obviously it needs to be suitably adapted.

The reader should not believe that the argument has to be applied, in the framework of the theory of $N-th$ gradient continua, to the surface density of $1-$ force. Instead it has to be applied to surface density of $N-$forces. It happens that $N-$force surface density depends only on the tangent planar shape to the contact surface and this dependence is a polynomial one: its grade is exactly N.

Theorem 14.8. *Let B a body occupying in one of its configurations the domain D. Let us consider a stress state $\mathfrak{S}$ having the form $\{\mathsf{F}_0^2, \mathsf{F}_1^2, \mathsf{F}_0^1\}$. Let us assume the regularity assumptions 10.1. If quasi balance of power holds then there exists a continuous three-tensor field C such that, at any point $x_0 \in D$ and for any plane shape n,*

$$\mathsf{F}_1^2(x_0, n) = (C(x_0).n).n. \tag{201}$$

Proof. We follow the tetrahedron construction due to Cauchy (see Figure 4 in dell'Isola and Seppecher (1997)). In an orthonormal coordinate system (x_0, e_1, e_2, e_3), we define the tetrahedron V whose faces S, S_1, S_2 and S_3 are respectively normal to n,$-e_1$,$-e_2$ and $-e_3$and whose height (perpendicular to S) is h. In our construction the origin of the coordinate system belongs to the face S. We denote respectively by $f_1 = (-e_2, -e_3, e_1)$, $f_2 = (-e_3, -e_1, e_2)$ and $f_3 = (-e_1, -e_2, e_3)$ the shapes of the edges L_1, L_2 and L_3. Let V_ε be the image of V under an homothetic transformation of ratio ε, we denote by S_ε, $S_{i\varepsilon}$ and $L_{i\varepsilon}$, and the faces and edges images of S, S_i and L_i $(i = 1, 2, 3)$ under this transformation. We consider the field $U : x \longrightarrow (x \cdot n)\, U_0$, where U_0 is any vector. As this field vanishes on S_ε, the inequality (174) applied to the domain V_ε implies

$$\begin{array}{r} \left| \displaystyle\sum_{i=1}^{3} \left\{ \int_{S_{i\varepsilon}} (x \cdot n)\, \mathsf{F}_0^2(x, -e_i).U_0 ds \right\} + \right. \\ + \displaystyle\sum_{i=1}^{3} \left\{ \int_{L_{i\varepsilon}} (x \cdot n)\, \mathsf{F}_0^1(x, f_i).U_0 dl \right\} + \\ + \displaystyle\sum_{i=1}^{3} \left\{ \int_{S_{i\varepsilon}} (-e_i \cdot n)\, \mathsf{F}_1^2(x, -e_i).U_0 ds \right\} + \\ \left. + \displaystyle\int_{S_\varepsilon} \mathsf{F}_1^2(x, n).U_0 ds \right| < K\varepsilon^3. \end{array} \tag{202}$$

Let us multiply this inequality by ε^{-2} and, changing variables in the integrals in order to transform them into integrals on the boundary of V, we obtain

$$\left|\varepsilon\sum_{i=1}^{3}\left\{\int_{S_i}(x\cdot n)\,\mathsf{F}_0^2(\varepsilon x,-e_i)\cdot U_0 ds\right\}\right.$$
$$+\sum_{i=1}^{3}\left\{\int_{L_i}(x\cdot n)\,\mathsf{F}_0^1(\varepsilon x,\ f_i)\cdot U_0 dl\right\}$$
$$\left.+\sum_{i=1}^{3}\left\{\int_{S_i}(-e_i\cdot n)\ \mathsf{F}_1^2(\varepsilon x,-e_i)\cdot U_0 ds\right\}+\int_S \mathsf{F}_1^2(\varepsilon x,n)\cdot U_0 ds\right|<K\varepsilon.$$

As the partial functions $\mathsf{F}(\cdot,\ f)$ are continuous, evaluating the limit as ε tends to 0, we get

$$\sum_{i=1}^{3}\left\{\mathsf{F}_0^1(x_0,\ f_i).U_0\int_{L_i}(x\cdot n)\,dl\right\}+ \tag{203}$$

$$+\sum_{i=1}^{3}\left\{|S_i|\,(-e_i\cdot n)\ \mathsf{F}_1^2(x_0,-e_i).U_0\right\}+ \tag{204}$$

$$+|S|\ \mathsf{F}_1^2(x_0,n).U_0=0. \tag{205}$$

This being valid for any U_0, we obtain

$$2\,|S|\ \mathsf{F}_1^2(x_0,n)=\sum_{i=1}^{3}\left\{\mathsf{F}_0^1(x_0,\ f_i)\,(n\cdot e^i)\,|L_i|^2\right\}+ \tag{206}$$

$$+2\sum_{i=1}^{3}\left\{\mathsf{F}_1^2(x_0,-e_i)\,|S_i|\,(n\cdot e_i)\right\}. \tag{207}$$

Using the geometrical relations

$$\begin{gathered} h=|L_1|\,(n\cdot e_1)=|L_2|\,(n\cdot e_2)=|L_3|\,(n\cdot e_3)\,, \\ 2\,|S|\,h=2\,|S_1|\,|L_1|=2\,|S_2|\,|L_2|=2\,|S_3|\,|L_3|=|L_1|\,|L_2|\,|L_3| \end{gathered} \tag{208}$$

and Theorem 14.6, we get

$$\begin{aligned} \mathsf{F}_1^2(x_0,n)&=\mathsf{F}_0^1(x_0,\ f_1)\,(n\cdot e_2)\,(n\cdot e_3)+\mathsf{F}_0^1(x_0,\ f_2)\,(n\cdot e_3)\,(n\cdot e_1)+ \\ &+\mathsf{F}_0^1(x_0,\ f_3)\,(n\cdot e_1)\,(n\cdot e_2)+\sum_{i=1}^{3}\mathsf{F}_1^2(x_0,e_i)\,(n\cdot e_i)^2\,. \end{aligned} \tag{209}$$

Thus we are led to define a three-tensor field C such that

$$\mathsf{F}_1^2(x_0,n)=(C(x_0).n).n=C(x_0)..n\otimes n \tag{210}$$

This tensor is not uniquely determined, as only its right-side products by symmetric two tensors are determined. We may impose its right side symmetry, by setting

$$\begin{aligned} C(x) = {} & \tfrac{1}{2}\mathsf{F}_0^1(x,\ f_1)\otimes(e_2\otimes e_3+e_3\otimes e_2)+ \\ & +\tfrac{1}{2}\mathsf{F}_0^1(x,\ f_2)\otimes(e_3\otimes e_1+e_1\otimes e_3)+ \\ & +\tfrac{1}{2}\mathsf{F}_0^1(x,\ f_3)\otimes(e_1\otimes e_2+e_2\otimes e_1)+ \\ & +\sum_{i=1}^{3}\left\{\mathsf{F}_1^2(x,e_i)\otimes e_i\otimes e_i\right\}, \end{aligned} \tag{211}$$

or its left side symmetry, by setting

$$\begin{aligned} C(x) = {} & \tfrac{1}{2}\mathsf{F}_0^1(x,\ f_1)\otimes(e_2\otimes e_3+e_3\otimes e_2)+ \\ & -\tfrac{1}{2}(e_2\otimes e_3+e_3\otimes e_2)\otimes\mathsf{F}_0^1(x,\ f_1)+ \\ & +\tfrac{1}{2}\left(e_2\otimes\mathsf{F}_0^1(x,\ f_1)\otimes e_3+e_3\otimes\mathsf{F}_0^1(x,\ f_1)\otimes e_2\right)+ \\ & +\tfrac{1}{2}\mathsf{F}_0^1(x,\ f_2)(e_3\otimes e_1+e_1\otimes e_3)+ \\ & -\tfrac{1}{2}(e_3\otimes e_1+e_1\otimes e_3)\otimes\mathsf{F}_0^1(x,\ f_2)+ \\ & +\tfrac{1}{2}\left(e_3\otimes\mathsf{F}_0^1(x,\ f_2)\otimes e_1+e_1\otimes\mathsf{F}_0^1(x,\ f_2)\otimes e_3\right)+ \\ & +\tfrac{1}{2}\mathsf{F}_0^1(x,\ f_3)(e_1\otimes e_2+e_2\otimes e_1)+ \\ & -\tfrac{1}{2}(e_1\otimes e_2+e_2\otimes e_1)\otimes\mathsf{F}_0^1(x,\ f_3)+ \\ & +\tfrac{1}{2}\left(e_1\otimes\mathsf{F}_0^1(x,\ f_3)\otimes e_2+e_2\otimes\mathsf{F}_0^1(x,\ f_3)\otimes e_1\right)+ \\ & +\sum_{i=1}^{3}\left\{\mathsf{F}_1^2(x,e_i)\otimes e_i\otimes e_i-e_i\otimes e_i\otimes\mathsf{F}_1^2(x,e_i)+e_i\otimes\mathsf{F}_1^2(x,e_i)\otimes e_i\right\}. \end{aligned} \tag{212}$$

□

Remark 14.9. The tensor C will be called the double-stress tensor.

Remark 14.10. Imposing the left side symmetry of C (as done in Seppecher (1987)) may seem complicated and artificial but the following Theorem 14.16 will show the advantage of such a choice.

The representation formula for the tensor C obtained in equation 211 allows us to prove also the following

Corollary 14.11. *For a generic dihedral shape (n_1, n_2, τ) the edge contact forces can be represented by*

$$\mathsf{F}_0^1(x,(n_1,n_2,\tau)) = (C(x)\cdot n_1)\cdot\nu_1+(C(x)\cdot n_2)\cdot\nu_2, \tag{213}$$

where

$$\nu_1=\tau\times n_1 \quad \nu_2=-\tau\times n_2. \tag{214}$$

Proof. Let us first consider the dihedral shapes $f_1 = (-e_2, -e_3, e_1)$, $f_2 = (-e_3, -e_1, e_2)$ and $f_3 = (-e_1, -e_2, e_3)$ where $\{e_i\}$ is an orthonormal basis. With simple calculations we get from 211

$$(C(x) . -e_2) . -e_3 = \tfrac{1}{2}\mathsf{F}_0^1(x, f_1) \quad (C(x) . -e_3) . -e_2 = \tfrac{1}{2}\mathsf{F}_0^1(x, f_1) \tag{215}$$

which implies 213 for dihedral shapes verifying the conditions

$$n_1 \cdot n_2 = 0; \quad \tau \cdot n_1 = 0; \quad \tau \cdot n_2 = 0. \tag{216}$$

For getting the statement for a generic dihedral shape it is enough to consider the ε–families of 3–prisms with height parallel to the vector τ the bases of which are triangles with two sides ortogonal to n_1 and n_2 and having a right angle. □

A first important consequence of the representation formula 211 can be obtained by means of a very simple reasoning. Indeed we can get the following lemma, which has been formulated already in dell'Isola and Seppecher (1995)-(1997):

Lemma 14.12. ***The form of surface 2–forces compatible with vanishing line 1–forces.*** *When F_0^1 is vanishing equations (210) and (211) imply that*

$$(C(x_0) .n) .n = \mathsf{F}_1^2(x_0, n) = \tag{217}$$

$$= \sum_{i=1}^{3} \mathsf{F}_1^2(x_0, e_i)(n \cdot e_i)^2 = \tag{218}$$

$$= \sum_{j=1}^{3} \mathsf{F}_1^2(x_0, e_j')(n \cdot e_j')^2 \tag{219}$$

where $\{e_i\}$ and $\{e_j'\}$ are two generic orthonormal bases. As the found expression is valid for every orthonormal vector basis, we can conclude that if F_0^1 is vanishing then F_1^2 does not depend on n: this is the only case in which the 1-normal surface distributions (i.e. surface 2–forces) can be nonzero with vanishing edge forces.

14.3 Representation theorems for contact 1–forces. Cauchy stress tensor.

To proceed we need to apply Gauss divergence Theorem to some tensor fields. Therefore we are guided to add some further regularity assumptions on the considered stress state $\mathfrak{S}$,which for second gradient continua has the form $\{\mathsf{F}_0^2, \mathsf{F}_1^2, \mathsf{F}_0^1\}$.

Although a careful analysis (which will be included in a forthcoming paper) is possible, leading us to the determination of the most adapted functional space to be chosen to "host" considered fields, for seek of simplicity and in order not to distract the reader with presently unessential mathematical formalism, we limit ourselves to the consideration of very regular stress states.

Hypothesis of C^1 regularity.

For every given regular shape f and every dihedral shape d, the partial functions

$$\mathsf{F}_1^2(\cdot, f), \quad \mathsf{F}_0^1(\cdot, d) \tag{220}$$

are C^1 functions.

As an obvious consequence of this assumption one easily gets that the tensor field C defined in 14.8 is also a field of class C^1.

Remark 14.13. We now can show how our hypothesis (174), which implies **but is not at all equivalent to** hypothesis (138) put forward by Noll and Virga (1990), allows us to prove the assumption III on page 21 in just mentioned paper, and to show that the example treated in its Section 9 actually is exhaustive of all possible cases.

Indeed the following two Theorems give the general representation of edge and surface contact forces. We prove them starting from the quasi balance of power and under the regularity assumptions 10.1 to which we add the last assumed hypothesis 14.3.

Theorem 14.14. *Let S be the contact surface of an admissible domain V. Let x be a regular point of an edge of S. Let (n_1, n_2, τ) be the tangent dihedral shape to S at x. Then the edge force density at x depends only on (n_1, n_2, τ) and is represented in terms of the second order stress tensor by*

$$\mathsf{F}_0^1(x, [(x, S)]) = (C(x) \cdot n_1) \cdot \nu_1 + (C(x) \cdot n_2) \cdot \nu_2, \tag{221}$$

where we have introduced the following notations:

$$\nu_1 = \tau \times n_1 \quad \nu_2 = -\tau \times n_2. \tag{222}$$

Remark 14.15. The observed arbitrariness regarding the tensor C has no influence on the representation formula (224) as the "shape" tensor

$$n_1 \otimes \nu_1 + n_2 \otimes \nu_2 \tag{223}$$

at any edge actually is a symmetric tensor.

Theorem 14.16. *At any regular point of the surface S it exists a continuous second order tensor field T such that*

$$\mathsf{F}_0^2(x,[(x,S)]) = T(x).n - \mathrm{div}_s((C(x).n).\Pi), \tag{224}$$

where Π denotes the projector on the tangent plane to the surface S

$$(\Pi = Id - n \otimes n). \tag{225}$$

Remark 14.17. The arbitrariness in C has an influence on T. With the choice (221), the tensor T is symmetric.

Proof. We prove at the same time both Theorems 14.14 and 14.16. Because of Theorem 14.8, inequality (174) may be written

$$\begin{aligned} &\left| \int_{L^r} \mathsf{F}_0^1(x,[(x,S)]).U(x)\,dl + \int_{S^r} \mathsf{F}_0^2(x,[(x,S)]).U(x)\,ds + \right. \\ &\left. + \int_{S^r} ((C(x).n).n).\tfrac{\partial U}{\partial n}(x)\,ds \right| < K_U\,|V|. \end{aligned} \tag{226}$$

Due to the equality

$$\left| \int_V \mathrm{div}(\nabla U..C)\,dv \right| = \left| \int_{S^r} \nabla U..(C.n)\,ds \right|, \tag{227}$$

the quantity

$$\int_{S^r} \nabla U..(C.n)\,ds \tag{228}$$

is quasi-balanced. Then there exists a scalar K'_U such

$$\begin{aligned} &\left| \int_{L^r} \mathsf{F}_0^1(x,[(x,S)]).U(x)\,dl + \int_{S^r} \mathsf{F}_0^2(x,[(x,S)]).U(x)\,ds + \right. \\ &\left. + \int_{S^r} \left\{((C(x).n).n).\tfrac{\partial U}{\partial n}(x) - \nabla U(x)..(C.n)\right\} ds \right| < K'_U\,|V|. \end{aligned} \tag{229}$$

Using now the decomposition

$$\nabla U..(C.n) = \partial U/\partial n.((C.n).n) + \nabla^s U..((C.n).\Pi)$$

where ∇^s denotes the surface gradient on S, we get

$$\begin{aligned} &\left| \int_{L^r} \mathsf{F}_0^1(x,[(x,S)]).U(x)\,dl + \int_{S^r} \mathsf{F}_0^2(x,[(x,S)]).U(x)\,ds + \right. \\ &\left. + \int_{S^r} \nabla^s U(x)..((C(x).n).\Pi)\,ds \right| < K'_U\,|V|. \end{aligned} \tag{230}$$

We apply the divergence theorem on every face of S, so obtaining

$$\left| \int_{L^r} \mathcal{F}'(x,[(x,S)]).U(x)\,dl + \int_{S^r} F'(x,[(x,S)]).U(x)\,ds \right| < K'_U\,|V|, \quad (231)$$

where we used the following definitions $\mathsf{F}_0^2, \mathsf{F}_1^2, \mathsf{F}_0^1$

$$\begin{aligned} \mathcal{F}'(x,[(x,S)]) &:= \mathsf{F}_0^1(x,[(x,S)]) - (C(x)\cdot n_1)\cdot \nu_1 + (C(x)\cdot n_2)\cdot \nu_2, \\ F'(x,[(x,S)]) &:= \mathsf{F}_0^2(x,[(x,S)]) + \nabla^s \cdot ((C(x)\cdot n)\cdot \Pi)\,. \end{aligned}$$

Because of the regularity hypotheses added in this section, F' and $\mathcal{F}'$ represent a stress state verifying the assumptions listed in 11. Because of inequality 231, we may apply to them our results found in 11. Theorem 11.7 implies that $\mathcal{F}'(x,[(x,S)])$ is vanishing and Theorem **??** establishes the existence of a continuous second order tensor field $T(\cdot)$ such that

$$F'(x,[(x,S)]) = T(x).n. \quad (232)$$

□

14.4 Surface N-th order Forces

In this section we indicate how the results obtained in the previous subsection devoted to surface double forces can be further generalized.

Dependence of n-th order stress functions on the shape of the contact surface. A theorem analogous to Noll Theorem.

We start with action-reaction theorem for surface $N-$forces and edge $N-1$ forces in $N-th$ gradient continua.

Theorem 14.18. *Let us consider a stress state $\mathfrak{S}$ of* **grade** *N. At every point x, for all plane shape n, for all dihedral shape (n,e_1,e_2) and for all polihedral shape $(n_1,n_2,....,n_k)$ we have*

$$\mathsf{F}_{N-1}^2(x,n) + (-1)^{N-1}\,\mathsf{F}_{N-1}^2(x_0,-n) = 0 \quad (233)$$

$$\begin{aligned} &\mathsf{F}_{N-2}^1(x,(n,-e_1,e_2)) + (-1)^{N-2}\,\mathsf{F}_{N-2}^1(x,(n,e_1,e_2)) = 0; \\ &\mathsf{F}_{N-2}^1(x,(-n,e_1,e_2)) + (-1)^{N-2}\,\mathsf{F}_{N-2}^1(x,(n,e_1,e_2)) = 0; \\ &\mathsf{F}_{N-2}^1(x,(n,e_1,-e_2)) + (-1)^{N-2}\,\mathsf{F}_{N-2}^1(x,(n,e_1,e_2)) = 0 \end{aligned} \quad (234)$$

$$\mathsf{F}_{N-3}^0\left(x,(-n_1,\widehat{-n_2,....},-n_k)\right) + \mathsf{F}_{N-3}^0\left(x,(n_1,\widehat{n_2,....},n_k)\right) = 0 \quad (235)$$

Proof. The reasoning proceeds exactly as in the case of stress states of grade 2,with a difference: while we still consider the coordinate system (x_0, e_1, e_2, n) and the domain

$$C_\varepsilon = [0,\varepsilon] \times [0,\varepsilon] \times \left[0,\varepsilon^2\right]$$

we use instead the vector field (where U_0 is a given vector)

$$U : x \mapsto (x_3)^{N-1} U_0 \tag{236}$$

which has as $(N-1)-th$ gradient the function

$$\nabla^{N-1} U : x \mapsto (N-1)! U_0 \otimes \underbrace{n ... \otimes n}_{(N-1)-times} =: (N-1)! U_0 \otimes n^{N-1} \tag{237}$$

The shapes of ∂C_ε are prescribed shapes therefore the functions F in the stress state $\mathfrak{S}$ are uniformly bounded in the set

$$\bigcup_{\varepsilon > 0} \left(C_\varepsilon \times \Phi\left(\partial C_\varepsilon\right)\right). \tag{238}$$

Considering the area or length of each face or edge, as in the previous theorem 14.6 the inequality 169 applied to C_ε implies together with the expression 168 (all other terms in 169 in the limit are vanishing faster than ε^2)

$$\lim_{\varepsilon \longrightarrow 0} \varepsilon^{-2} \int_{S_\varepsilon^+} \mathsf{F}_{N-1}^2 (x,n) . U_0 \, ds + \tag{239}$$

$$+ \lim_{\varepsilon \longrightarrow 0} \varepsilon^{-2} \int_{S_\varepsilon^-} (-1)^{N-1} \mathsf{F}_{N-1}^2 (x,-n) . U_0 \, ds = 0, \tag{240}$$

where S_ε^+ and S_ε^- denote the upper and lower faces of C_ε i.e. in formulas

$$S^+ = \left\{x : x \cdot e_3 = +\varepsilon^2\right\}, S^- = \{x : x \cdot e_3 = 0\}$$

The regularity assumptions accepted for F_{N-1}^2 and the arbitrariness of U_0 finally imply

$$\mathsf{F}_{N-1}^2 (x,n) + (-1)^{N-1} \mathsf{F}_{N-1}^2 (x_0,-n) = 0.$$

Repeating then the construction conceived for dihedral shapes in Theorem 14.6 we easily get also 234. Finally in order to get 235 one needs to consider the ε^2-translation of the considered polihedral shape and construct the $\varepsilon-$family of volumes so obtained. □

We are now able to prove that in $N-th$ gradient continua contact surface $N-$ forces and edge $(N-1)$-forces on any couple of contact surface which are tangent assume the same value.

Theorem 14.19. *Let us consider a $N-th$ gradient continuum. Under regularity assumptions 10.1 and having postulated the quasi-balance of contact actions, we have that: i) at every point x and for every regular shape f tangent to the plane shape n we have*

$$\mathsf{F}^2_{N-1}(x,f) = \mathsf{F}^2_{N-1}(x,n) \tag{241}$$

that is F^2_1 at surface regular points depends on the shape of the contact surface only through its normal ii) at every point x and for every regular edge shape f tangent to the dihedral shape (n_1, n_2, e_1) we have

$$\mathsf{F}^1_{N-2}(x,f) = \mathsf{F}^1_{N-2}(x,(n_1,n_2,e_1)). \tag{242}$$

that is F^1_{N-2} at edge points depends on the shape of the contact surface only through the tangent dihedral shape; iii) at every point x and for every wedge shape f tangent to the nondegenerate Polihedral Wedge Shape $(n_1,...,n_k)$ we have

$$\mathsf{F}^0_{N-3}(x,f) = \mathsf{F}^0_{N-3}(x,(\widehat{n_1,...,n_k})). \tag{243}$$

that is F^0_{N-3} at wedge points depends on the shape of the contact surface only through the tangent polihedral shape.

Proof. The proof of 241 is very close to that we have given for the Theorem 12.5. The $\varepsilon-$family to be considered is exactly the same: the only difference to be adopted in the demonstration consists in the need of using as test function in the quasi-balance of power the vector field

$$U : x \longrightarrow (x \cdot n_0)^{N-1} U_0 \tag{244}$$

where n_0 denotes the normal to S at x_0 and U_0 is a generic vector. On the other hand slightly different constructions are needed to prove 242 and 243: these constructions require the $\varepsilon-$translation of a neighborhood of the contact surface keeping fixed the tangent dihedral or polihedral shape. □

14.5 A Representation Theorem for surface $(N-1)$ −forces generalizing the Cauchy tetrahedron Theorem.

Celebrated Cauchy tetrahedron argument deserves to be admired. It allows us to prove also for $N-th$ gradient continua a representation theorem for contact $(N-1)$-forces in terms of a tensor of order $N+1$. Indeed

$N-$force surface density (which has been seen to depend only on the tangent planar shape to the contact surface) can be obtained by calculating a vector-valued polynomial function having grade exactly N.

Theorem 14.20. *Let B a body occupying in one of its configurations the domain D. Let us consider a physically admissible stress state $\mathfrak{S}$ of grade N. Let us assume the regularity assumptions 10.1. If quasi balance of power holds then there exists a continuous tensor field C_N of order $N+1$ such that, at any point $x_h \in D$ and for any plane shape n,*

$$\mathsf{F}^2_{N-1}(x_h, n) = (C_N(x_h)\underbrace{.n)....).n}_{N \quad times} \tag{245}$$

Proof. We adapt the Cauchy tetrahedron construction. We consider the point x_0 and the three unit vectors (e_1, e_2, e_3). We define the tetrahedron V as follows i) its face S opposed to the vertex in x_0 is normal to the unit vector n; ii) its edges

$$L_i = \overline{x_0 x_i} = l_i e_i \qquad (i = 1, 2, 3) \tag{246}$$

are such that $x_i \in S$; iii) its height H parallel to n has length h: if we call x_h the second endpoint of the segment H we have

$$H = \overline{x_0 x_h}; \qquad x_h - x_0 = hn. \tag{247}$$

It is easy to check that: i) the equation of the plane π_S orthogonal to n and passing through the point x_h is given by

$$(x - x_h) \cdot n = 0 \tag{248}$$

ii) as $S \subset \pi_S$ and $x_i \in S$ then

$$(x_0 + l_i e_i - x_h) \cdot n = 0 \Rightarrow -h + l_i e_i \cdot n = 0 \Rightarrow h = l_i e_i \cdot n \Rightarrow \frac{h}{e_i \cdot n} = l_i \tag{249}$$

iii) the edges of the tetrahedron V lying in the plane π_S are

$$\begin{aligned} \overrightarrow{x_1 x_2} &= l_2 e_2 - l_1 e_1 = h\left(\tfrac{e_2}{e_2 \cdot n} - \tfrac{e_1}{e_1 \cdot n}\right) \\ \overrightarrow{x_2 x_3} &= l_3 e_3 - l_2 e_2 = h\left(\tfrac{e_3}{e_3 \cdot n} - \tfrac{e_2}{e_2 \cdot n}\right) \\ \overrightarrow{x_3 x_1} &= l_1 e_1 - l_3 e_3 = h\left(\tfrac{e_1}{e_1 \cdot n} - \tfrac{e_3}{e_3 \cdot n}\right) \end{aligned} \tag{250}$$

iv) the triangles S_3, S_1 and S_2 having a vertex in x_0 and respectively $\overline{x_1 x_2}$, $\overline{x_2 x_3}$, $\overline{x_3 x_1}$ as side opposed to x_0 have as outward pointing unit normals respectively the vectors

$$n_3 = \frac{e_2 \times e_1}{|e_2 \times e_1|}, \quad n_1 = \frac{e_3 \times e_2}{|e_3 \times e_2|}, \quad n_2 = \frac{e_1 \times e_3}{|e_1 \times e_3|}; \tag{251}$$

It is important to establish the relationship between the reciprocal basis (e^j) of the introduced basis (e_i) and the three normals (n_j). Using the notation $v := e_1 \cdot (e_2 \times e_3)$ we have:

$$ve^1 = e_2 \times e_3; \quad ve^2 = e_3 \times e_1; \qquad ve^3 = e_1 \times e_2; \qquad -n_j = \frac{e^j}{|e^j|} \tag{252}$$

v) the areas of the faces S, S_3, S_1 and S_2 can be easily calculated. We can verify some simple relationships:

$$\begin{aligned}
2|S| &= \overrightarrow{x_1x_2} \times \overrightarrow{x_1x_3} \cdot n = h^2 \left(\frac{e_2}{e_2 \cdot n} - \frac{e_1}{e_1 \cdot n} \right) \times \left(-\frac{e_1}{e_1 \cdot n} + \frac{e_3}{e_3 \cdot n} \right) \cdot n = \\
&= h^2 \left(\frac{e_2}{e_2 \cdot n} \times \left(-\frac{e_1}{e_1 \cdot n} + \frac{e_3}{e_3 \cdot n} \right) - \frac{e_1}{e_1 \cdot n} \times \left(\frac{e_3}{e_3 \cdot n} \right) \right) \cdot n = \\
&= h^2 \left(\frac{-(e_2 \times e_1) \cdot n}{(e_2 \cdot n)(e_1 \cdot n)} + \frac{(e_2 \times e_3) \cdot n}{(e_2 \cdot n)(e_3 \cdot n)} + \frac{-(e_1 \times e_3) \cdot n}{(e_1 \cdot n)(e_3 \cdot n)} \right) = \\
&= h^2 v \left(\frac{e^3 \cdot n}{(e_2 \cdot n)(e_1 \cdot n)} + \frac{e^1 \cdot n}{(e_2 \cdot n)(e_3 \cdot n)} + \frac{e^2 \cdot n}{(e_1 \cdot n)(e_3 \cdot n)} \right) = \\
&= \frac{h^2 v}{(e_3 \cdot n)(e_2 \cdot n)(e_1 \cdot n)} \left(\sum_{i=1}^{3} (e^i \cdot n)(e_i \cdot n) \right) \\
&= \frac{h^2 (e_1 \cdot (e_2 \times e_3))}{(e_3 \cdot n)(e_2 \cdot n)(e_1 \cdot n)}
\end{aligned} \tag{253}$$

$$2|S| = \frac{vh^2}{(e_2 \cdot n)(e_1 \cdot n)(e_3 \cdot n)} \tag{254}$$

$$2|S_3| = (l_2 e_2 \times l_1 e_1) \cdot n_3 = h^2 \left(\frac{e_2}{e_2 \cdot n} \times \frac{e_1}{e_1 \cdot n} \right) \cdot n_3 \tag{255}$$

$$= h^2 \frac{|e_2 \times e_1|}{(e_2 \cdot n)(e_1 \cdot n)} = \frac{h^2 v |e^3|}{(e_2 \cdot n)(e_1 \cdot n)} \tag{256}$$

$$2|S_1| = \frac{vh^2 |e^1|}{(e_2 \cdot n)(e_3 \cdot n)}; \quad 2|S_2| = \frac{vh^2 |e^2|}{(e_1 \cdot n)(e_3 \cdot n)} \tag{257}$$

so that we have

$$|S| = \frac{|S_i|}{(e_i \cdot n)|e^i|}; \quad (i = 1, 2, 3) \tag{258}$$

vi) the shapes of the edges L_1, L_2 and L_3 are respectively $f_1 := (n_2, n_3, e_1)$, $f_2 := (n_3, n_1, e_2)$ and $f_3 := (n_1, n_2, e_3)$. Let V^ε be the image of V under an homothetic transformation of ratio ε and center x_h, we denote by L^ε and S^ε (with eventually suitable subscripts) the images of all edges and surfaces previously introduced. We will use now the quasi-balance of contact power for the fixed (it does not vary with ε!) test field

$$U : x \longrightarrow ((x - x_h) \cdot n)^{N-1} U_0 \tag{259}$$

where U_0 is a generic vector and for the varying family of domains V^ε, its contact surface, having the edges and wedges we have just described. The equation 168 applied to state the quasi-balance of power for a physically admissible stress state $\mathfrak{S}$ in the domain V_ε becomes

$$|\mathcal{P}_U(\mathfrak{S}, V_\varepsilon)| := \left| \int_{\mathcal{S}_r^\varepsilon} \mathsf{F}_0^2 \cdot U + \int_{\mathcal{L}_r^\varepsilon} \mathsf{F}_0^1 \cdot U + \sum_{k=1}^{N-1} \int_{\mathcal{S}_r^\varepsilon} \mathsf{F}_k^2 \cdot \left(\nabla^k U\right)_\perp + \right. \tag{260}$$

$$\left. + \sum_{k=1}^{N-2} \int_{\mathcal{L}_r^\varepsilon} \mathsf{F}_k^1 \cdot \left(\nabla^k U\right)_\perp + \sum_{k=0}^{N-3} \int_{\mathcal{W}_S^\varepsilon} \mathsf{F}_k^0 \cdot \nabla^k U \right| \leq K_U \varepsilon^3 \tag{261}$$

where with the symbols $\mathcal{S}_r^\varepsilon$, $\mathcal{L}_r^\varepsilon$ and $\mathcal{W}_S^\varepsilon$ we have denoted respectively the set of regular surface points, regular edges points and wedges belonging to the contact surface of the domain V_ε.We now start by remarking that the field U vanishes on the surface S^ε together with all its derivatives up to the order $N-2$.Indeed:

$$\nabla^k U = \frac{(N-1)!}{(N-1-k)!} ((x - x_h) \cdot n)^{N-1-k} U_0 \otimes n^k$$

$$\left(\nabla^k U\right)_\perp = \frac{(N-1)!}{(N-1-k)!} ((x - x_h) \cdot n)^{N-1-k} U_0 \otimes (n - (n \cdot e_i) e_i)^k$$

We now consider one by one all the addends in 275 in which the previous equalities have been replaced

$$\left|\int_{\mathcal{S}_r^\varepsilon} \mathsf{F}_0^2 \cdot U\right| = \tag{262}$$

$$= \left|\sum_{i=1}^{3} \int_{\mathcal{S}_i^\varepsilon} \mathsf{F}_0^2 \cdot \left((x - x_h) \cdot n\right)^{N-1} U_0\right| \leq K_0^2 \left\|U_0\right\| \varepsilon^2 \varepsilon^{N-1} \tag{263}$$

$$(\forall k < N-1) \left(\begin{array}{c} \left|\int_{\mathcal{S}_r^\varepsilon} \mathsf{F}_k^2 \cdot \left(\nabla^k U\right)_\perp\right| = \\ = \left|\sum_{i=1}^{3} \int_{S_i^\varepsilon} \mathsf{F}_k^2 \cdot \frac{(N-1)!}{(N-1-k)!} \left((x - x_h) \cdot n\right)^{N-1-k} U_0 \left(n \cdot n_i\right)^k\right| \\ \leq K_k^2 \left\|U_0\right\| \varepsilon^2 \varepsilon^{N-1-k} \end{array} \right) \tag{264}$$

$$\left|\int_{\mathcal{S}_r^\varepsilon} \mathsf{F}_{N-1}^2 \cdot \left(\nabla^{N-1} U\right)_\perp\right| = \left|\sum_{i=1}^{3} \int_{S_i^\varepsilon} \mathsf{F}_{N-1}^2 \cdot (N-1)! U_0 \left(n \cdot n_i\right)^{N-1} \tag{265}$$

$$+ \int_{S^\varepsilon} \mathsf{F}_{N-1}^2 \cdot (N-1)! U_0 \leq K_{N-1}^2 \left\|U_0\right\| \varepsilon^2 \tag{266}$$

$$\left|\int_{\mathcal{L}_r^\varepsilon} \mathsf{F}_0^1 \cdot U\right| = \left|\sum_{i=1}^{3} \int_{L_i^\varepsilon} \mathsf{F}_0^1 \cdot \left((x - x_h) \cdot n\right)^{N-1} U_0\right| \leq K_0^1 \left\|U_0\right\| \varepsilon^1 \varepsilon^{N-1} \tag{267}$$

$$(\forall k < N-2) \left(\left|\int_{\mathcal{L}_r^\varepsilon} \mathsf{F}_k^1 \cdot \left(\nabla^k U\right)_\perp\right| = \right. \tag{268}$$

$$\left. = \left|\sum_{i=1}^{3} \int_{L_i^\varepsilon} \mathsf{F}_k^1 \cdot \left(\frac{(N-1)!}{(N-1-k)!} \left((x - x_h) \cdot n\right)^{N-1-k} U_0 \otimes \left(n - \left(n \cdot e_i\right) e_i\right)^k\right)\right| \leq \atop \leq K_k^1 \left\|U_0\right\| \varepsilon^1 \varepsilon^{N-1-k} \right) \tag{269}$$

$$\left|\int_{\mathcal{L}_r^\varepsilon} \mathsf{F}_{N-2}^1 \cdot \left(\nabla^{N-2} U\right)_\perp\right| = \tag{270}$$

$$= \left|\sum_{i=1}^{3} \int_{L_i^\varepsilon} \mathsf{F}_{N-2}^1 \cdot \left(\left((N-1)! \left((x - x_h) \cdot n\right) U_0 \otimes \left(n - \left(n \cdot e_i\right) e_i\right)^{N-2}\right)\right)\right|$$

$$\leq K_{N-2}^1 \left\|U_0\right\| \varepsilon^2 \tag{271}$$

$$(\forall k \leq N-3) \left(\begin{array}{c} \left|\int_{\mathcal{W}_S^\varepsilon} \mathsf{F}_k^0 \cdot \nabla^k U\right| = \\ \left|\mathsf{F}_k^0(x_0, [(x_0, S^\varepsilon)]) \cdot \left(\frac{(N-1)!((x - x_h) \cdot n)^{N-1-k}}{N-1-k!}\right) U_0 \otimes n^k\right| \leq \\ \leq K_0^0 \left\|U_0\right\| \varepsilon^{N-1-k} \end{array} \right) \tag{272}$$

We now simply multiply both terms in 261 (in which we have replaced 259) times ε^{-2}. Then we take into account all inequality listed from 262 to 272 and we remark that the shapes on every face and edge of the boundary ∂V^{ε} are spatially constants and, together with wedge shapes, also independent of ε. We are thus able to calculate the limit for ε tending to zero and finally get (remark that having already treated the case $N = 2$, we can consider $N \geq 3$)

$$\lim_{\varepsilon \to 0} \varepsilon^{-2} \left(\sum_{i=1}^{3} \int_{S_i^{\varepsilon}} \mathsf{F}_{N-1}^{2} \cdot (N-1)! U_0 \left(n \cdot n_i\right)^{N-1} + \right. \tag{273}$$

$$+ \int_{S^{\varepsilon}} \mathsf{F}_{N-1}^{2} \cdot (N-1)! U_0 + \tag{274}$$

$$+ \sum_{i=1}^{3} \int_{L_i^{\varepsilon}} \mathsf{F}_{N-2}^{1} \cdot \left(\left((N-1)! \left((x - x_h) \cdot n \right) U_0 \otimes \left(n - (n \cdot e_i) \, e_i \right)^{N-2} \right) \right) + \tag{275}$$

$$\left. + \mathsf{F}_{N-3}^{0}(x_h - \varepsilon h n, [(x_h - \varepsilon h n, S^{\varepsilon})]) \cdot \left(\frac{(N-1)! \, (\varepsilon h)^2}{2!} \right) U_0 \otimes n^{N-3} \right) = 0 \tag{276}$$

Before starting to calculate the indicated limit we remark that the set of shapes $\bigcup \Phi(\partial V^{\varepsilon})$ is a prescribed set of shapes, so that all the continuous functions F are equi-uniformly bounded. The previous equality, where we have now explicitly indicated the argument of the functions F and once we apply the first mean value theorem for integration, becomes

$$\begin{aligned} \lim_{\varepsilon \to 0} \varepsilon^{-2} & \left(\sum_{i=1}^{3} |S_i^{\varepsilon}| \, \mathsf{F}_{N-1}^{2}(x_i^{\varepsilon}, n_i) \cdot (N-1)! U_0 \left(n \cdot n_i\right)^{N-1} \right. \\ & + |S^{\varepsilon}| \, \mathsf{F}_{N-1}^{2}(x_{\varepsilon}, n) \cdot (N-1)! U_0 + \\ & + \sum_{i=1}^{3} \left(\int_{L_i^{\varepsilon}} \left((x - x_h) \cdot n \right) \right) \mathsf{F}_{N-2}^{1} \left(\tilde{x}_i^{\varepsilon}, f_i \right) \\ & \cdot \left(\left((N-1)! U_0 \otimes \left(n - (n \cdot e_i) \, e_i \right)^{N-2} \right) \right) + \\ & \left. + \mathsf{F}_{N-3}^{0}(x_h - \varepsilon h n, (\widehat{n_1, n_2, n_3})) \cdot \left(\frac{(N-1)! \, (\varepsilon h)^2}{2!} \right) U_0 \otimes n^{N-3} \right) = 0 \end{aligned} \tag{277}$$

where $x_i^{\varepsilon} \in S_i^{\varepsilon}, \quad x_{\varepsilon} \in S^{\varepsilon}, \quad \tilde{x}_i^{\varepsilon} \in L_i^{\varepsilon}$. We now calculate the integral

$$\int_{L_i^{\varepsilon}} \left((x - x_h) \cdot n \right) = \frac{1}{2} \left(-e_i \cdot n \right) \varepsilon^2 l_i^2 \tag{278}$$

and, as the following limit is finite

$$\lim_{\varepsilon\to 0} \varepsilon^{-2} |S^\varepsilon| = \frac{vh^2}{2\,(e_2\cdot n)\,(e_1\cdot n)\,(e_3\cdot n)}, \tag{279}$$

we factorize $|S^\varepsilon|$ in equality 277, to get

$$\lim_{\varepsilon\to 0}\left(\sum_{i=1}^{3} |S_i^\varepsilon|\,|S^\varepsilon|^{-1}\,\mathsf{F}^2_{N-1}(x_i^\varepsilon, n_i)\cdot (N-1)!U_0\,(n\cdot n_i)^{N-1} + \right.$$

$$+\mathsf{F}^2_{N-1}(x_\varepsilon, n)\cdot (N-1)!U_0+ \tag{280}$$

$$+\sum_{i=1}^{3}\left(\frac{1}{2}\,(-e_i\cdot n)\,\varepsilon^2 l_i^2\right)|S^\varepsilon|^{-1}\,\mathsf{F}^1_{N-2}\,(\tilde{x}_i^\varepsilon, f_i)\cdot \tag{281}$$

$$\cdot\left((N-1)!U_0\otimes (n-(n\cdot e_i)\,e_i)^{N-2}\right)+ \tag{282}$$

$$+\mathsf{F}^0_{N-3}(x_h-\varepsilon hn, (\widehat{n_1, n_2, n_3}))\cdot \tag{283}$$

$$\left.\cdot\left(\frac{(N-1)!\,(\varepsilon h)^2}{2!}\right)|S^\varepsilon|^{-1}\left(U_0\otimes n^{N-3}\right)\right)=0$$

To proceed we need i) to use the following formulas $(i=1,2,3)$

$$(e_i\cdot n)\,|e^i| = |S_i|\,|S|^{-1} = |S_i^\varepsilon|\,|S^\varepsilon|^{-1};\qquad \frac{h}{(e_i\cdot n)} = l_i \tag{284}$$

$$\varepsilon^2 l_i^2\,|S^\varepsilon|^{-1} = l_i^2\,|S|^{-1} = v^{-1}\left(\frac{2\,(e_2\cdot n)\,(e_1\cdot n)\,(e_3\cdot n)}{(e_i\cdot n)^2}\right) \tag{285}$$

$$|S^\varepsilon|^{-1}\,(\varepsilon h)^2 = h^2\,|S^\varepsilon|^{-1}\,\varepsilon^2 = 2v^{-1}\,(e_2\cdot n)\,(e_1\cdot n)\,(e_3\cdot n) \tag{286}$$

which are implied by 254 and 258, ii) to simplify the nonvanishing factor $(N-1)!$ and iii) to use simple algebra. We get:

$$\lim_{\varepsilon\to 0}\left(\sum_{i=1}^{3}(e_i\cdot n)\,|e^i|\,\mathsf{F}^2_{N-1}(x_i^\varepsilon, n_i)\cdot U_0\,(n\cdot n_i)^{N-1} + \right.$$

$$+\mathsf{F}^2_{N-1}(x_\varepsilon, n)\cdot U_0 \tag{287}$$

$$+\sum_{i=1}^{3}\left(-v^{-1}\right)\frac{(e_2\cdot n)\,(e_1\cdot n)\,(e_3\cdot n)}{(e_i\cdot n)}\mathsf{F}^1_{N-2}\,(\tilde{x}_i^\varepsilon, f_i)\cdot \tag{288}$$

$$\cdot\left(U_0\otimes (n-(n\cdot e_i)\,e_i)^{N-2}\right)+ \tag{289}$$

$$+\mathsf{F}^0_{N-3}(x_h-\varepsilon hn, (\widehat{n_1, n_2, n_3}))\cdot \tag{290}$$

$$\left.\cdot v^{-1}\,(e_2\cdot n)\,(e_1\cdot n)\,(e_3\cdot n)\left(U_0\otimes n^{N-3}\right)\right)=0$$

It is now possible to calculate the limit and factorize the arbitrary vector U_0

$$\left(\sum_{i=1}^{3}(e_i\cdot n)\left|e^i\right|\mathsf{F}^2_{N-1}(x_h,n_i)\left(n\cdot n_i\right)^{N-1}+\mathsf{F}^2_{N-1}(x_h,n)+\right.$$

$$+\sum_{i=1}^{3}\left(-v^{-1}\right)\frac{(e_2\cdot n)(e_1\cdot n)(e_3\cdot n)}{(e_i\cdot n)}\mathsf{F}^1_{N-2}\left(x_h,f_i\right)\cdot\left(n-\left(n\cdot e_i\right)e_i\right)^{N-2}+ \tag{291}$$

$$\left.+v^{-1}(e_2\cdot n)(e_1\cdot n)(e_3\cdot n)\,\mathsf{F}^0_{N-3}(x_h,(\widehat{n_1,n_2,n_3}))\cdot n^{N-3}\right).U_0=0$$

As the last equality holds for every nonvanishing vector U_0 it implies

$$\sum_{i=1}^{3}(e_i\cdot n)\left|e^i\right|\mathsf{F}^2_{N-1}(x_h,n_i)\left(n\cdot n_i\right)^{N-1}+\mathsf{F}^2_{N-1}(x_h,n)+$$

$$+\sum_{i=1}^{3}\left(-v^{-1}\right)\frac{(e_2\cdot n)(e_1\cdot n)(e_3\cdot n)}{(e_i\cdot n)}\mathsf{F}^1_{N-2}\left(x_h,f_i\right)\cdot\left(n-\left(n\cdot e_i\right)e_i\right)^{N-2}+ \tag{292}$$

$$+v^{-1}(e_2\cdot n)(e_1\cdot n)(e_3\cdot n)\,\mathsf{F}^0_{N-3}(x_h,(\widehat{n_1,n_2,n_3}))\cdot n^{N-3}=0$$

The formula which we have just obtained generalizes equation 209. As the following implication holds

$$(\forall\,(e_i))\left((e_i\cdot e_j=\delta_{ij})\Rightarrow(\forall i)\,(n_i=-e_i)\wedge(v=1)\right) \tag{293}$$

then 292, together with theorem 14.18 obviously implies that there exists a tensor of order N which allows for the representation of surface contact $(N-1)$ –forces given in formula 245. □

Remark 14.21. When re-considering the flow of the demonstration of the previous theorem the reader will be persuaded that the tetrahedron argument can be used to get an alternative proof of Theorems 13.6 and 13.4.

The theorem which we have just proven leads us to define a tensor field C_N of order $N+1$ such that

$$\mathsf{F}^2_{N-1}\left(x_h,n\right)=\underbrace{(C_N(x_h).n)....).n}_{N\quad times}\;=C_N\left(x_0\right)\cdot n^N \tag{294}$$

This tensor is not uniquely determined, as only its right-side products by symmetric $(N+1)-th$ order tensors are determined. We may impose its right side symmetry and by using 292 we can easily prove the following

Lemma 14.22. *The only tensor field C_N verifying the equation 294 and completely symmetric with respect its last N indices has the following form*

$$C_N(x) = \sum_{i=1}^{3} \left(\mathsf{F}^2_{N-1}(x, n_i) \otimes (n_i)^{N-2} \right) \otimes \frac{1}{2} \left(e^i \otimes e_i + e_i \otimes e^i \right) + \tag{295}$$

$$+ \sum_{i=1}^{3} \mathsf{F}^1_{N-2} \left(x, f_i\right)_{\perp} \otimes \left(v^{-1}\right) \frac{1}{2} \left(e_{i+1} \otimes e_{i+2} + e_{i+2} \otimes e_{i+1}\right) + \tag{296}$$

$$-\mathsf{F}^0_{N-3}(x, (\widehat{n_1, n_2, n_3})) \otimes v^{-1} \check{S} \tag{297}$$

where we have introduced the third order tensor S as follows ($\sigma(\{1,2,3\})$ denotes the set of all permutations in the set $\{1,2,3\}$)

$$\sum_{\pi \in \sigma(\{1,2,3\})} e_{\pi(1)} \otimes e_{\pi(2)} \otimes e_{\pi(3)} =: 6\check{S} \tag{298}$$

The previous Lemma allows us to state the following

Proposition 14.23. *Given a physically admissible stress state $\mathfrak{S}$ of grade N. The tensor field C_N the existence of which has been proven in the previous Theorem 14.20 allows for the representation of the highest order nonvanishing contact actions in $\mathfrak{S}$*

$$\mathsf{F}^2_{N-1}, \mathsf{F}^1_{N-2}, \mathsf{F}^0_{N-3} \tag{299}$$

This representation is exactly the one given in equations 101,102 and 103.

14.6 Representation theorem for contact $k-$forces with $k < N$.

To proceed we will need to apply Gauss divergence Theorem to a sequence of tensor fields defined by recursively applying the previously presented tetrahedron argument. Indeed these tensor fields will be considered as arguments of the functionals introduced by formulas 105.

Therefore we are naturally induced to add some further regularity assumptions on the considered physically admissible stress state $\mathfrak{S}$ of grade N. In the present work we limit ourselves to the consideration of very regular stress states.

Hypothesis of C^{K+1} regularity.

For every given regular shape f the partial functions

$$\mathsf{F}^2_K(\cdot,\ f) \in \mathfrak{S} \tag{300}$$

are C^{K+1} functions.

As an obvious consequence of this assumption one easily gets that the tensor field C_N defined in 14.20 is a field of class C^N.

We can now easily prove the following

Proposition 14.24. *Let us consider a physically admissible stress state* $\mathfrak{S}$ *of grade N. The stress state* $\hat{\mathfrak{S}} := \left\{ \hat{\mathsf{F}} \right\}$ *defined by*

$$\begin{array}{l} \hat{\mathsf{F}}^2_J := \mathsf{F}^2_J - \mathfrak{F}(\partial B, \{0, ...0, C_N\}, J) \\ \hat{\mathsf{F}}^1_J := \mathsf{F}^1_J - \mathfrak{F}(\partial\partial B, \{0, ...0, C_N\}, J) \\ \hat{\mathsf{F}}^0_J := \mathsf{F}^0_J - \mathfrak{F}(\partial\partial\partial B, \{0, ...0, C_N\}, J) \end{array} \tag{301}$$

is physically admissible, verifies the Hypothesis 14.6 and is a stress state of grade not greater than $N-1$.

Proof. The demonstration is obtained by remarking that because of 14.23 the definition 301 has all vanishing the actions $\left\{ \hat{\mathsf{F}}^2_{N-1}, \hat{\mathsf{F}}^1_{N-2}, \hat{\mathsf{F}}^0_{N-3} \right\}$. Therefore the stress state $\hat{\mathfrak{S}}$ cannot have a grade greater than $N-1$. Then the definition itself of the functional $\mathfrak{F}$ assures that $\hat{\mathfrak{S}}$ verifies the regularity condition 14.6. Finally a simple integration by parts argument allows us to conclude that $\hat{\mathfrak{S}}$ is physically admissible. □

Definition 14.25. It is possible to apply recursively the tetrahedron argument and the previous proposition to defined the set of tensors C_K where $1 < k \leq N$.

Remark 14.26. Because of Hypothesis 14.6 the tensor fields C_K have all the $k-th$ order derivatives and these derivatives are continuous.

With a simple recursive argument it is then possible to prove the following

Theorem 14.27. ***Existence of Generalized Cauchy Stress Tensors.*** *Given a physically admissible stress state* $\mathfrak{S} = \{\mathsf{F}^2_J, \mathsf{F}^1_J, \mathsf{F}^0_J\}$ *of grade N. Then the following representation formula holds*

$$\begin{array}{l} \mathsf{F}^2_J = \mathfrak{F}(\partial B, \{C_2, C_3..., C_N\}, J) \\ \mathsf{F}^1_J = \mathfrak{F}(\partial\partial B, \{C_2, C_3..., C_N\}, J) \\ \mathsf{F}^0_J = \mathfrak{F}(\partial\partial\partial B, \{C_2, C_3..., C_N\}, J) \end{array} \tag{302}$$

In other words: for every physically admissible stress state of grade N there exists a N – tuple of stress tensors which allow for the representation of all admissible contact actions.

15 Conclusions

In this work it is proven, for the class of $N - th$ gradient continua, that the approach *à la Cauchy* and the approach *à la d'Alembert* are absolutely equivalent.

There are several aspects of the studied theory which deserve further investigations. We list here few of them.

1. It is needed to study contact actions on contact surfaces in which there are present curves of discontinuity of Gaussian curvature or its derivatives.
2. Cauchy tetrahedron argument needs to be extended to Cosserat-type continua, where further kinematical descriptors, in addition to placement, are introduced.
3. The weakest regularity assumptions for introduced fields which allow for a Cauchy type representation theorem need to be found.
4. More singular types of wedge shapes need to be introduced and studied as parts of admissible Cauchy cuts.

Acknowledgements

The authors are fully responsible for the opinions expressed and results presented in the present paper. However they must, also in this occasion, thank G. Maugin for the interest he has always shown in their research. Moreover the interesting discussions with A. Di Carlo were really important for a true understanding of the problems presented by an approach à la Cauchy to second gradient theories: he also attracted their attention to the novel by Borges quoted in one of the previous sections and cited in the references. Indeed the role of A.Di Carlo as "ideas spreader" in the interest of the advancement of science, must be recognized. His activity personifies the metaphor of the Bacon's bee (see e.g. Muntersbjom (2003)), or (using an expression from the ancient greek) of the true scientific "melitta" i.e. the honey bee who spreads the pollen of science from flower to flower.

15.1 REFERENCES

Bibliography

[1] R. K.Abu Al-Rub **Modeling the interfacial effect on the yield strength and flow stress of thin metal films on substrates** *Mechanics Research Communications* 35 65–72 (2008)

[2] R. Abeyaratne,N. Triantafyllidis, **An investigation of localization in a porous elastic material using homogenization theory.** *Trans. ASME J. Appl. Mech.* 51, no. 3, 481–486 (1984)

[3] R. Abraham, J.E. Marsden, and T. Ratiu, '**Manifolds,Tensor Analysis, and Applications**', Applied Mathematical Sciences, 75, Springer Verlag, (1988).

[4] Alibert, J.-J., Seppecher, P., dell'Isola, F. **Truss modular beams with deformation energy depending on higher displacement gradients.** *Math. Mech. Solids* 8 , no. 1, 51–73 (2003).

[5] V.I. Arnold, **Mathematical Methods of Classical Mechanics** Springer Verlag (1979)-(1989)

[6] C. Banfi, A. Marzocchi and A. Musesti **On the principle of virtual powers in continuum mechanics** *Ricerche di Matematica* 55: 299–310 (2006)

[7] S.Bardenhagen, N. Triantafyllidis **Derivation of higher order gradient continuum theories in 2,3-D nonlinear elasticity from periodic lattice models.** *J. Mech. Phys. Solids* 42 , no. 1, 111–139 (1994)

[8] E. Benvenuto **La scienza delle costruzioni e il suo sviluppo storico** Sansoni, Firenze, (1981)

[9] V. Berdichevsky, **Variational Principles of Continuum Mechanics**, Springer, (2009).

[10] J.L. Bleustein, **A note on the boundary conditions of Toupin's strain-gradient theory** *International Journal of Solids and Structures*, 3 (6), pp. 1053-1057. (1967)

[11] B. Bourdin, G.A. Francfort and J.-J. Marigo, **The variational approach to fracture**, *J. Elasticity*, 91, 1-3, 2008, 1-148 (also appeared as a Springer book: ISBN: 978-1-4020-6394-7).

[12] J.L.Borges, **Pierre Menard Author of the Quixote** (translation by James E. Irby) http://www.coldbacon.com/writing/borges-quixote.html

[13] P. Casal, et H.Gouin, '**Relation entre l'équation de l'énergie et l'équation du mouvement en théorie de Korteweg de la capillaritè**', *C. R. Acad. Sci. Paris*, t. 300, Série II, N. 7 231–233 (1985).

[14] P.Casal, '**La théorie du second gradient et la capillarité**', *C. R. Acad. Sci. Paris*, t. 274, Série A 1571–1574 (1972).

[15] P.Casal, **La capillarité interne**, Cahier du groupe Français de rhéologie, CNRS VI, 3, pp. 31-37 (1961).

[16] P. Casal, et H. Gouin, '**Relation entre l'équation de l'énergie et l'équation du mouvement en théorie de Korteweg de la capillarité**', *C. R. Acad. Sci. Paris*, t. 300, Série II, N. 7 231–233 (1985).

[17] A. Carcaterra , A. Akay , I.M. Koc **Near-irreversibility in a conservative linear structure with singularity points in its modal density** *Journal of the Acoustical Society of America* 119 4 2141-2149 (2006)

[18] A. Carcaterra **Ensemble energy average and energy flow relationships for nonstationary vibrating systems** *Journal of Sound and Vibration* 288, 3, 751-790 (2005)

[19] F.Collin, R.Chambon and R.Charlier **A finite element method for poro mechanical modelling of geotechnical problems using local second gradient models.** *Int. J. Num. Meth. Engng.* 65, 1749–1772 (2006).

[20] G. Colonnetti, **Scienza delle costruzioni**, Torino, Edizioni scientifiche Einaudi, 3° ed., (1953-57).

[21] E. Cosserat and F. Cosserat **Note sur la théorie de l'action euclidienne.** Paris, Gauthier-Villars, (1908).

[22] E. Cosserat, and F. **Cosserat Sur la Théorie des Corps Déformables**, Herman, Paris, (1909).

[23] N. Daher, G. A.Maugin **Virtual power and thermodynamics for electromagnetic continua with interfaces.** *J. Math. Phys.* 27 , no. 12, 3022–3035 (1986).

[24] N.Daher, G. A. Maugin **The method of virtual power in continuum mechanics. Application to media presenting singular surfaces and interfaces.** *Acta Mech.* 60, no. 3-4, 217–240, (1986)

[25] F. dell'Isola and W. Kosinski "**Deduction of thermodynamic balance laws for bidimensional nonmaterial directed continua modelling interphase layers,**" *Archives of Mechanics*, vol. 45, , pp. 333-359 (1993).

[26] F. dell'Isola and P. Seppecher, "**The relationship between edge contact forces, double force and interstitial working allowed by the principle of virtual power**", *Comptes Rendus de l'Academie de Sciences* - Serie IIb: Mecanique, Physique, Chimie, Astronomie, vol. 321, , pp. 303-308 (1995).

[27] F. dell'Isola and P. Seppecher, "**Edge Contact Forces and Quasi-Balanced Power**", *Meccanica*, vol. 32, pp. 33-52 (1997)

[28] F. dell'Isola, G. Sciarra, and S.Vidoli, **Generalized Hooke's law for isotropic second gradient materials**. *Proc. R. Soc. Lond. Ser. A Math. Phys. Eng. Sci.* 465, no. 2107, 2177–2196 (2009)

[29] F.dell'Isola, G. Sciarra, R. C.Batra, **Static deformations of a linear elastic porous body filled with an inviscid fluid**. Essays and papers dedicated to the memory of Clifford Ambrose Truesdell III. Vol. III. *J. Elasticity* 72 , no. 1-3, 99–120 (2003)

[30] F.dell'Isola, M. Guarascio, K. Hutter **A variational approach for the deformation of a saturated porous solid. A second-gradient theory extending Terzaghi's effective stress principle.** *Arch. Appl. Mech.* 70, 323–337 (2000)

[31] A. Di Carlo e A. Tatone **(Iper-)Tensioni & Equi-Potenza** AIMETA'01 XV Congresso AIMETA di Meccanica Teorica e Applicata 15th AIMETA Congress of Theoretical and Applied Mechanics 2001

[32] M. Degiovanni, A. Marzocchi and A. Musesti **Edge-force densities and second-order powers** *Annali di Matematica* 185, 81–103 (2006)

[33] M. Degiovanni, A. Marzocchi, and A. Musesti, A., '**Cauchy fluxes associated with tensor fields having divergence measure**', *Arch. Ration. Mech. Anal.* 147 197–223 (1999).

[34] J.E. Dunn, and J.Serrin, '**On the thermomechanics of interstitial working**', *Arch. Rational Mech. Anal.*, 88 (2) 95–133 (1985).

[35] J.E. Dunn '**Interstitial working and a non classical continuum thermodynamics**', In: J. Serrin (Ed), New Perspectives in Thermodynamics, Springer Verlag, Berlin, , pp. 187–222 (1986).

[36] G. E. Exadaktylos, I. Vardoulakis, **Microstructure in linear elasticity and scale effects: a reconsideration of basic rock mechanics and rock fracture mechanics**, *Tectonophysics*, Volume 335, Issues 1-2, 25 June 2001, Pages 81-109,

[37] A.C. Fannjiang, Y.S. Chan, and G.H. Paulino **Strain gradient elasticity for antiplane shear cracks: a hypersingular integrodifferential equation approach**. *SIAM J. Appl. Math.* 62 , no. 3, 1066–1091 (electronic) (2001/02).

[38] S. Forest, M. Amestoy, S. Cantournet, G. Damamme, S. Kruch *Mécanique des Milieux Continus* ECOLE DES MINES DE PARIS Année 2005–2006

[39] S. Forest **Mechanics of generalized continua: construction by homogenization** *J.Phys.* IV France 8 1998

[40] S. Forest **Homogenization methods and the mechanics of generalized continua** - part 2 Theoretical and applied Mechanics vol. 28-29, pp. 113-143 (2002)

[41] S. Forest. **Milieux continus généralisés et matériaux hétérogènes**. Les Presses de l'Ecole des Mines de Paris, ISBN : 2-911762-67-3, 200 pages, (2006).

[42] S. Forest. **Generalized continua.** In K.H.J. Buschow, R.W. Cahn, M.C. Flemings, B. Ilschner, E.J. Kramer, and S. Mahajan, editors, Encyclopedia of Materials : Science and Technology updates, pages 1–7. Elsevier, Oxford, (2005).

[43] S. Forest and M. Amestoy. **Mécanique des milieux continus.** Cours de l'Ecole des Mines de Paris n25e6 3121, 264 pages, (2004, 2005, 2006).

[44] S. Forest. **Milieux continus généralisés et matériaux hétérogènes**. Mémoire d'habilitation à diriger des recherches, (2004).

[45] E. Fried and M.E. Gurtin, **Tractions, balances, and boundary conditions for nonsimple materials with application to liquid flow at small length scales**. *Archive for Rational Mechanics and Analysis* 182 , 513-554 (2006).

[46] E. Fried and M.E. Gurtin, **A continuum mechanical theory for turbulence: a generalized Navier -Stokes-equation with boundary conditions**. *Theoretical and Computational Fluid Dynamics* 182 , 513-554 (2008)

[47] P. Germain **Cours de Mècanique des Milieux Continus**, tome I, Masson, Paris, (1973).

[48] P.Germain **La méthode des puissances virtuelles en mécanique des milieux continus. Première partie. Théorie du second gradient**. *J. Mécanique* 12, 235–274 (1973)

[49] P. Germain **The method of virtual power in continuum mechanics. Part 2: Microstructure.** *SIAM J. Appl. Math.* 25, 556–575 (1973)

[50] P. Germain '**Sur l'application de la méthode des puissances virtuelles en mécanique des milieux continus**', *C. R. Acad. Sci. Paris* Série A-B 274 A1051–A1055.(1972)

[51] A. E. Green, R. S. Rivlin, , **Multipolar continuum mechanics,** *Arch. Rational Mech. Anal.*, 17, 113–147 (1964)

[52] A. E. Green, R. S.Rivlin, **Simple force and stress multipoles,** *Arch. Rational Mech. Anal.*,16, 325–353 (1964)

[53] A. E. Green, R. S. Rivlin, .**On Cauchy's equations of motion**, *Z. Angew. Math. Phys.,ZAMP.* 15,, 290–292, (1964)

[54] A. E. Green, R. S. Rivlin **Multipolar continuum mechanics: Functional theory. I**, *Proc. Roy. Soc. Ser. A*, 284, 303–324 (1965)

[55] M.E. Gurtin, **A gradient theory of single-crystal viscoplasticity that accounts for geometrically necessary dislocations**, *International Journal of Plasticity* 50 , 809-819 (2002).

[56] N. Kirchner, P. Steinmann, **On the material setting of gradient hyperelasticity.** (English summary) *Math. Mech. Solids* 12 (2007)

[57] O.D. Kellogg, , **Foundations of Potential Theory** Springer, Berlin, (1929)

[58] W. Kosinski, **Field Singularities and Wave Analysis in Continuum Mechanics.** Ellis Horwood Series: Mathematics and Its Applications, Wiley & Sons, PWN - Polish Scientific Publishers, Warsaw 1986,

[59] R. Larsson, S. A Diebels, **Second-order homogenization procedure for multi-scale analysis based on micropolar kinematics.** *Internat. J. Numer. Methods Engrg.* 69 , no. 12, 2485–2512 (2007).

[60] M. Lazar , G. A. Maugin **A note on line forces in gradient elasticity** *Mechanics Research Communications* 33 674–680 (2006)

[61] M. Lucchesi ·M. Silhavý ·N. Zani **On the Balance Equation for Stresses Concentrated on Curves** *J Elasticity* 90:209–223 (2008).

[62] A. Madeo, F. dell'Isola, N. Ianiro, G. Sciarra, "**A second gradient poroelastic model of consolidation**", SIMAI 2008, Rome 15 – 19 September (2008).

[63] A. Madeo, F. dell'Isola, N. Ianiro, and G. Sciarra, "**A variational deduction of second gradient poroelasticity II: An application to the consolidation problem,**" *Journal of Mechanics of Materials and Structures*, vol. 3, , pp. 607-625 (2008)

[64] A.Marzocchi,, A.Musesti, **Balanced virtual powers in Continuum Mechanics.** *Meccanica* 38, 369–389 (2003)

[65] A. Marzocchi, A.Musesti,, '**Decomposition and integral representation of Cauchy interactions associated with measures**', *Cont. Mech. Thermodyn.* 13 149–169 (2001) .

[66] G. Maugin MathReview MR1437786 (98d:73003) 73A05 (73B18 73S10) on the paper F.dell'Isola P. Seppecher (1997).

[67] G. Maugin MathReview MR1600928 (99e:73005) on the paper G. Capriz and G. Mazzini Invariance and balance in continuum mechanics.Nonlinear analysis and continuum mechanics (Ferrara,1992), 27–35, Springer, New York, 1998.

[68] G. A.Maugin A.V. Metrikine Editors **Mechanics of Generalized Continua, One Hundred Years After the Cosserats**, Springer (2010)

[69] R.D. Mindlin, **Second gradient of strain and surface tension in linear elasticity** *Int. J. Solids and Struct.* 1,4,417-438 (1965)

[70] R.D. Mindlin, **Micro-structure in linear elasticity**, *Arch. Rat. Mech. Analysis*, , vol. 16, pp. 51-78 (1964)

[71] R. D. Mindlin, **Complex representation of displacements and stresses in plane strain with couple-stresses**. 1965 Appl. Theory of Functions in Continuum Mechanics (Proc. Internat. Sympos ., Tbilisi), Vol. I, *Mechanics of Solids* (Russian) pp. 256–259 Izdat. "Nauka", Moscow (1963)

[72] R. D. Mindlin, H. F.Tiersten, **Effects of couple-stresses in linear elasticity**. *Arch. Rational Mech. Anal.* 11 415–448 (1962)

[73] R. D. Mindlin **Influence of couple-stresses on stress concentrations Main features of cosserat theory are reviewed by lecturer and some recent solutions of the equations, for cases of stress concentration around small holes in elastic solids, are described** EXPERIMENTAL MECHANICS 3, 1, 1-7, THE WILLIAM M. MURRAY LECTURE, (1962)

[74] R. D. Mindlin N. N. Eshel **On first strain-gradient theories in linear elasticity** *International Journal of Solids and Structures* 4, 1 1968, 109-124

[75] R.D. Mindlin, .**Stress functions for a Cosserat continuum** *International Journal of Solids and Structures*, 1 (3), pp. 265-271 (1965)

[76] R.D.Mindlin **On the equations of elastic materials with microstructure** *International Journal of Solids and Structures*, 1 (1), pp. 73-78 (1965).

[77] M. Muntersbjom, **Francis Bacon's Philosophy of Science: Machina intellectus and Forma indita** *Philosophy of Science*, 70 (December 2003) pp. 1137-1148.

[78] W. Noll, E.G.,Virga, '**On edge interactions and surface tension**', *Arch. Rational Mech. Anal.,* 111 (1) 1–31 (1990).

[79] W. Noll '**The foundations of classical mechanics in the light of recent advances in continuum mechanics**' Proceeding of the Berkeley Symposium on the Axiomatic Method, Amsterdam, , pp. 226–281 (1959).

[80] W. Noll '**Lectures on the foundations of continuum mechanics and thermodynamics**', *Arch. Rational Mech. Anal.* 52 62–92 (1973).

[81] W. Noll '**The geometry of contact separation and reformation of continuous bodies**', Arch. Rational Mech. Anal., 122 (3) 197–212 (1993).

[82] C. Pideri and P. Seppecher **A second gradient material resulting from the homogenization of an heterogeneous linear elastic medium**. *Contin. Mech. Thermodyn.* 9 , no. 5, 241–257 (1997).

[83] P. Podio-Guidugli **A virtual power format for thermomechanics** *Continuum Mech. Thermodyn.* 20: 479–487 (2009)

[84] P. Podio-Guidugli, **Contact interactions, stress, and material symmetry, for nonsimple elastic materials.** (English, Serbo-Croatian summary) Issue dedicated to the memory of Professor Rastko Stojanovic (Belgrade, 2002). *Theoret. Appl. Mech.* 28/29 (2002), 261–276.

[85] P. Podio-Guidugli and M. Vianello **Hypertractions and hyperstresses convey the same mechanical information** *Continuum Mech. Thermodyn.* 22:163–176 (2010)

[86] C. Polizzotto, **Strain-gradient elastic-plastic material models and assessment of the higher order boundary conditions.** *Eur. J. Mech. A Solids* 26 , no. 2, 189–211 (2007).

[87] Rorres C. **Completing Book II of Archimedes's On Floating Bodies** *THE MATHEMATICAL INTELLIGENCER* 26,3 , Pages 32-42 (2004)

[88] L. Russo **The Forgotten Revolution** Springer Verlag (2003)

[89] J. Salençon **Mécanique des milieux continus** Ed. Ellipses (1988)-(1995) **Handbook of Continuum Mechanics** Ed. Springer (Berlin, 2001) Mécanique des milieux continus. Tome I. Éd. École polytechnique, Palaiseau ; Ellipses, Paris, (2002)-(2005)

[90] L. Schwartz , **Théorie des Distributions**, Hermann Paris, (1973).

[91] G. Sciarra, F. dell'Isola and O. Coussy **Second gradient poromechanics.** *Internat. J. Solids Structures* 44 , no. 20, 6607–6629 (2007).

[92] G. Sciarra, F. dell'Isola and K. Hutter **A solid-fluid mixture model allowing for solid dilatation under external pressure.** *Contin. Mech. Thermodyn.* 13 , no. 5, 287–306 (2001).

[93] G. Sciarra, F. dell'Isola, N. Ianiro, and A. Madeo, "**A variational deduction of second gradient poroelasticity part I: General theory**", *Journal of Mechanics of Materials and Structures,* vol. 3, , pp. 507-526 (2008).

[94] P. Seppecher '**Etude des conditions aux limites en théorie du second gradient: cas de la capillarité**, *C. R. Acad. Sci. Paris*, t. 309, Série II 497–502 (1989).

[95] P. Seppecher **Etude d'une Modélisation des Zones Capillaires Fluides: Interfaces et Lignes de Contact**, Thèse de l'Université Paris VI, Avril (1987).

[96] M. Šilhavý '**The existence of the flux vector and the divergence theorem for general Cauchy fluxes**', *Arch.Ration. Mech. Anal.* 90 195–211 (1985).

[97] M. Šilhavý '**Cauchy's stress theorem and tensor fields with divergences in Lp**', *Arch. Ration. Mech. Anal.* 116 223–255 (1991).

[98] M. Sokolowski, **Theory of couple-stresses in bodies with constrained rotations.** In CISM courses and lectures, vol. 26. Berlin, Germany: Springer (1970).

[99] M. Spivak **A comprehensive introduction to differential geometry.** Voll. I and II. Second edition. Publish or Perish, Inc., Wilmington, Del. (1979)

[100] A. S. J. Suiker and C. S. Chang, **Application of higher-order tensor theory for formulating enhanced continuum models.** *Acta Mech.* 142, 223–234 (2000).

[101] R.A. Toupin **Elastic Materials with couple-stresses**, *Arch. Rat. Mech. Analysis*, , vol. 11, pp. 385-414 (1962)

[102] R. A. Toupin, **Theories of elasticity with couple-stress.** *Arch. Rational Mech. Anal.* 17 85–112 (1964).

[103] N. Triantafyllidis and S. Bardenhagen, **On higher order gradient continuum theories in 1-D nonlinear elasticity. Derivation from and comparison to the corresponding discrete models.** *J. Elasticity* 33 (1993)

[104] N. Triantafyllidis and S. Bardenhagen **The influence of scale size on the stability of periodic solids and the role of associated higher order gradient continuum models.** *J. Mech. Phys. Solids* 44 , no. 11, 1891–1928 (1996)

[105] C.A.Truesdell **A First Course in Rational Continuum Mechanics**, Vol. I General Concepts, Academic Press, New York, 1977.

[106] N. Triantafyllidis and E.C. Aifantis **A gradient approach to localization of deformation. I. Hyperelastic materials.** *J. Elasticity* 16 , no. 3, 225–237 (1986).

[107] Y.Yang, and A. Misra "**Higher-order stress-strain theory for damage modeling implemented in an element-free Galerkin formulation,**" *Computer Modeling in Engineering and Sciences*, Vol. 64, No. 1, 1-36 (2010)

[108] Y. Yang, W.Y. Ching and Misra, A.,"**Higher-order continuum theory applied to fracture simulation of nano-scale intergranular glassy film,**" *Journal of Nanomechanics and Micromechanics*, (in print). (2011)

[109] G. Vailati, **Il principio dei lavori virtuali da Aristotele a Erone d'Alessandria**, Scritti (Bologna, Forni, 1987), vol. II, pp. 113-128, *Atti della R. Accademia delle Scienze di Torino*, vol. XXXII, adunanza del 13 giugno 1897, quaderno IG (091) 75 I - III (1897).

Fracture

B. Bourdin [*] and G.A. Francfort [†]

[*] Louisiana State University, Baton Rouge, LA 70803, USA

[†] Université Paris-Nord, 93430 Villetaneuse, France

Part of these notes has been extracted verbatim from Bourdin et al. (2008)

Abstract. These notes begin with a review of the mainstream theory of brittle fracture, as it has emerged from the works of Griffith and Irwin. We propose a re-formulation of that theory within the confines of the calculus of variations, focussing on crack path prediction. We then illustrate the various possible minimality criteria in a simple 1d-case as well as in a tearing experiment and discuss in some details the only complete mathematical formulation so far, that is that where global minimality for the total energy holds at each time. Next we focus on the numerical treatment of crack evolution and detail crack regularization which turns out to be a good approximation from the standpoint of crack propagation. This leads to a discussion of the computation of minimizing states for a non-convex functional. We illustrate the computational issues with a detailed investigation of the tearing experiment.

1 From Griffith to the variational

In this section, the starting premise is Griffith's model for crack evolution, as presented in Griffith (1920). Of course, continuum mechanics has evolved and it would make little sense to present fracture exactly as in Griffith (1920). The reader will find below what we believe to be a very classical introduction to brittle fracture within a rational mechanical framework.

Our starting assumptions are two-fold. First we restrict our focus to quasi-static evolution, a huge restriction: At each time, the investigated sample is in static equilibrium with the loads that are applied to it at that time. We use the blanket label "loads" for both hard devices (displacement type boundary conditions) and soft devices (traction type boundary conditions and/or body forces). Then, we do not concern ourselves with changes in temperature, implicitly assuming that those will not impact upon the mechanics of the evolution: in particular, thermal expansion is not covered by this model, at least to the extent that it couples thermal and mechanical effects. However, thermal stresses induced by a known temperature field fall squarely within the scope of the forthcoming analysis.

Also, we only discuss the 2d-case in this section. However, it will be clear that the resulting formulation applies as well to dimensions 1 and 3.

We consider Ω, a bounded open domain of $\mathbb{R}^2$. That domain is filled with a brittle elastic material. At this level of generality, the type of elastic behavior matters little, as long as it is represented by a bulk energy $F \mapsto W(F)$ which will be assumed to be a function of the gradient of the deformation field φ; in linearized elasticity W will become a function of $e(u) := \frac{1}{2}(\nabla u + \nabla u^t)$ with $\varphi(x) = x + u(x)$.

Time dependent loads are applied to Ω. We will assume that the force part of the load is given in the reference configuration (that is defined on $\overline{\Omega}$). Those are

- body forces denoted by $f_b(t)$ and defined on Ω;
- surface forces denoted by $f_s(t)$ and defined on $\partial_s\Omega \subset \partial\Omega$;
- boundary displacements denoted by $g(t)$ and defined on $\partial_d\Omega := \partial\Omega \setminus \partial_s\Omega$. Precisely, we assume throughout that $g(t)$ is defined and smooth enough on all of $\mathbb{R}^2$ and that the boundary displacement is the trace of $g(t)$ on $\partial_d\Omega$.

1.1 Griffith's theory

Griffith's theory is purely macroscopic. The crack or cracks are discontinuity surfaces for the deformation field of the continuum under investigation. If that continuum behaves elastically, material response under external loading will be unambiguous once the laws that preside over the onset and propagation of the crack(s) are specified. The construction of such laws – the goal of Griffith's theory – requires three foundational ingredients,

1. A surface energy associated to the surfaces where the deformation is discontinuous;
2. A propagation criterion for those surfaces;
3. An irreversibility condition for the cracking process.

The surface energy adopted by Griffith is simple. Throughout the cracking process, a(n isotropic) homogeneous material spends an energy amount which remains proportional to the area of the surface of discontinuity. We call fracture toughness of the material the proportionality factor, and denote it by k.

A simple counting argument demonstrates that, if inter-atomic bonding is ruled by a Lennard-Jones type interaction potential, then the add-energy spent in moving two atoms apart while the remaining atoms stay put is additive, which ultimately yields a total (macroscopic) energy proportional to the separation area. In the absence of contact the crack lips do not interact and cohesiveness is prohibited.

The propagation criterion is energy based. The test is a balance between the potential energy released through a virtual increase of the crack length (area) and the energy spent in creating additional length (area). The crack will extend only if the balance favors creation.

Finally a crack will form where and at the time at which the displacement field becomes discontinuous. It will then stay so forever, oblivious to the actual state of displacement at any posterior time.

We now formulate Griffith's view of the crack evolution problem in a(n isotropic) homogeneous elastic material. For now, the crack path $\hat{\Gamma}$ is assumed to be known *a priori.* We wish to include partial debonding as a possible crack behavior, so that $\hat{\Gamma} \subset \overline{\Omega} \setminus \partial_s\Omega$. The crack at time t is assumed to be a time increasing connected subset of $\hat{\Gamma}$; it can thus live partially, or totally on $\partial\Omega$. It is therefore completely determined by its length l and denoted by $\Gamma(l)$.

By the quasi-static assumption, the cracked solid (see Figure 1.1.1) is, at each time, in elastic equilibrium with the loads that it supports at that time; in other words, if the crack length at that time is l, then the kinematic unknown at that time, $\varphi(t,l)$ (the transformation, or displacement) satisfies

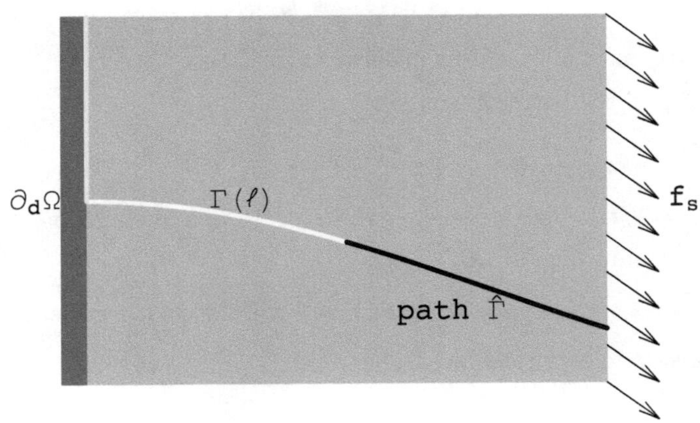

Figure 1.1.1. The cracked solid.

$$\begin{cases} -\text{div } \dfrac{\partial W}{\partial F}(\nabla\varphi(t,l)) = f_b(t) \text{ in } \Omega\setminus\Gamma(l) \\ \varphi(t,l) = g(t) \text{ on } \partial_d\Omega\setminus\Gamma(l) \\ \dfrac{\partial W}{\partial F}(\nabla\varphi(t,l))n = f_s(t) \text{ on } \partial_s\Omega\setminus\Gamma(l) \\ \dfrac{\partial W}{\partial F}(\nabla\varphi(t,l))n = 0 \quad \text{ on } \overline{\Omega}\cap\Gamma(l) \end{cases} \tag{1.1.1}$$

where n denotes the appropriate normal vector.

The last relation in (1.1.1) calls for several comments. In an anti-plane shear setting, it merely states, in accord with Griffith's premise, the absence of cohesive forces along the crack lips. In a planar situation, it implicitly assumes separation of the crack lips, hence non-interpenetration.

The system (1.1.1) assumes that the crack length is known. Griffith's contribution is to propose the following criteria for the determination of that length. At time t, compute the potential energy associated to the crack of length l, that is

$$\mathcal{P}(t,l) := \int_{\Omega\setminus\Gamma(l)} W(\nabla\varphi(t,l))\,dx - \mathcal{F}(t,\varphi(t,l)) \tag{1.1.2}$$

with

$$\mathcal{F}(t,\varphi) := \int_\Omega f_b(t).\varphi\,dx + \int_{\partial_s\Omega} f_s(t).\varphi\,ds. \tag{1.1.3}$$

Then, $l(t)$ must be such that it obeys

- `The Griffith's criterion:`

i. $l \overset{t}{\nearrow}$ (the crack can only grow);

ii. $-\dfrac{\partial\mathcal{P}}{\partial l}(t,l(t)) \le k$ (the energy release rate is bounded from above by the fracture toughness);

iii. $\left(\dfrac{\partial\mathcal{P}}{\partial l}(t,l(t)) + k\right)\dot{l}(t) = 0$ (the crack will not grow unless the energy release rate is critical).

Remark 1.1.1 *From a thermodynamical viewpoint, Griffith's criterion should be interpreted as follows. The crack length is a global internal variable, and its variation induces a dissipation which must in turn satisfy Clausius–Duhem's inequality.*

A convenient enforcement of Clausius–Duhem's inequality is provided through the introduction of a convex dissipation potential $\mathcal{D}(\dot{l})$, further satisfying $\mathcal{D}(0) = 0$. Then, the inequality reduces to

$$-\frac{\partial \mathcal{P}}{\partial l}(t, l(t)) \in \partial \mathcal{D}(\dot{l}(t)). \tag{1.1.4}$$

The correct dissipation potential in Griffith's setting is denoted by $\mathcal{D}_G$ and given by (see Figure 1.1.2)

$$\mathcal{D}_G(\dot{l}) := \begin{cases} k\dot{l}, & \dot{l} \geq 0 \\ \infty, & \dot{l} < 0, \end{cases} \tag{1.1.5}$$

and (1.1.4) then yields precisely Griffith's criteria. So, summing up, Griffith's modeling of crack evolution reduces to (1.1.1), (1.1.4) with (1.1.5) as dissipation potential.

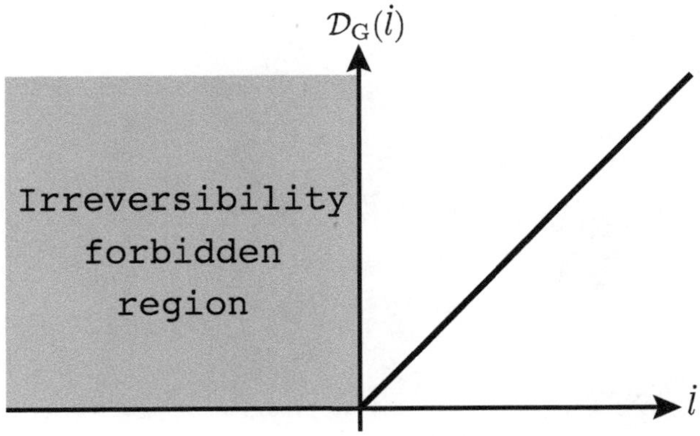

Figure 1.1.2. Griffith dissipation potential.

Note that the the dissipation potential is positively 1-homogeneous. This is an essential feature if adopting a variational viewpoint.

We are now ready to explore the system (1.1.1), (1.1.4). For completeness, we should add an initial condition to (1.1.4); we will thus assume that

$$l(0) = l_0, \tag{1.1.6}$$

and denote, from now onward, any pair-solution $(l(t), \varphi(t, l(t)))$, if it exists, by $(l(t), \varphi(t))$.

1.2 A variational equivalence

Assuming suitable – and unstated – smoothness of all relevant quantities, we propose to establish the equivalence between the original system (1.1.1), (1.1.4), (1.1.6) and a formulation which states that a certain energy must remain stationary at every time among all virtual admissible crack-displacement pairs at that time, and that an energy conservation statement must be satisfied throughout the time evolution. This is the object of the following

Proposition 1.2.1 *The pair $(l(t), \varphi(t))$ (satisfying (1.1.6)) satisfies (1.1.1), (1.1.4) (with appropriate smoothness) on $[0,T]$ iff, for every $t \in [0,T]$, it satisfies (with that same smoothness)*

(Ust) *$(l(t), \varphi(t))$ is a stationary point – in the sense of (1.2.5) below – of*

$$\mathcal{E}(t;\varphi,l) := \int_{\Omega\setminus\Gamma(l)} W(\nabla\varphi)\,dx - \mathcal{F}(t,\varphi) + kl, \tag{1.2.1}$$

among all $l \geq l(t)$ and $\varphi = g(t)$ on $\partial_d\Omega \setminus \Gamma(l)$;

(Ir)

$$\dot{l}(t) \geq 0\,; \tag{1.2.2}$$

(Eb) $$\frac{dE}{dt}(t) = \int_{\partial_d\Omega\setminus\Gamma(l(t))} \frac{\partial W}{\partial F}(\nabla\varphi(t))n.\dot{g}(t)\,ds - \dot{\mathcal{F}}(t,\varphi(t))$$

with

$$\dot{\mathcal{F}}(t,\varphi) := \int_\Omega \dot{f}_b(t).\varphi\,dx + \int_{\partial_s\Omega} \dot{f}_s(t).\varphi\,ds \tag{1.2.3}$$

$$\begin{aligned} E(t) &:= \int_{\Omega\setminus\Gamma(l(t))} W(\nabla\varphi(t))\,dx - \mathcal{F}(t,\varphi(t)) + kl(t) \\ &= \mathcal{P}(t,l(t)) + kl(t). \end{aligned} \tag{1.2.4}$$

The unilateral stationarity statement (Ust) is rather unusual because the functional $\mathcal{E}(t;\cdot)$ that should be stationary at $(l(t), \varphi(t))$ explicitly depends on $l(t)$; hence the label unilateral. The energy balance (Eb) can be turned, through various integration by parts in time, into what is referred to in the literature as the mechanical form of the second law of thermodynamics; see e.g. Gurtin (2000).

Proof. First we should clearly articulate what is meant by (Ust). To this effect, we introduce a one-parameter family of variations of the kinematic variable $\varphi(t)$ and of the crack length $l(t)$ as follows. We set

$$l(t,\varepsilon) := l(t) + \varepsilon\hat{l};\ \hat{l} \geq 0; \quad \varphi(t,\varepsilon,l) := \varphi(t,l) + \varepsilon\psi(t,l),$$

where $\psi(t,l) = 0$ on $\partial_d\Omega \setminus \Gamma(l)$ and $\varphi(t,l(t)) = \varphi(t)$. Then, unilateral stationarity is meant as

$$\frac{d}{d\varepsilon}\mathcal{E}(t;\varphi(t,\varepsilon,l(t,\varepsilon)),l(t,\varepsilon))\Big|_{\varepsilon=0} \geq 0. \tag{1.2.5}$$

Recall the expression (1.2.1) for $\mathcal{E}$. Then, the above also reads as

$$\int_{\Omega\setminus\Gamma(l(t))} \frac{\partial W}{\partial F}(\nabla\varphi(t)).\nabla\psi\, dx - \mathcal{F}(t,\psi) + \frac{\partial \mathcal{P}}{\partial l}(t,l(t))\,\hat{l} + k\hat{l} \geq 0,$$

where we recall that $\mathcal{P}$ was defined in (1.1.2). Consequently, through integration by parts, `(Ust)` is equivalent to

$$(1.1.1) \qquad \text{and} \qquad \frac{\partial \mathcal{P}}{\partial l}(t,l(t)) + k \geq 0. \tag{1.2.6}$$

Then, assume that `(Ust)`, `(Ir)`, `(Eb)` hold. In view of the above, (1.1.1) is satisfied, so that `(Eb)` reduces to

$$(\frac{\partial \mathcal{P}}{\partial l}(t,l(t)) + k)\dot{l} = 0. \tag{1.2.7}$$

Conversely, if (1.1.1) holds true, then

$$\begin{aligned}\frac{dE}{dt}(t) &= \int_{\partial_d\Omega\setminus\Gamma(l(t))} \frac{\partial W}{\partial F}(\nabla\varphi(t)).\dot{g}(t)\, ds - \dot{\mathcal{F}}(t,\varphi(t)) \\ &\quad + \{\frac{\partial \mathcal{P}}{\partial l}(t,l(t)) + k)\dot{l}(t)\},\end{aligned} \tag{1.2.8}$$

and, in view of the third item in Griffith's criterion, the term in brackets in (1.2.8) cancels out and `(Eb)` is established.

The second item in Griffith's criterion, together with (1.1.1), implies `(Ust)`. □

Remark 1.2.2 *Elimination of the kinematic field in the variational formulation leads to the sometimes more convenient equivalent formulation for* `(Ust)`*.*
`(Ust)` *$l(t)$ is a stationary point of $\mathcal{P}(t,l) + kl$, among all $l \geq l(t)$.*

Modulo smoothness, Griffith's formulation and the variational formulation obtained in Proposition 1.2.1 are strictly *one and the same* and cannot be opposed on mechanical grounds anymore than the original formulation. Pre-assuming smoothness is universal practice in deriving a notion of weak solution, so that we feel perfectly justified in doing so, and will be quite qualify as "weak" the solutions of what we will, from now onward, label the "variational evolution".

Remark 1.2.3 *Consider the case of a $p > 1$-homogeneous elastic energy density and of a monotonically increasing load, that is*

$$W(tF) = t^p W(F), \qquad \mathcal{F}(t, tu) = t^p \mathcal{F}(1, u), \quad g(t) = t\, g,$$

where u is the displacement field. Then, by homogeneity,

$$u(t, l) = t\bar{u}(\ell), \qquad \mathcal{P}(t, l) = t^p \bar{\mathcal{P}}(l),$$

where $\bar{\mathcal{P}}(l)$ is the potential energy associated to a crack of length l and loads corresponding to the value $t = 1$. We assume that $\bar{\mathcal{P}}$ is a sufficiently smooth function of l. Then it can be shown that that energy is a strictly convex function of l on $[l_0, l_1]$, if, and only if Griffith's criterion is satisfied by a unique smooth crack propagation $l(t)$ on $[t_0, t_1]$ given by

$$l(t) = (\bar{\mathcal{P}}')^{-1}\left(-\frac{k}{t^p}\right), \qquad t_1 = \sqrt[p]{\frac{k}{-\bar{\mathcal{P}}'(l_1)}}. \tag{1.2.9}$$

Then, at each time t, $-t^p \bar{\mathcal{P}}'(l(t)) = k$.

Thus, smoothness of the propagation leads to a reinforcement of the unilateral stationarity principle (Ust)*. The crack length $l(t)$ must actually be a minimizer for $\mathcal{P}(t, l) + kl$, because of the necessary convexity of $\mathcal{P}$.*

So Griffith's criterion, because it assumes smoothness of the crack evolutions, implicitly pre-supposes the global convexity of the potential energy as a function of the crack length.

Stationarity is not a very pleasant mathematical notion from the standpoint of existence and it it is tempting to somewhat strengthen (Ust). Observe that (Ust) amounts to a first order optimality condition for $(l(t), \varphi(t))$ to be a local unilateral minimizer – in any reasonable topology – of $\mathcal{E}(t; \cdot)$.

The preceding analysis suggests the adoption of some kind of minimality principle. Consequently, we propose the following two levels of departure from Griffith's classical theory:

- Local level – (Ust) is replaced by
  (Ulm) $(l(t), \varphi(t))$ is a local minimizer (in a topology that remains to be specified) for $\mathcal{E}(t; \varphi, l)$ among all $l \geq l(t)$ and all $\varphi = g(t)$ on $\partial_d\Omega \setminus \Gamma(l)$;
- Global level – (Ust) is replaced by
  (Ugm) $(l(t), \varphi(t))$ is a global minimizer for $\mathcal{E}(t; \varphi, l)$ among all $l \geq l(t)$ and all $\varphi = g(t)$ on $\partial_d\Omega \setminus \Gamma(l)$.

Those criteria are common in mechanics, but can never be justified on mechanical grounds, at least when departing from a purely convex setting. Local minimality is equivalent to Lyapunov stability for systems with a finite number of degrees of freedom, while the modern treatment of finite elasticity usually resorts to global minimality of the potential energy. Of course, in our setting, the minimization criterion, be it global or local, must also accommodate irreversibility, hence the already mentioned notion of *unilaterality*.

We now return to the time-continuous variational evolution and recast it in a more suitable functional framework.

1.3 Functional framework– A weak variational evolution

Using a minimality criterion immediately frees the crack path. Indeed, the minimality-modified Griffith variational evolution states that the actual length $l(t)$ of the crack is a local (or global) minimum among all lengths l greater than, or equal to $l(t)$ along the pre-determined crack path $\hat{\Gamma}$. There is then no point in restricting the future evolution precisely to the curve $\hat{\Gamma}$. We may as well let the crack choose which future path it wishes to borrow, according to the minimality principle.[1] Thus, denoting by $\Gamma(t)$ the crack at time t, we replace `(Ulm)`, resp. `(Ugm)` by

`(Ulm)` $(\Gamma(t), \varphi(t))$ is a local minimizer (in a topology that remains to be specified) for

$$\mathcal{E}(t; \varphi, \Gamma) := \int_{\Omega\setminus\Gamma} W(\nabla\varphi)\, dx - \mathcal{F}(t, \varphi) + k\mathcal{H}^1(\Gamma), \qquad (1.3.1)$$

among all $\Gamma \supset \Gamma(t)$ and all $\varphi = g(t)$ on $\partial_d\Omega \setminus \Gamma$; or, resp.,

`(Ugm)` $(\Gamma(t), \varphi(t))$ is a global minimizer for $\mathcal{E}(t; \varphi, \Gamma)$ among all $\Gamma \supset \Gamma(t)$ and all $\varphi = g(t)$ on $\partial_d\Omega \setminus \Gamma$.

Note that the test φ's depend on the test Γ's. Correspondingly, we also replace (1.1.6) by

`(Ic)` $\quad \Gamma(0) = \Gamma_0$,

(1.2.2) by

`(Ir)` $\Gamma(t) \nearrow_t$,

and the definition (1.2.4) of $E(t)$ in (Eb) by

$$\begin{aligned} E(t) &:= \int_{\Omega\setminus\Gamma(t)} W(\nabla\varphi(t))\, dx - \mathcal{F}(t, \varphi(t)) + k\mathcal{H}^1(\Gamma(t)) \\ &= \mathcal{P}(t, \Gamma(t)) + k\mathcal{H}^1(\Gamma(t)), \end{aligned} \qquad (1.3.2)$$

[1] It is precisely that freedom which truly distinguishes our approach from those usually adopted in mechanics; it is also precisely that freedom which confuses the mechanician, enslaved from a very early stage to a preconceived notion of what the crack can or cannot do.

with an obvious extension of the definition (1.1.2) of the potential energy $\mathcal{P}$.

We allow the test cracks Γ to be pretty much any *closed* set in $\overline{\Omega} \setminus \partial_s\Omega$ with finite Hausdorff measure. This allows us to envision very rough cracks, with length that coincide with the usual length when the crack is a rectifiable curve. We do not allow for the crack to lie on $\partial_s\Omega$ because the crack cannot live where soft devices are applied.

We shall refer to the above formulation, that is `(Ic)`, `(Ulm)` or `(Ugm)`, `(Ir)`, `(Eb)`, as *the strong variational evolution.*

Local minimality directly refers to a topology, whereas global minimality is topology-independent. But, even then, we need to impart upon test cracks a decent topology. A natural candidate is the Hausdorff metric, defined for two closed sets A, B as

$$d_H(A, B); = \max\{\sup_{a\in A} d(a, B), \sup_{b\in B} d(b, A)\}.$$

Examine for instance the initial time in the global minimality context with $\Gamma_0 = \emptyset$, $f_b(0) = f_s(0) = 0$. Then, we should minimize

$$\int_{\Omega\setminus\Gamma} W(\nabla\varphi)\, dx + k\mathcal{H}^1(\Gamma)$$

among all pairs (Γ, φ) with $\varphi = g(0)$ on $\partial_d\Omega \setminus \Gamma$. The direct method of the calculus of variations would have us take an infimizing sequence $\{(\Gamma_n, \varphi_n)\}$. In particular, we are at liberty to assume that $\mathcal{H}^1(\Gamma_n) \leq \mathcal{C}$. Say that the sequence Γ_n converges in the Hausdorff metric to some Γ; this is not a restriction, thanks to Blaschke's compactness theorem Rogers (1970). Then we would like to have

$$\mathcal{H}^1(\Gamma) \leq \liminf_n \mathcal{H}^1(\Gamma_n).$$

But, this is generically false, except in 2d and for, say, connected Γ_n's! That topology has been used with success to prove existence, in the global minimality framework, for the 2d variational evolution restricted to connected cracks in Dal Maso and Toader (2002). We shall come back to this point in Section 3.

In the context of image segmentation, D. Mumford and J. Shah proposed to segment image through the following algorithm: Find a pair K, compact of $\Omega \subset \mathbb{R}^2$ (the picture) representing the contours of the image in the picture, and φ, the true pixel intensity at each point of the picture, an element of $C^1(\Omega \setminus K)$, which minimizes

$$\int_{\Omega\backslash K} |\nabla\varphi|^2 \, dx + k\mathcal{H}^1(K) + \int_\Omega |\varphi - g|^2 \, dx, \tag{1.3.3}$$

where g is the measured pixel intensity. The minimization proposed in Mumford and Shah (1989) was then shown in De Giorgi et al. (1989) to be equivalent to a well-posed one-field minimization problem on a subspace $SBV(\Omega)$ of the space $BV(\Omega)$ of functions with bounded variations on Ω, namely,

$$\int_\Omega |\nabla\varphi|^2 \, dx + k\mathcal{H}^1(S(\varphi)) + \int_\Omega |\varphi - g|^2 \, dx, \tag{1.3.4}$$

where $\nabla\varphi$ represents the absolutely continuous part of the weak derivative of φ (a measure), and $S(\varphi)$ the set of jump points for φ.

We recall that a function $\varphi : \Omega \mapsto \mathbb{R}$ is in $BV(\Omega)$ iff $\varphi \in L^1(\Omega)$ and its distributional derivative $D\varphi$ is a measure with bounded total variation. Then, the theory developed by E. De Giorgi (see e.g. Evans and Gariepy (1992)) implies that

$$D\varphi = \nabla\varphi(x) \, dx + (\varphi^+(x) - \varphi^-(x))\nu(x)\mathcal{H}^1 \llcorner S(\varphi) + C(\varphi),$$

with $\nabla\varphi$, the approximate gradient, $\in L^1(\Omega)$ ($\nabla\varphi$ is no longer a gradient), $S(\varphi)$ the complement of the set of Lebesgue points of φ, a $\mathcal{H}^1$ σ–finite and countably 1-rectifiable set (a countable union of compacts included in C^1–hypersurfaces, up to a set of 0 $\mathcal{H}^1$–measure), $\nu(x)$ the common normal to all those hypersurfaces at a point $x \in S(\varphi)$, $\varphi^\pm(x)$ the values of $\varphi(x)$ "above and below" $S(\varphi)$, and $C(\varphi)$ a measure (the Cantor part) which is mutually singular with dx and with $\mathcal{H}^1$ (it only sees sets that have 0 Lebesgue–measure and infinite $\mathcal{H}^1$–measure). The subspace $SBV(\Omega)$ is that of those $\varphi \in BV(\Omega)$ such that $C(\varphi) \equiv 0$. It enjoys good compactness properties established in Ambrosio (1990), namely

$$\varphi_n \in SBV(\Omega) \text{ with} \begin{cases} \varphi_n \text{ bounded in } L^\infty(\Omega) \\ \nabla\varphi_n \text{ bounded in } L^q(\Omega;\mathbb{R}^2), \; q > 1 \\ \mathcal{H}^1(S(\varphi_n)) \text{ bounded in } \mathbb{R} \end{cases}$$

$$\Downarrow$$

$$\exists\{\varphi_{k(n)}\} \subset \{\varphi_n\}, \exists\varphi \in SBV(\Omega) \text{ s.t.}$$

$$\begin{cases} \varphi_{k(n)} \to \varphi, \text{ strongly in } L^p(\Omega), p < \infty \\ \nabla\varphi_{k(n)} \rightharpoonup \nabla\varphi, \text{ weakly in } L^q(\Omega;\mathbb{R}^2) \\ \mathcal{H}^1(S(\varphi)) \leq \liminf_n \mathcal{H}^1(S(\varphi_{k(n)}) \end{cases} \tag{1.3.5}$$

Thanks to Ambrosio's compactness result, a simple argument of the direct method applied to (1.3.4) establishes existence of a minimizer φ_g for that functional. The further result that the pair $\left(\varphi_g, \overline{(S(\varphi_g))}\right)$ is a minimizer for (1.3.3) is highly non-trivial and makes up the bulk of De Giorgi et al. (1989).

In De Giorgi's footstep, we thus reformulate the variational evolution in the weak functional framework of SBV, or rather of those functions that have all their components in SBV, the jump set $S(\varphi)$ becoming the union of the jump set of each component of φ. To do this, it is more convenient to view the hard device $g(t)$ as living on all of $\mathbb{R}^2$ and to integrate by parts the boundary term involving $\dot{g}(t)$ in `(Eb)`. So, after elementary integrations by parts, we propose to investigate

- `The weak variational evolution :` Find $(\varphi(t), \Gamma(t))$ satisfying

`(Ic)` $\Gamma(0) = \Gamma_0$;

`(Ulm)` $(\Gamma(t), \varphi(t))$ is a local minimizer (in a topology that remains to be specified) for

$$\mathcal{E}(t;\varphi,\Gamma) := \int_\Omega W(\nabla\varphi)\,dx - \mathcal{F}(t,\varphi) + k\mathcal{H}^1(\Gamma), \qquad (1.3.6)$$

among all $\overline{\Omega}\setminus\partial_s\Omega \supset \Gamma \supset \Gamma(t)$ and all $\varphi \equiv g(t)$ on $\mathbb{R}^2\setminus\overline{\Omega}$ with $S(\varphi) \subset \Gamma$; or, resp.,

`(Ugm)` $(\Gamma(t), \varphi(t))$ is a global minimizer for $\mathcal{E}(t;\varphi,\Gamma)$ among all $\overline{\Omega}\setminus\partial_s\Omega \supset \Gamma \supset \Gamma(t)$ and all $\varphi \equiv g(t)$ on $\mathbb{R}^2\setminus\overline{\Omega}$ with $S(\varphi) \subset \Gamma$;

`(Ir)` $\Gamma(t) \nearrow_t$;

`(Eb)` $\dfrac{dE}{dt}(t) = \displaystyle\int_\Omega \frac{\partial W}{\partial F}(\nabla\varphi(t)).\nabla\dot{g}(t)\,dx - \dot{\mathcal{F}}(t,\varphi(t)) - \mathcal{F}(t,\dot{g}(t))$
with

$$E(t) = \mathcal{E}(t;\varphi(t),\Gamma(t)). \qquad (1.3.7)$$

Remark that, in spite of the previous considerations on SBV, we have not explicitly indicated where φ (or $\varphi(t)$) should live. This is because, when dealing with vector-valued SBV-functions (the case of plane (hyper)elasticity, for example), that space – that is the Cartesian product of SBV for each of the component – is not quite sufficient. One should really work in $GSBV$ Dal Maso et al. (2005). Forget about that here.

Likewise, it is not so that the crack should belong to $\overline{\Omega}\setminus\partial_s\Omega$. Any rigorous analysis will actually require $\partial_s\Omega$, the site of application of the surface forces, to be part of the boundary of a non-brittle piece of the material. In other words, we should single out a thin layer around $\partial_s\Omega$ with infinite fracture toughness. This also will be overlooked in the sequel.

Also the test cracks Γ do not have to be even essentially, *i.e.*, up to a set of $\mathcal{H}^1$-measure 0, closed subsets of $\overline{\Omega} \setminus \partial_s\Omega$, but only countably 1-rectifiable curves. Whether the actual crack $\Gamma(t)$ that could be produced through the weak variational evolution is closed or not will be deemed a question of regularity and briefly commented upon in Paragraph 3.4 in the setting of global minimization.

Finally, as before, the same labels have been kept. The context will clearly indicate if the relevant formulation is weak or strong.

Remark 1.3.1 *The crack* $\Gamma(t)$ *can be identified with* $\Gamma_0 \cup \left[\bigcup_{s\le t} S(\varphi(s))\right]$*; see Dal Maso et al. (2009).*

The recasting of Griffith's evolution model in a variational framework is now complete.

2 Stationarity versus local or global minimality – a comparison

We wish to explore the consequences of minimality. The adopted setting, or rather settings, for such an analysis are designed so that the "crack path" is not at stake. Nor is irreversibility a concern here because the monotonicity of the loads combined with the geometry of the problems result in an increase of both the measure of the discontinuity set and the magnitude of the discontinuities on that set. The focus is squarely on minimality, although, at times energy balance (`Eb`) will also be invoked.

The two settings are

1. A 1d-traction experiment under a hard or a soft device;
2. A 2d-tearing experiment.

In the first setting, cracks are merely points of discontinuity along the bar; in the second setting, symmetry of the geometry and of the loads suggests a straight crack path in mode III. In both settings, we assess the potential existence of weak variational evolutions satisfying unilateral stationarity (`Ust`), unilateral minimality (`Ulm`), or still unilateral global minimality (`Ugm`), together with energy balance (`Eb`).

2.1 1d traction

A "crack-free" homogeneous linearly elastic bar of length L, cross-sectional area Σ, Young's modulus E, toughness k is clamped at $x = 0$ and subject to a displacement load εL, $0 \le \varepsilon \nearrow$ (hard device), or to a force load $\sigma\Sigma$, $0 \le \sigma \nearrow$ (soft device) at $x = L$. The parameters σ, ε play the role of the time variable. Thus, all evolutions will be parameterized by either σ, or ε. The results are concatenated in Conclusions 2.1.1, 2.1.4.

The soft device Assume that u is an admissible displacement field for a value σ of the loading parameter; that field may have jumps $S(u) \subset [0, L]$, or it may correspond to the elastic state, in which case it lies in $W^{1,2}(0, L)$. In any case we view it as a field defined in $SBV(\mathbb{R})$ and such that $u \equiv 0$ on $(-\infty, 0)$. Its associated energy is

$$\mathcal{E}(\sigma, u) = \frac{1}{2} \int_{(0,L)} E\Sigma(u')^2 \, dx - \sigma\Sigma u(L+) + k\Sigma\#(S(u)), \tag{2.1.1}$$

and that energy will only be finite if $S(u)$ is finite and $u' \in L^2(0, L)$, which we assume from now onward. This in turn implies that we may as well restrict the admissible fields to be in $SBV(\mathbb{R}) \cap L^\infty(\mathbb{R})$. Then,

Conclusion 2.1.1 *In a 1d traction experiment with a soft device, the elastic evolution is the only one that satisfies the weak variational evolution with either* (`Ust`), *or* (`Ulm`), *and* (`Eb`). *There is no solution to the weak variational evolution with* (`Ugm`) *and* (`Eb`).

Remark 2.1.2 *Testing the elastic solution against non-interpenetrating jumps is easy, since it suffices to restrict test jumps to be non-negative. In this context, the elastic solution is checked to be a global minimum for $\sigma < 0$, if non-interpenetration is imposed.*

Remark 2.1.3 *The above result demonstrates that soft devices prohibit global minimality. This is a significant drawback of global minimality and it clearly militates for a more local criterion. This has to be somewhat tempered, since one can build a reasonable class of non linear soft devices for which global minimality works; see Dal Maso et al. (2005).*

The hard device The admissible deformations are still in $SBV(\mathbb{R})$ and they satisfy $u \equiv 0$ on $(-\infty, 0)$ and $u \equiv \varepsilon L$ on (L, ∞). The associated energy is

$$\mathcal{E}(\varepsilon, u) = \frac{1}{2} \int_{(0,L)} E\Sigma(u')^2 \, dx + k\Sigma\#(S(u)), \tag{2.1.2}$$

and, once again it is only finite if $\#(S(u))$ is finite and $u' \in L^2(0, L)$, which we assume. Then,

Conclusion 2.1.4 *In a 1d traction experiment with a hard device, the elastic evolution, and all admissible evolutions with a set finite number of jumps satisfy the weak variational evolution with* (Ulm) *– and also* (Ust) *– and* (Eb)*. Only* $u_g(\varepsilon)$ *defined as*

$$u_g(\varepsilon) = \begin{cases} \varepsilon x \text{ if } 0 < \varepsilon \le \sqrt{2k/EL} \\ 0,\ x \le a;\ \varepsilon L,\ x > a \text{ if } \varepsilon \ge \sqrt{2k/EL} \end{cases}$$

satisfies the weak variational evolution with (Ugm) *and* (Eb)*.*

Also, all evolutions that are elastic, up to $\varepsilon = \sqrt{2ik/EL}$*, then have* i *jumps satisfy* (Ulm) *– and also* (Ust) *– and* (Eb)*.*

2.2 A tearing experiment

Consider a thin semi-infinite homogeneous, linearly elastic slab of thickness $2H$, $\Omega = (0, +\infty) \times (-H, +H)$. Its shear modulus is μ and its toughness k. Tearing amounts to a displacement load $tH\mathbf{e}_3$ on $\{0\} \times (0, +H)$ and $-tH\mathbf{e}_3$ on $\{0\} \times (-H, 0)$. The upper and lower edges are traction free and no forces are applied.

We assume throughout that all solutions respect geometric symmetry, although doing so cannot be justified; see in this respect the numerical experiment in Subsection 4.3. The symmetry assumption permits to look for anti-plane shear solution, anti-symmetric with respect to $y = 0$ and for a crack along that axis. We seek a displacement solution field of the form

$$\mathbf{u}(x, y, t) = \text{sign}(y) u(t, x) \mathbf{e}_3 \qquad \text{with} \qquad u(t, 0) = tH \tag{2.2.1}$$

and note that such a displacement cannot be the exact solution, because it fails to ensure the continuity of the normal stress at the points $(l(t), y)$, $y \neq 0$ (see (2.2.3)). The true symmetric solution can only be evaluated numerically, but it will be close to the proposed approximate solution as H becomes large.

The field $\mathbf{u}(t)$ will be discontinuous at the points x on the $y = 0$ - axis where $u(x, t) \neq 0$, that is $S(\mathbf{u}(t)) = \{x \ge 0 \ : \ u(t, x) \neq 0\}$. Then, the energy reads as

$$\mathcal{E}(\varphi) = \int_0^\infty \mu H (u'(x))^2 dx + k \int_0^\infty \text{sg}^+(|u(x)|)\, dx,$$

with $\text{sg}^+(z) := \begin{cases} 0, z \le 0 \\ 1, z > 0 \end{cases}$.

The kinematically admissible test fields u at time t are elements of $W^{1,2}(0, +\infty)$ and satisfy $u(t, 0) = tH$. A global minimum for $\mathcal{E}$ exists for

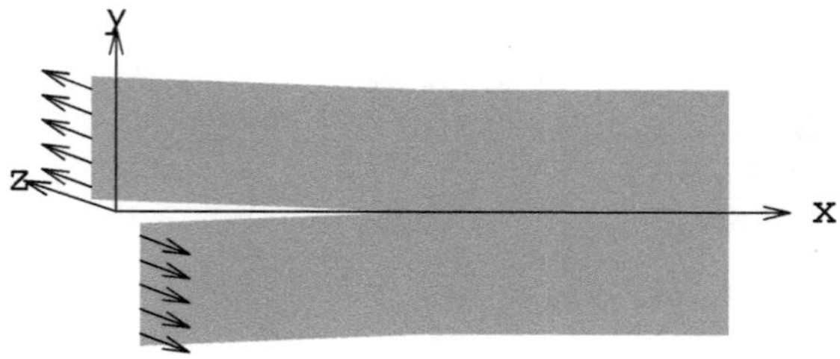

Figure 2.2.1. Tearing.

each t by elementary lower semi-continuity properties. We propose to show that (Ust), (Eb) has a unique solution, which identifies with the global minimum for $\mathcal{E}$ at t, which is thus unique.

Fix t. First, if $u(t)$ is solution to (Ust), then it is such that, once it reaches 0, it stays equal to 0. Indeed, assume that a, with $0 \le a$, is such that $u(t,a) = 0$. Take v with $v = -u(t,x)$ in (a,∞) and $v = 0$ otherwise. For $h \in (0,1)$, $u + hv$ is an admissible test and

$$\begin{aligned}\mathcal{E}(u+hv) - \mathcal{E}(u) &= (-2h+h^2)\int_a^\infty u'(t,x)^2\,dx \\ &\quad + k\int_a^\infty \Big(\operatorname{sg}^+(|(1-h)u(t,x)|) - \operatorname{sg}^+(|u(t,x)|)\,dx\Big) \\ &= (-2h+h^2)\int_a^\infty u'(t,x)^2\,dx.\end{aligned}$$

Thus, invoking (Ust),

$$0 \le \frac{d}{dh}\mathcal{E}(u+hv)\bigg|_{h=0+} \le -2\int_a^\infty u'(t,x)^2 dx \le 0,$$

so that $u(t) = 0$ in (a,∞). But $u(t)$ is continuous in x; thus, there exists $\infty \ge l(t) > 0$ such that $S(u(t)) = [0, l(t))$ with $u(t,0) = tH$ and $u(t,l(t)) = 0$.

We now perform an inner variation in $\mathcal{E}$. Take v be in $C_0^\infty(0,\infty)$. When $|h|$ is sufficiently small, $x \mapsto \phi_h(x) = x + hv(x)$ is a direct diffeomorphism onto $\mathbb{R}^+$. Moreover, if $u(0) = tH$, $u_h(0) = tH$ and u_h converges to u when h goes to 0. The change of variables $y = (\phi_h)^{-1}(x)$ in the energy yields

$$\mathcal{E}(u(t)\circ\phi_h^{-1}) = \int_0^\infty \Big(\mu H \frac{u'(t,x)^2}{\phi_h'(x)} + \phi_h'(x) k \ \operatorname{sg}^+(u(t,x))\Big) dx,$$

which in turn leads to

$$\frac{d}{dh}\mathcal{E}(u(t)\circ\phi_h^{-1})\Big|_{h=0} = \int_0^\infty \Big(-\mu H u'(t,x)^2 + k\ \ \mathrm{sg}^+(u(t,x))\Big) v'(x)dx.$$

Thus, provided that (Ust) holds for this kind of variations,

$$\mu H u'(t,x)^2 - k\ \ \mathrm{sg}^+(u(t,x)) = c, \tag{2.2.2}$$

for some constant c.

Now, take $v \geq 0$ be in $\mathcal{C}_0^\infty(0, l(t))$. Then,

$$\frac{d}{dh}\mathcal{E}(u(t)+hv)\Big|_{h=0} = -2\mu H \int_0^{l(t)} u''(t,x)v(x)dx.$$

Thus, invoking (Ust) again, we get that, on $(0, l(t))$, $u'' \leq 0$, that is that u' is monotonically decreasing there.

Now, if $l(t)$ were infinite, we would have from (2.2.2) that $u' \equiv d$, some negative constant, which contradicts the convergence of u to 0 at infinity. Thus $l(t)$ is finite and $u' \equiv 0$ on $(l(t), \infty)$, hence $c = 0$, and

$$u'(t,x) = -\sqrt{\frac{k\ \ \mathrm{sg}^+(u(t,x))}{\mu H}}, x > 0,\ u(0) = tH. \tag{2.2.3}$$

We then conclude that the solution $u(t)$ to (Ust) is unique and that it is given by

$$S(u) = [0, l(t)) \quad \text{with} \quad l(t) = tH\sqrt{\frac{\mu H}{k}},$$

while

$$u(t,x) = tH\left(1 - \frac{x}{l(t)}\right)^+.$$

Also note that $l(t)$ and $u(t)$ increase with t, so that irreversibility is automatic, while energy balance is guaranteed by the evoked smoothness of $u(t)$.

Here, in contrast with the setting of Subsection 2.1, unilateral stationarity, unilateral local, or unilateral global minimality are indistinguishable, at least for an increasing load.

Remark 2.2.1 *Note the surreptitious assumption that inner variations are valid tests for stationarity. In the presentation of Section 1, stationarity was introduced in the form of a combination of outer and inner variation (see*

(1.2.5)). It is in that sense that the re-formulated problem of Proposition 1.2.1 was equivalent to the original problem (1.1.1), (1.1.4) and an investigation of possible additional constraints on that problem resulting from the introduction of inner variations should be undertaken.

For a given length l of the tear (crack), the total energy at time t is immediately seen to be $\mu H^3 t^2/l + kl$, hence strictly convex in l, so that, according to Remark 1.2.3, the smoothness of the evolution $l(t)$ is hardly surprising.

3 Global minimality

In a Griffith setting, irreversibility is a simple notion: the crack can only extend with time,

$$\Gamma(t) \supset \Gamma(s), s < t.$$

With that notion in mind, we now discuss the variational evolution in a global minimality setting, noting that existence in such a context will automatically provide existence of that evolution for any kind of local minimality criterion. Once again, the argument put forth in Paragraph 2.1 (see Remark 2.1.3) essentially prohibits force loads. We thus assume throughout this subsection that the only load is a displacement $g(t)$ defined on $\partial_d\Omega$, or rather, as we saw earlier in Subsection 1.3, on $\mathbb{R}^2 \setminus \overline{\Omega}$.

3.1 Discrete evolution

As mentioned in the Introduction, the basic tool is also the natural computational tool: time discretization over the interval $[0, T]$. We thus consider

$$t_0 = 0 < t_1^n < ...... < t_{k(n)}^n = T \text{ with } k(n) \overset{n}{\nearrow} \infty, \; \Delta_n := t_{i+1}^n - t_i^n \overset{n}{\searrow} 0.$$

Time-stepping the strong or weak minimality condition (Ugm), we obtain

(Sde) The strong discrete evolution: Find $(\Gamma_{i+1}^n, \varphi_{i+1}^n)$ a minimizer for

$$\min_{\varphi,\Gamma} \left\{ \int_{\Omega\setminus\Gamma} W(\nabla\varphi)\, dx + k\mathcal{H}^1(\Gamma \setminus \partial_s\Omega) : \right.$$
$$\left. \varphi = g(t_{i+1}^n) \text{ on } \partial_d\Omega \setminus \Gamma;\; \Gamma \supset \Gamma_i^n \right\};$$

resp.

(Wde) The weak discrete evolution: Find φ^n_{i+1} a minimizer for

$$\min_{\varphi}\left\{\int_\Omega W(\nabla\varphi)\,dx + k\mathcal{H}^1(S(\varphi)\setminus(\Gamma^n_i \cup \partial_s\Omega)) : \varphi = g(t^n_{i+1}) \text{ on } \partial_d\Omega \setminus S(\varphi)\right\};$$

then, $\Gamma^n_{i+1} = \Gamma^n_i \cup (S(\varphi^n_{i+1}) \setminus \partial_s\Omega)$.

The balance (Eb) seems to have been forgotten all together in the discrete evolution, yet it will reappear in the time-continuous limit of those evolutions.

The first mathematical issue to confront is the existence of a solution to those discrete evolutions. As we mentioned before in Subsection 1.3, we cannot expect, even in 2d, a direct existence proof for the *strong discrete evolution* without imposing further restrictions on the class of admissible cracks. This is easily understood through the Neumann sieve example Murat (1985).

A Neumann sieve situation occurs when boundaries close up at a critical speed that creates channels of non–zero capacity in the domain. For example, consider $\Omega = (-1,1)^2$ and assume, in a linear anti-plane shear setting, that the crack Γ_n is given as $\{0\}\times[-1,1]\setminus\left(\bigcup_{p=-n,\ldots,n}\left(\frac{p}{n}-e^{-n},\frac{p}{n}+e^{-n}\right)\right)$

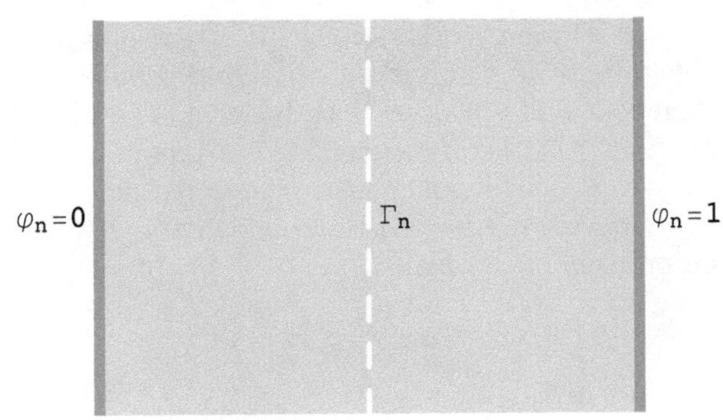

Figure 3.1.1. The Neumann sieve.

with

$$\varphi_n = \begin{cases} 0 \text{ on } \{x_1=-1\},\\ 1 \text{ on } \{x_1=1\}.\end{cases}$$

Then φ_n satisfies

$$-\Delta\varphi_n = 0 \text{ on } \Omega_n := (-1,1)^2 \setminus \Gamma_n,$$

with $\dfrac{\partial \varphi_n}{\partial \nu} = 0$ on all boundaries of $\Omega \setminus \Gamma_n$, except $\{x_1 = \mp 1\}$. According to the results in Murat (1985) $\varphi_n \to \varphi$ strongly in $L^2(\Omega)$, with $\Omega = [(-1,0) \cup (0,1)] \times (-1,1)$ and φ is the solution, for some $\mu \neq 0$ of

$$-\Delta\varphi = 0 \text{ on } \Omega,$$

with

$$\begin{cases} \dfrac{\partial \varphi}{\partial x_2} & = & 0 & \text{on} & \partial\Omega \cap \{x_2 = \pm 1\} \\ \varphi & = & 0, \text{ resp. } 1 & \text{on} & \{-1\} \times (-1,1), \text{ resp. } \{1\} \times (-1,1) \\ \dfrac{\partial \varphi}{\partial x_1} & = & \mu[\varphi] & \text{on} & \{0\} \times (-1,1). \end{cases}$$

Hence φ_n does not converge to the solution

$$\hat{\varphi} = 0 \text{ on } (-1,0) \times (-1,1);\ 1 \text{ on } (0,1) \times (-1,1)$$

of the Neumann problem on $\Omega \setminus \Gamma$, with $\Gamma = \{0\} \times (-1,1)$.

The Neumann sieve must thus be prevented so as to ensure the very existence of a pair-solution to the strong discrete evolution at each time step. A possible exit strategy consists in "prohibiting" disconnected cracks. A result of Chambolle and F. Doveri Chambolle and Doveri (1997) (see also Bucur and Varchon (2000)) states that, if Ω is a Lipschitz two dimensional domain and $\{\Gamma_n\}$ is a sequence of compact connected sets with $\mathcal{H}^1(\Gamma_n) \leq \mathcal{C}$ and such that it converges – for the Hausdorff metric – to Γ, the solution to a Neumann problem of the form

$$\begin{cases} -\Delta\varphi_n + \varphi_n & = g \text{ in } \Omega \setminus \Gamma^n \\ \dfrac{\partial \varphi_n}{\partial \nu} & = 0 \text{ on } \partial[\Omega \setminus \Gamma^n] \end{cases}$$

is such that $\varphi_n, \nabla\varphi_n \overset{n}{\longrightarrow} \varphi, \nabla\varphi$, strongly in $L^2(\Omega)$, with φ solution to

$$\begin{cases} -\Delta\varphi + \varphi & = g \text{ in } \Omega \setminus \Gamma \\ \dfrac{\partial \varphi}{\partial \nu} & = 0 \text{ on } \partial[\Omega \setminus \Gamma] \end{cases}$$

An adaptation of that result by Dal Maso and Toader Dal Maso and Toader (2002) proves the existence of a minimizer to the strong discrete evolution at each time step under the restriction that the cracks have an a priori set number of connected components. In turn, Chambolle in Chambolle (2003) proves an analogous result for plane elasticity. We will not discuss the strong evolution any further in these notes.

Note that the connectedness restriction can be weakened to include cracks with an a priori set number of connected components Dal Maso and Toader (2002).

The discrete weak evolution behaves better as far as existence is concerned. Indeed, existence is a direct consequence of Ambrosio's compactness result (1.3.5), together with the following lower semi-continuity result which applies to the kind of elastic energy under consideration and to the sequence φ_n in (1.3.5) (see Ambrosio (1994))

$$\int_\Omega W(\nabla\psi)dx \le \liminf_n \int_\Omega W(\nabla\varphi_n)\, dx. \tag{3.1.1}$$

(There is a slight modification in (1.3.5) which consists in replacing $\mathcal{H}^1$ by $\mathcal{H}^1 \llcorner (\Gamma_i^n \cup \partial_s\Omega)^c$.)

To be precise, existence is established in

- The anti-plane shear case: φ is scalar-valued and W is convex and has p-growth, $p > 1$;
- The non-linear elasticity case: φ is vector-valued and W is quasi-convex with p-growth, $p > 1$. We refer the reader to the abundant literature on quasi-convexity (see e.g. Ball and Murat (1984)) for details on that notion; for our purpose, it suffices to remark that quasi-convexity, plus growth implies sequential weak lower semi-continuity on the Sobolev space $W^{1,p}(\Omega;\mathbb{R}^2)$ Ball and Murat (1984), but also, see Ambrosio (1994), on

  $$L^\infty(\Omega;\mathbb{R}^2) \cap \left\{\varphi \in SBV(\Omega;\mathbb{R}^2) : \nabla\varphi \in L^p(\Omega;\mathbb{R}^{2\times 2})\right\}.$$

  It should be noted that the growth assumption prevents the energy density $W(F)$ from blowing up as $\det F \searrow 0$, a desirable feature in hyperelasticity; and, most recently,
- The case of finite elasticity: φ is vector-valued and W is poly-convex with blow-up as $\det F \searrow 0$ and a weak form of non-interpenetration is imposed Dal Maso and Lazzaroni (2010). We will not consider this case in the sequel.

Existence will not however be achieved in the setting of linearized elasticity which thus seems confined, for the time being, to the strong formulation.

Consider any setting for which the discrete evolution is meaningful. Then, for a given n (a given time step), we define the piecewise in time fields

$$\begin{cases} \varphi^n(t) := \varphi_i^n \\ \Gamma^n(t) := \Gamma_i^n \qquad \text{on } [t_i^n, t_{i+1}^n), \quad \text{and, for } i = -1, \Gamma_{-1}^n := \Gamma_0. \\ g^n(t) = g(t_i^n) \end{cases}$$

Remark that irreversibility is guaranteed at the discrete level because of the definition of Γ_i^n in terms of its predecessors. In other words, $\Gamma^n(t) \nearrow$ with t.

Summing up, we have constructed, for each time $t \in [0,T]$, a pair $(\Gamma^n(t), \varphi^n(t))$ such that

(Wde) The weak discrete evolution: $\varphi^n(t)$ is a minimizer for

$$\min_{\varphi} \left\{ \int_\Omega W(\nabla\varphi)\, dx + k\mathcal{H}^1(S(\varphi)\backslash(\Gamma^n(t-\Delta_n) \cup \partial_s\Omega)) : \right.$$
$$\left. \varphi = g^n(t) \text{ on } \partial_d\Omega\backslash S(\varphi) \right\}$$

and $\Gamma^n(t) = \Gamma^n(t-\Delta_n) \cup S(\varphi^n(t))$.

Here again, the functional dependence of $\varphi^n(t)$ is not specified because it depends upon the scalar/vectorial nature of the specific problem, as well as on the coercivity/growth properties of the bulk energy density W.

At time $t = 0$, generically, it is not true that Γ_0^n (independent of n) $\equiv \Gamma_0$, but merely that $\Gamma_0^n \supset \Gamma_0$. There is an increase in the initial condition.

The goal is to pass to the limit in n and hope that the limit fields will be solutions to the strong/weak variational evolutions. As will be seen below, this is not a straightforward proposition.

3.2 Global minimality in the limit

A usual first step in a limit process is to obtain n-independent a priori estimates on the fields. This will be obtained here upon testing the weak discrete evolution (Wde) at each time by appropriate test fields. The two choice test fields are $g^n(t)$ and $\varphi^n(t-\Delta_n)+g^n(t)-g^n(t-\Delta_n)$ (the addition of the terms involving g^n are so that the test deformations satisfy the boundary conditions at time t).

Then, provided we impose decent regularity on g, namely

$$g \in W^{1,1}(0,T; W^{1,p}(\Omega(;\mathbb{R}^2)) \cap L^\infty((0,T) \times \Omega(;\mathbb{R}^2)), \qquad (3.2.1)$$

for an energy with $p > 1$-growth, we obtain the following a priori bounds:

$$\begin{cases} \|\nabla\varphi^n(t)\|_{L^p(\Omega(;\mathbb{R}^2))} \le \mathcal{C} \\ \mathcal{H}^1(S(\varphi^n(t)) \le \mathcal{C}, \end{cases} \tag{3.2.2}$$

and

$$\mathcal{H}^1(\Gamma^n(t)) \le \mathcal{C}, \tag{3.2.3}$$

together with the following upper bound on the total energy

$$\begin{aligned} E^n(t) &:= \int_\Omega W(\nabla\varphi^n(t))\,dx + k\mathcal{H}^1(\Gamma^n(t)\setminus\partial_s\Omega) \\ &\le E^n(0) + \int_0^{\tau^n(t)}\int_\Omega \frac{\partial W}{\partial F}(\nabla\varphi^n(s)).\nabla\dot{g}(s)\,dx\,ds, \end{aligned} \tag{3.2.4}$$

where $\tau^n(t) := \sup\{t_i^n \le t\}$. Remark that the derivation of (3.2.4) actually requires a bit of care; see Dal Maso et al. (2005), Section 6.

It remains to pass to the n–limit in the minimality statement (Wde) under the above convergences and to avoid a Neumann sieve phenomenon as more and more crack components accumulate at a given time when $n \nearrow$.

To this effect, remark that the circumstances that presided over the appearance of the Neumann sieve phenomenon in Paragraph 3.1 were deceiving, for they failed to account for the role played by the surface energy. Indeed, consider n large enough; the pair φ_n, Γ_n considered in that example cannot be a joint minimizer of

$$\frac{1}{2}\int_{\Omega\setminus\Gamma} |\nabla\varphi|^2\,dx + \mathcal{H}^1(\Gamma),\ \Gamma \supset \Gamma_n$$

with the same boundary conditions. By lower semi-continuity,

$$\liminf_n \frac{1}{2}\int_{\Omega\setminus\Gamma} |\nabla\varphi_n|^2\,dx + \mathcal{H}^1(\Gamma_n) \ge \frac{1}{2}\int_{\Omega\setminus\Gamma} |\nabla\varphi|^2\,dx + 1,$$

with φ, the solution to the Neumann sieve. Now, φ has non zero bulk energy $\frac{1}{2}\int_{\Omega\setminus\Gamma} |\nabla\varphi|^2\,dx$, say $\mathcal{C}$, so that, for n large enough,

$$\frac{1}{2}\int_{\Omega\setminus\Gamma} |\nabla\varphi_n|^2\,dx + \mathcal{H}^1(\Gamma_n) \ge 1 + \frac{\mathcal{C}}{2}.$$

But the energy associated to the pair $(\{0\}\times[-1,1], \hat{\varphi})$ is exactly 1, a strictly smaller value, while $\{0\}\times[-1,1] \supset \Gamma_n$. For n large enough, closing the holes of the sieve and taking the crack to be $\{0\}\times[-1,1]$ is the energetically sound choice. There is thus hope for a derivation of the global minimality condition (`Ugm`) in the weak variational evolution from (`Wde`) under refinement of the time step.

That this is by no means a trivial endeavor can be illustrated as follows. We note first that, since $\mathcal{H}^1(B \setminus A) \geq \mathcal{H}^1(B) - \mathcal{H}^1(A)$, (`Wde`) implies in particular that $\varphi^n(t)$ is *a minimizer for its own jump set*, that is

$$\frac{1}{2}\int_\Omega W(\nabla\varphi^n(t))\,dx \leq \frac{1}{2}\int_\Omega W(\nabla\varphi)\,dx + k\mathcal{H}^1(S(\varphi)\setminus(S(\varphi^n(t))\cup\partial\Omega_s)). \tag{3.2.5}$$

If (`Ugm`) is to be obtained in the limit, then $\varphi(t)$ should also in particular be a minimizer for its own jump set. In view of (3.2.2) and of the already quoted lower semi-continuity result of Ambrosio (1994), the left hand side of (3.2.5) is well behaved and the result would follow easily, provided that

$$\limsup_n \mathcal{H}^1(S(\varphi)\setminus S(\varphi^n(t))) \leq \mathcal{H}^1(S(\varphi)\setminus S(\varphi(t))).$$

Consider however φ such that $S(\varphi)\subset S(\varphi(t))$, while the jump set of $\varphi^n(t)$ does not intersect that of $\varphi(t)$ (which would surely happen if $S(\varphi^n(t))\subset K_n$, with $K_n\cap K=\emptyset$ and the Hausdorff distance from K_n to K goes to 0). Then $\mathcal{H}^1(S(\varphi))$ must be 0!

The stability of the own jump set minimality condition cannot be established without a modification of the test fields φ. This is the essence of the jump transfer Theorem Francfort and Larsen (2003), Section 2. We now quote it without proof in its simplest version.

Theorem 3.2.1 *Let $\varphi^n, \varphi \in SBV(\Omega)$ with $\mathcal{H}^1(S(\varphi))<\infty$, be such that*
- *$|\nabla\varphi^n|$ weakly converges in $L^1(\Omega)$;and*
- *$\varphi^n \to \varphi$ in $L^1(\Omega)$.*

Then, for every $\zeta\in SBV(\Omega)$ with $\nabla\zeta\in L^p(\Omega)$, $1\leq p<\infty$, and $\mathcal{H}^1(S(\zeta))<\infty$, there exists a sequence $\{\zeta^n\}\subset SBV(\Omega)$ with $\nabla\zeta^n\in L^p(\Omega)$, such that
- *$\zeta^n\to\zeta$ strongly in $L^1(\Omega)$;*
- *$\nabla\zeta^n\to\nabla\zeta$ strongly in $L^q(\Omega)$; and*
- *$\limsup_n \mathcal{H}^1\llcorner A\,(S(\zeta^n)\setminus S(\varphi^n)) \leq \mathcal{H}^1\llcorner A\,(S(\zeta)\setminus S(\varphi))$, for any Borel set A.*

We fix a time t and recall (3.2.2). Ambrosio's compactness result permits to assert the existence of a t-dependent subsequence $\{\varphi^{n_t}(t)\}$ of $\{\varphi^n(t)\}$ and of $\varphi(t)$ such that the assumptions of Theorem 3.2.1 – or rather of a corollary of Theorem 3.2.1 which takes into account the boundary conditions on the test fields at t, namely $\varphi^{n_t}(t)=g^{n_t}(t)$ on $\mathbb{R}^2\setminus\Omega$ – are met. The conclusion of that theorem then allows for a corresponding sequence $\{\zeta^{n_t}\}$ that is an admissible test in (Wde), so that

$$\int_\Omega W(\nabla\varphi^{n_t}(t))\,dx \le \int_\Omega W(\nabla\zeta^{n_t})\,dx + k\mathcal{H}^1\left(S(\zeta^{n_t}) \setminus (S(\varphi^{n_t}(t)) \cup \partial_s\Omega)\right),$$

and then, from the convergences obtained in the theorem, together with the assumed p-growth of the energy, we pass to the limit in n_t and obtain that the limit $\varphi(t)$ is a minimizer for its own jump set, that is

$$\int_\Omega W(\nabla\varphi(t))\,dx \le \int_\Omega W(\nabla\zeta)\,dx + k\mathcal{H}^1\left(S(\zeta) \setminus (S(\varphi(t)) \cup \partial_s\Omega)\right).$$

We are inching ever closer to the global minimality statement (Ugm) in the weak variational evolution, but are not quite there yet, because we would like to remove not only $S(\varphi(t))\cup\partial_s\Omega$ but $\Gamma(t)\cup\partial_s\Omega$ in the minimality statement above. To do this, we need to define the limit crack $\Gamma(t)$. There are various setting-dependent paths to a meaningful definition of the limit crack. An encompassing view of that issue is provided by the notion of σ_p-convergence introduced in Dal Maso et al. (2005), Section 4, a kind of set convergence for lower dimensional sets.

Definition 3.2.2 *Γ^n σ_p-converges to Γ if $\mathcal{H}^1(\Gamma^n)$ is bounded uniformly with respect to n, and*

(1) Whenever $\varphi^j, \varphi \in SBV(\mathbb{R}^2)$ are such that

$$\begin{cases} \varphi^j \overset{weak-*}{\rightharpoonup} \varphi, \ \text{in } L^\infty(\mathbb{R}) \\ \nabla\varphi^j \overset{weak}{\rightharpoonup} \nabla\varphi, \ \text{in } L^p(\mathbb{R}^2) \\ S(\varphi^j) \subset \Gamma^{n_j} \end{cases}$$

for some sequence $n_j \nearrow \infty$, then $S(\varphi) \subset \Gamma$;

(2) there exist a function $\varphi \in SBV^p(\mathbb{R}^2)$ with $S(\varphi) = \Gamma$ and a sequence φ^n with the properties of item (1).

The following compactness result proved in Dal Maso et al. (2005), Section 4.2, holds true:

Theorem 3.2.3 *Let $\Gamma^n(t)$ be a sequence of increasing sets defined on $[0,T]$ and contained in a bounded set B. Assume that $\mathcal{H}^1(\Gamma^n(t))$ is bounded uniformly with respect to n and t. Then there exist a subsequence Γ^{n_j} and $\Gamma(t)$ defined on $[0,T]$ such that*

$$\Gamma^{n_j}(t) \quad \sigma_p\text{-converges to } \Gamma(t),\ \forall t \in [0,T].$$

The estimate (3.2.3) permits to apply the theorem above and thus to define a meaningful crack $\Gamma(t)$ such that, for a subsequence still labeled $\Gamma^n(t)$, $\Gamma^n(t)$ σ_p–converges to $\Gamma(t)$, hence also $\Gamma^{n_t}(t)$. Thanks to item (2) in Definition 3.2.2, we can construct φ with $S(\varphi) = \Gamma(t)$ and φ^{n_t} satisfying the assumptions of Theorem 3.2.1 with $S(\varphi^{n_t}) \subset \Gamma^{n_t}(t)$. But (Wde) implies in particular that

$$\begin{aligned}\int_\Omega W(\nabla\varphi^{n_t}(t))\,dx &\le \int_\Omega W(\nabla\zeta)\,dx + k\mathcal{H}^1(S(\zeta)\backslash(\Gamma^{n_t}(t)\cup\partial_s\Omega))\\ &\le \int_\Omega W(\nabla\zeta)\,dx + k\mathcal{H}^1(S(\zeta)\backslash(S(\varphi^{n_t})\cup\partial_s\Omega)).\end{aligned}$$

and the jump transfer Theorem 3.2.1 delivers the minimality property (Ugm).

Having obtained global minimality, we still have to derive energy conservation (Eb). This is the object of the next paragraph.

3.3 Energy balance in the limit

Inequality (3.2.4) derived at the onset of Paragraph 3.2 hints at the possibility of an energy inequality. To obtain such an inequality in the limit, one should ensure that, as $n_t \nearrow \infty$,

$$\int_\Omega W(\nabla\varphi^{n_t}(t))\,dx \to \int_\Omega W(\nabla\varphi(t))\,dx \tag{3.3.1}$$

and that

$$\limsup_{n_t} \int_0^{\tau^n(t)} \int_\Omega \frac{\partial W}{\partial F}(\nabla\varphi^{n_t}(s)).\ \nabla\dot g(s)dx\,ds \le \int_0^t \int_\Omega \frac{\partial W}{\partial F}(\nabla\varphi(s)).\nabla\dot g(s)\,dx\,ds. \tag{3.3.2}$$

Equality (3.3.1) is nearly immediate; one inequality holds true by lower semi-continuity as seen several times before. The other is obtained upon applying the jump transfer Theorem 3.2.1 to $\varphi(t)$ itself and inserting the resulting test sequence in (3.2.5). This yields the other inequality, namely

$$\limsup_{n_t} \int_\Omega W(\nabla\varphi^{n_t}(t))\,dx \le \int_\Omega W(\nabla\varphi(t))\,dx.$$

The derivation of (3.3.2) is more involved in the non-quadratic case. Indeed, it amounts, modulo application of Fatou's lemma for the time integral, to showing that the stresses $\dfrac{\partial W}{\partial F}(\nabla\varphi^{n_t}(t))$ converge weakly to the limit stress

$\frac{\partial W}{\partial F}(\nabla\varphi(t))$. Although a surprising result, this is indeed the case in view of the convergences announced for $\varphi^{n_t}(t)$ to $\varphi(t)$ and of (3.3.1); we omit the proof and refer the interested reader to Dal Maso et al. (2005), Section 4.3.

The following energy inequality is established:

$$\begin{aligned} E(t) &:= \int_\Omega W(\nabla\varphi(t))\,dx + k\mathcal{H}^1(\Gamma(t)\setminus\partial_s\Omega) \\ &\leq E(0) + \int_0^t\int_\Omega \frac{\partial W}{\partial F}(\nabla\varphi(s)).\nabla\dot{g}(s)\,dx\,ds, \end{aligned} \tag{3.3.3}$$

The other energy inequality is a byproduct of the minimality statement (Ugm). Simply test global minimality at time s by $\varphi(t)+g(s)-g(t)$, $t>s$. Then, since $S(\varphi(t) \subset \Gamma(t)$,

$$\begin{aligned} \int_\Omega W(\nabla\varphi(s))\,dx &\leq \int_\Omega W(\nabla\varphi(t)+g(s)-g(t))\,dx \\ &\qquad + \mathcal{H}^1(S(\varphi(t))\setminus(\Gamma(s)\cup\partial_s\Omega)) \\ &\leq \int_\Omega W(\nabla\varphi(t)+g(s)-g(t))\,dx \\ &\qquad + \mathcal{H}^1(\Gamma(t)\setminus(\Gamma(s)\cup\partial_s\Omega)) \\ &= \int_\Omega W(\nabla\varphi(t))\,dx + \mathcal{H}^1(\Gamma(t)\setminus(\Gamma(s)\cup\partial_s\Omega)) \\ &\qquad - \int_\Omega \frac{\partial W}{\partial F}\left(\nabla\varphi(t)+\rho(s,t)\int_s^t \nabla\dot{g}(\tau)\,d\tau\right). \\ &\qquad \int_s^t \nabla\dot{g}(\tau)\,d\tau\,dx, \end{aligned}$$

for some $\rho(s,t)\in[0,1]$. Hence

$$E(t)-E(s) \geq \int_\Omega \frac{\partial W}{\partial F}\left(\nabla\varphi(t)+\rho(s,t)\int_s^t \nabla\dot{g}(\tau)\,d\tau\right).\int_s^t \nabla\dot{g}(\tau)\,d\tau\,dx.$$

We then choose a partition $0 < s_1^n < .... < s_{k(n)}^n = t$ of $[0,t]$, with $\Delta_n' := s_{i+1}^n - s_i^n \searrow 0$; summing the contributions, we get

$$\begin{aligned} E(t)-E(0) \geq \sum_{i=0}^{k(n)} \int_\Omega \frac{\partial W}{\partial F}\left(\nabla\varphi(s_{i+1}^n)+\rho(s_i^n,s_{i+1}^n)\int_{s_i^n}^{s_{i+1}^n} \nabla\dot{g}(\tau)\,d\tau\right). \\ \int_{s_i^n}^{s_{i+1}^n} \nabla\dot{g}(\tau)d\tau dx. \end{aligned}$$

A uniform continuity type result – already implicitly used in the derivation of (3.2.4) – permits to drop the term depending on $\rho(s_i^n, s_{i+1}^n)$ in the previous inequality in the limit $\Delta'_n \searrow 0$; see Dal Maso et al. (2005), Section 4.3. Thus

$$E(t) - E(0) \geq \limsup_n \left\{ \sum_{i=0}^{k(n)} \int_{s_i^n}^{s_{i+1}^n} \int_\Omega \frac{\partial W}{\partial F} \left(\nabla\varphi(s_{i+1}^n)\right).\nabla\dot{g}(\tau)\, dx\, d\tau \right\}.$$

The expression on the right hand-side of the previous inequality looks very much like a Riemann sum. A not so well-known result in integration asserts that Riemann sums of a Lebesgue integrable function do converge to the integral of that function, but only for carefully chosen partitions Hahn (1914). Since we enjoy complete liberty in our choice of the partition $\{s_j^n\}$ of $[0,t]$, we conclude that

$$E(t) - E(0) \geq \int_0^t \int_\Omega \frac{\partial W}{\partial F} \left(\nabla\varphi(s)\right).\nabla\dot{g}(\tau)\, dx\, d\tau,$$

which, together with (3.3.3), provides the required equality `(Eb)`.

3.4 The time-continuous evolution

Here, the results obtained in the previous paragraphs are coalesced into an existence statement to the weak variational evolution. The result is expressed in a 2d setting, but it applies equally in a 3d setting, upon replacing $\mathcal{H}^1$ by $\mathcal{H}^2$. We also recall similar existence results obtained in Dal Maso and Toader (2002),Chambolle (2003) in the 2d connected case.

In what follows, the energy density W is a nonnegative convex – in the anti-plane shear setting – or quasiconvex – in the plane setting – C^1 function on $\mathbb{R}^2$ with

$$(1/\mathcal{C})|F|^p - \mathcal{C} \leq W(F) \leq C|F|^p + C, \quad \forall F,\ 1 < p < \infty.$$

Note that the assumptions on W immediately imply that (see, e.g., Dacorogna (1989))

$$|DW(F)| \leq \mathcal{C}(1 + |F|^{p-1}).$$

The domain Ω under consideration is assumed throughout to be Lipschitz and bounded , and the function g, which appears in the boundary condition on $\partial\Omega_d$, is assumed to be defined on all of $\mathbb{R}^2$; actually, each of its components is taken to be in $W^{1,1}_{loc}([0,\infty); W^{1,p}(\mathbb{R}^2))$.

The traction-free part $\partial_s\Omega$ of the boundary is assumed to be closed. Finally, the pre-existing crack Γ_0 is a closed set in Ω, with $\mathcal{H}^1(\Gamma_0) < \infty$.

Theorem 3.4.1 *$\exists \Gamma(t) \subset \overline{\Omega}$ and φ such that*

- *each component of $\varphi(t) \in SBV(\mathbb{R}^2)$, with $\nabla\varphi$ p-integrable;*
- *$\Gamma(t) \supset \Gamma_0$ increases with t and $\mathcal{H}^1(\Gamma(t)) < +\infty$;*
- *$S(\varphi(t)) \subset \Gamma(t) \cup \partial_s\Omega$ and $\varphi(t) = g(t)$ a.e. on $\mathbb{R}^2 \setminus \overline{\Omega}$;*
- *For every $t \geq 0$ the pair $(\varphi(t), \Gamma(t))$ minimizes*

$$\int_\Omega W(\nabla\varphi)\, dx + k\mathcal{H}^1(\Gamma \setminus \partial_s\Omega)$$

  among all $\Gamma \supset \Gamma(t)$ and φ with components in $SBV(\mathbb{R}^2)$ s.t. $S(\varphi) \subset \Gamma$ and $\varphi = g(t)$ a.e. on $\mathbb{R}^2 \setminus \overline{\Omega}$;
- *the total energy*

$$E(t) := \int_\Omega W(\nabla\varphi(t))\, dx + k\mathcal{H}^1(\Gamma(t) \setminus \partial_s\Omega)$$

  is absolutely continuous, $DW(\nabla\varphi) \cdot \nabla\dot{g} \in L^1_{loc}([0,\infty); L^1(\mathbb{R}^2))$, and

$$E(t) = E(0) + \int_0^t \int_\Omega DW(\nabla\varphi(s)) \cdot \nabla\dot{g}(s)\, dx\, ds.$$

We could incorporate body or surface loads, provided they belong to a certain class of soft devices Dal Maso et al. (2005), Section 3.

Note that, in the vector-valued setting, it is assumed that somehow, the deformations are always capped in sup-norm by some set number. This is an a-priori assumption which can be verified for certain classes of quasi-convex energies Leonetti and Siepe (2005). There is no need for such an assumption in the anti-plane shear case, provided that the displacement load g is also bounded in sup-norm.

In 2d only and in the case where the cracks are assumed a priori to be connected – or to have a pre-set number of connected components – then the same existence result for the strong variational evolution is obtained in Dal Maso and Toader (2002) in the quadratic case and in Chambolle (2003) in the case of linearized elasticity. The statement is identical to that of Theorem 3.4.1 at the expense of replacing $\int_\Omega$ by $\int_{\Omega\setminus\Gamma}$, and considering φ's with components in $L^{1,2}(\Omega\setminus\Gamma) := \{f \in L^2_{loc}(\Omega\setminus\Gamma) : \nabla f \in L^2(\Omega\setminus\Gamma)\}$, resp. $\varphi \in LD(\Omega\setminus\Gamma) := \{\zeta \in L^2_{loc}(\Omega\setminus\Gamma; \mathbb{R}^2) : e(\zeta) \in L^2(\Omega\setminus\Gamma; \mathbb{R}^4)\}$. in the case of linear elasticity.

This existence result calls for comments. First and foremost, it is an existence result, not a uniqueness result. As in other non-convex problems in mechanics, uniqueness should not be expected.

Then the lack of regularity of the field $\varphi(t)$ indicates that time jumps could appear in the various fields. Indeed, still referring to that same example, we witness there a brutal decrease to 0 at time t_i of the bulk energy with a corresponding increase of the surface energy. The total energy (Eb) will remain impervious to those jumps.

Third, an implicit change of initial conditions may occur, since it might happen that $\Gamma(0)$ contains, but does not equal Γ_0.

Finally, the weak evolution might just turn out to be a strong evolution, as was the case for image segmentation thanks to De Giorgi et al. (1989), in which case there would be no need for the strong variational evolution. Recent results of J. F. Babadjian and A. Giacomini seem to confirm this in a 2d setting.

Remark 3.4.2 *The unilateral global minimality condition (item 4. in Theorem 3.4.1) can actually be strengthened as follows:*
For every $t \geq 0$ the pair $(\varphi(t), \Gamma(t))$ minimizes

$$\int_\Omega W(\nabla\varphi)\, dx + k\mathcal{H}^1(\Gamma \setminus \partial_s\Omega)$$

among all $\Gamma \supset \cup_{s<t}\Gamma(s)$ and φ with components in $SBV(\mathbb{R}^2)$ s.t. $S(\varphi) \subset \Gamma$ and $\varphi = g(t)$ a.e. on $\mathbb{R}^2 \setminus \overline{\Omega}$.

This states that the admissible test cracks do not have to contain the current crack, but only those up to, but not including the current time, a clearly stronger minimality condition. The two conditions are actually equivalent because, for $s < t$, unilateral global minimality implies in particular that

$$\int_\Omega W(\nabla\varphi(s))\, dx + k\mathcal{H}^1(\Gamma(s) \setminus \partial_s\Omega) \leq \int_\Omega W(\nabla\varphi + \nabla g(s) - \nabla g(t))\, dx + k\mathcal{H}^1(\Gamma \setminus \partial_s\Omega),$$

for any φ with components in $SBV(\mathbb{R}^2)$ s.t. $\varphi = g(t)$ a.e. on $\mathbb{R}^2\setminus\overline{\Omega}$, $S(\varphi) \subset \Gamma$ and any $\Gamma \supset \cup_{s<t}\Gamma(s)$. Let $s \nearrow t$ and use item 5. (the continuity of the total energy) to pass to the limit in the left hand-side of the inequality above. The stronger minimality result is then obtained by dominated convergence (since W has p-growth).

4 Numerics

At first glance, numerical implementation of the variational approach advocated in these notes is hopeless because the "classical" numerical methods dealing with discontinuous displacement fields rely on some non-negligible amount of *a priori* knowledge of that path. This includes the extended finite element method and other enrichment-based variants. A proper discretization scheme for the total energy needs to both approximate potentially discontinuous displacement fields –and thus the position of their discontinuity sets – and to lead to an accurate and isotropic approximation of the surface energy. Such a scheme does not easily accommodate cohesive finite element methods or discontinuous Galerkin methods. Note that this is partially addressed by a careful estimate of the anisotropy induced by the mesh in Negri (1999), Negri (2003) or still through the use of adaptive finite element methods Bourdin and Chambolle (2000).

Further, if the variational framework contends that it addresses crack initiation and crack propagation in a unified framework, the same should be true of the numerical method. In particular, methods based on considering energy restitution caused by small increments of existing cracks are ruled out. It can actually be shown (see Chambolle et al. (2008)) that "small" cracks will never lead to descent directions for the global minimization of the total energy in the absence of strong singularities in the elastic field.

Non-convexity of the total energy is yet another major obstacle to overcome. The typical size of the discrete problems prohibits appeal to global or non-deterministic optimization techniques. Global minimization of the energy is an arguable postulate, but it is at present the only one theoretically suitable for a thorough investigation of any numerical implementation. So far, the only exception to that would be the 1d setting where one can establish convergence of the critical points of the approximating functional discussed later on in this section Francfort et al. (2009).

Since, as already noted at the onset of Section 3, global minimality deals badly with force loads, the only loads considered throughout this section are displacement loads.

The numerical method that will be described below finds, once again, its inspiration in the Mumford-Shah functional for image segmentation (see Subsection 1.3). The main ingredients were first introduced in the latter context in Ambrosio and Tortorelli (1990), Ambrosio and Tortorelli (1992), Bellettini and Coscia (1994), Bourdin (1998), Bourdin (1999), Negri and Paolini (2001) and later adapted to fracture in Bourdin et al. (2000), Giacomini and Ponsiglione (2003), Chambolle (2004), Chambolle (2005), Giacomini (2005), Giacomini and Ponsiglione (2006).

The method allows for an isotropic and mesh independent approximation of the total energy. It copes rather successfully with both initiation and propagation as seen through the various numerical experiments presented in Subsection 4.3. Like the actual variational model, it applies to the one, two, or three dimensional cases without alteration.

Finally, time dependence will be approached through time discretization, and all computations will be performed for a sequence of times $t_0 = 0 < t_1^n < \dots\dots < t_{k(n)}^n = T$ with $k(n) \overset{n}{\nearrow} \infty$, $\Delta_n := t_{i+1}^n - t_i^n \overset{n}{\searrow} 0$. We will mostly drop the n-dependence, unless explicitly referring to the putative convergence of the time-discrete evolution to the time-continuous evolution.

4.1 Numerical approximation of the energy

The essence of the numerical implementation is to be found in the concept of variational convergence. Specifically, the first step consists in devising a good approximation of the total energy in the sense of Γ–convergence. We refer the reader to Dal Maso (1993), Braides (2002) for a complete exposition of the underlying theory.

Consider a $\overline{\mathbb{R}}$-valued functional $\mathcal{F}$ defined over, say a metrizable topological space X, and a sequence $\mathcal{F}_\varepsilon$ of the same type. Then, $\mathcal{F}_\varepsilon$ Γ–converges to $\mathcal{F}$ as $\varepsilon \searrow 0$ iff the following two conditions are satisfied for any $u \in X$:

1. *lower–inequality:* for any sequence $(u_\varepsilon)_\varepsilon \in X$ converging to u,

$$\liminf_{\varepsilon \to 0} \mathcal{F}_\varepsilon(u_\varepsilon) \geq \mathcal{F}(u); \tag{4.1.1}$$

2. *existence of a recovery–sequence:* there exists a sequence $(u_\varepsilon)_\varepsilon \in X$ converging to u, such that

$$\limsup_{\varepsilon \to 0} \mathcal{F}_\varepsilon(u_\varepsilon) \leq \mathcal{F}(u). \tag{4.1.2}$$

The interest of Γ–convergence from the standpoint of numerics lies in the following elementary theorem in Γ-convergence:

Theorem 4.1.1 *If $\mathcal{F}_\varepsilon$ Γ–converges to $\mathcal{F}$ and u_ε^* is a minimizer of $\mathcal{F}_\varepsilon$ and if, further, the sequence u_ε^* is compact in X, then there exists $u^* \in X$ such that $u_\varepsilon^* \to u$, u^* is a global minimizer for $\mathcal{F}$, and $\mathcal{F}_\varepsilon(u_\varepsilon^*) \to \mathcal{F}(u^*)$.*

Stability of global minimizers under Γ–convergence is indeed a powerful numerical tool. Rather than attempting to minimize the total energy – thus having to reconcile discretization and discontinuous functions – we propose to construct, at each time step t_i, a family of regularized energies $\mathcal{E}_\varepsilon^i$ that Γ-converge to $\mathcal{E}^i$, the energy for the weak variational evolution at that time step (see (1.3.6)). In the footstep of Ambrosio and Tortorelli (1990), Ambrosio and Tortorelli (1992), we will approximate the potentially discontinuous field φ^i and its crack set Γ^i by two smooth functions. The implementation of the first time step, which is very close to that of the original approximation in the context of the Mumford-Shah functional, is presented in Paragraph 4.1. while Paragraph 4.1 shows how to account for irreversibility and approximate the weak discrete time evolution (Wde).

The first time step Consider the first time step of the weak discrete evolution under the unilateral global minimality condition (Ugm). The irreversibility condition is trivially satisfied, so that it suffices to minimize the total energy

$$\mathcal{E}(\varphi) = \int_\Omega W(\nabla\varphi)dx + k\mathcal{H}^1(S(\varphi))$$

with respect to any kinematically admissible φ. In all that follows, $\widetilde{\Omega}$ denotes a "large enough" open bounded set such that $\Omega \subset \widetilde{\Omega}$, and the Dirichlet boundary conditions are enforced on $\widetilde{\Omega} \setminus \bar{\Omega}$, not on $\mathbb{R}^2 \setminus \bar{\Omega}$ because, as will be seen below, the computations are performed on that larger domain, and not only on Ω.

Following Ambrosio and Tortorelli (1990), Ambrosio and Tortorelli (1992), we introduce a secondary variable $v \in W^{1,2}(\widetilde{\Omega} \setminus \partial_s\Omega)$ and two small positive parameters ε, and $\eta_\varepsilon = o(\varepsilon)$, and define, for any kinematically admissible φ,

$$\mathcal{F}(\varphi, v) = \begin{cases} \displaystyle\int_\Omega W(\nabla\varphi)\,dx + k\mathcal{H}^{N-1}(S(\varphi) \setminus \partial_S\Omega) & \text{if } v = 1 \text{ a.e.} \\ +\infty & \text{otherwise,} \end{cases} \tag{4.1.3}$$

and

$$\mathcal{F}_\varepsilon(\varphi, v) = \int_\Omega (v^2 + \eta_\varepsilon)W(\nabla\varphi)\,dx + k\int_{\widetilde{\Omega}\setminus\partial_s\Omega} \left[\frac{(1-v)^2}{4\varepsilon} + \varepsilon|\nabla v|^2\right] dx. \tag{4.1.4}$$

In the anti-plane case, proving the Γ–convergence of $\mathcal{F}_\varepsilon$ to $\mathcal{F}$ is a simple adaptation of Ambrosio and Tortorelli's result (see Bourdin (1998)) while it is more involved in that of linearized elasticity Chambolle (2004). We limit the analysis to the former case assuming that the energy is quadratic in the field, *i.e.*, $W(F) := 1/2\mu|F|^2$. The proof of the lower inequality of Theorem 4.1.1 is technical and does not shed much light on the proposed numerical method. Instead, we present a simpler but weaker inequality *i.e.* a version of (4.1.1) with a "wrong" constant that still highlights the link between the regularized and weak energies. The construction of the recovery sequence in Theorem 4.1.1 provides valuable insight and we propose to detail it, at least when the target is a mildly regular kinematically admissible field for $\mathcal{E}$. Actually, deriving the lim-sup inequality (4.1.2) for minimizers can easily be seen to be no restriction. But for those, the mild regularity assumption below holds true, at least in anti-plane shear and energy densities of the form $|F|^p$ with $p > 1$.

We start with the compactness of minimizing sequences. Let $(\varphi_\varepsilon, v_\varepsilon)$ be a sequence of minimizers for $\mathcal{F}_\varepsilon$. We show that $v_\varepsilon \to 1$ almost everywhere and that there exists $\varphi \in SBV(\Omega)$ such that $\varphi_\varepsilon \to \phi$ in L^2.

A simple truncation argument shows that we are at liberty to assume that

$$\|\varphi_\epsilon\|_{L^\infty} \leq C \tag{4.1.5}$$

and

$$0 \leq v_\epsilon \leq 1. \tag{4.1.6}$$

Using the classical inequality $a^2 + b^2 \geq 2ab$, we also have that

$$\frac{(1-v_\varepsilon)^2}{4\varepsilon} + \varepsilon^2 |\nabla v_\varepsilon|^2 \geq (1-v_\varepsilon)|\nabla v_\varepsilon|. \tag{4.1.7}$$

Finally, testing with $v = 1$ and φ kinematically admissible, we get

$$\mathcal{F}_\varepsilon(\varphi_\varepsilon, v_\varepsilon) \leq C. \tag{4.1.8}$$

From (4.1.8), $v_\varepsilon \to 1$ *a.e.* in Ω. In order to obtain the compactness of the sequence φ_ε, we consider the function $\omega_\varepsilon := (2v_\varepsilon - v_\varepsilon^2)\varphi_\varepsilon$ and note that ω_ε is uniformly bounded in $L^\infty(\Omega)$. We have that

$$\nabla\omega_\varepsilon = (2v_\varepsilon - v_\varepsilon)^2 \nabla\varphi_\varepsilon + 2(1-v_\varepsilon)\nabla v_\varepsilon \varphi_\varepsilon.$$

From (4.1.6) and (4.1.8), we easily obtain that $(2v_\varepsilon - v_\varepsilon)^2 \nabla\varphi_\varepsilon$ is uniformly bounded in L^1 and from (4.1.7), (4.1.5), and (4.1.8), that $2(1-v_\varepsilon)\nabla v_\varepsilon \varphi_\varepsilon$ is also bounded in L^1. Thus ω_ε is uniformly bounded in $L^\infty(\Omega) \cap BV(\Omega)$, so that there exists ω such that

$$\omega_\epsilon \to \omega \text{ } a.e. \text{ in } \Omega.$$

Finally, remark that $\varphi_\varepsilon = \omega_\varepsilon / 2v_\varepsilon - v_\varepsilon^2$ and that, since $\omega_\varepsilon \to \omega$ and $v_\varepsilon \to 1$ *a.e.* in Ω, $\varphi := \omega$ is such that

$$\varphi_\varepsilon \to \varphi \text{ in } L^2(\Omega).$$

We now focus on the lower inequality of Γ–convergence, or rather on a weaker inequality. Take any $(\varphi_\varepsilon, v_\varepsilon)$ such that $(\varphi_\varepsilon, v_\varepsilon) \to (\varphi, 1)$ in $\left(L^2(\Omega)\right)^2$. Then,

$$\liminf_{\varepsilon \to 0} \mathcal{F}_\varepsilon(\varphi_\varepsilon, v_\varepsilon) \geq \int_\Omega W(\nabla\varphi)\, dx + \frac{k}{2}\mathcal{H}^{N-1}(S(\varphi)). \tag{4.1.9}$$

We may as well assume that (4.1.5), (4.1.6), (4.1.8) hold true, in which case $\varphi \in BV(\Omega)$.

Using (4.1.7) again,

$$\mathcal{F}_\varepsilon(\varphi_\varepsilon, v_\varepsilon) \geq \int_\Omega v_\varepsilon^2 |\nabla \varphi_\varepsilon|^2 \, dx + k \int_{\widetilde{\Omega} \setminus \partial_s \Omega} (1 - v_\varepsilon) |\nabla v_\varepsilon| \, dx.$$

By increasing rearrangement,

$$\int_\Omega v_\varepsilon^2 |\nabla \varphi_\varepsilon|^2 \, dx = \int_0^1 2s \int_{\{v_\varepsilon > s\}} |\nabla \varphi_\varepsilon|^2 \, dx \, ds,$$

and using the co-area formula for BV-functions (see for instance Ambrosio et al. (2000); Evans and Gariepy (1992)),

$$\int_{\widetilde{\Omega} \setminus \partial_s \Omega} (1 - v_\varepsilon) |\nabla v_\varepsilon| \, dx = \int_0^1 (1 - s) \mathcal{H}^{N-1} \left(\partial^* \{v_\varepsilon > s\} \setminus \partial_s \Omega \right) ds.$$

Putting the above expressions together, we obtain

$$\begin{aligned} \mathcal{F}_\varepsilon(\varphi_\varepsilon, v_\varepsilon) \geq & \int_0^1 2s \int_{\{v_\varepsilon > s\}} |\nabla \varphi_\varepsilon|^2 \, dx \, ds \\ & + k \int_0^1 (1 - s) \mathcal{H}^{N-1} \left(\partial^* \{v_\varepsilon > s\} \setminus \partial_s \Omega \right) ds. \end{aligned} \tag{4.1.10}$$

Consider any $s \neq 1$ and define $\omega_\varepsilon^s := \varphi_\varepsilon \chi_{\{v_\varepsilon > s\}}$. Since $\varphi_\varepsilon \in W^{1,2}$ and $\{v_\varepsilon > s\}$ is a set of finite perimeter, $\omega_\varepsilon^s \in SBV$,

$$\partial^* \{v_\varepsilon > s\} = S(\omega_\varepsilon^s),$$

the jump set (set of non Lebesgue points) of ω_ε^s, and

$$\int_{\{v_\varepsilon > s\}} |\nabla \varphi_\varepsilon|^2 \, dx = \int_\Omega |\nabla \omega_\varepsilon^s|^2 \, dx.$$

Since $v_\varepsilon \to 1$ *a.e.*, we also obtain that $\omega_\varepsilon^s \to \varphi$ in L^2.

Combining (4.1.10) and the two identities above and using Fatou's lemma, we obtain

$$\begin{aligned} \liminf_{\varepsilon \to 0} \mathcal{F}_\varepsilon(\varphi_\varepsilon, v_\varepsilon) \geq & \int_0^1 2s \liminf_{\varepsilon \to 0} \int_\Omega |\nabla \omega_\varepsilon^s|^2 \, dx \, ds \\ & + k \int_0^1 (1 - s) \liminf_{\varepsilon \to 0} \mathcal{H}^{N-1} \left(S(\omega_\varepsilon^s) \setminus \partial_s \Omega \right) ds. \end{aligned}$$

By Ambrosio's Compactness and lower semi-continuity theorems in SBV, we obtain

$$\liminf_{\varepsilon\to 0} \mathcal{F}_\varepsilon(\varphi_\varepsilon, v_\varepsilon) \geq \int_0^1 2s \int_\Omega |\nabla\varphi|^2\, dx\, ds + k\int_0^1 (1-s)\mathcal{H}^{N-1}\left(S(\varphi)\setminus\partial_s\Omega\right)\, ds.$$

Integrating in s yields (4.1.9).

We finally turn to the construction of the recovery sequence (4.1.2). The following construction does not account for the Dirichlet boundary condition and the interested reader is referred to Bourdin (1998) for the corresponding technicalities. As a corollary, we may as well take $\widetilde{\Omega} \equiv \Omega$ in the construction of the attainment sequence that follows. In the case of interest to us, *i.e.*, that with Dirichlet boundary conditions on a part $\partial_d\Omega = \partial\Omega\setminus\partial_s\Omega$ it will be enough to re-introduce $\widetilde{\Omega}\setminus\partial\Omega_s$ in lieu of Ω in the second integral in (4.1.4).

We also assume that φ is a solution to the minimization of $\mathcal{E}$ that satisfies

$$\mathcal{H}^1(\overline{S(\varphi)}) = \mathcal{H}^1(S(\varphi)). \tag{4.1.11}$$

For minimizers of the Mumford-Shah functional, this mild, albeit difficult regularity property was established in De Giorgi et al. (1989). In the scalar-valued setting, the case of a certain class of convex bulk energies which includes $p > 1$-homogeneous energies was investigated in Fonseca and Fusco (1997). The regularity result was generalized to our setting in Bourdin (1998), at least for minimizers in anti-plane shear with a quadratic elastic energy density. The closure property (4.1.11) is not so clearly true in more general settings, and different approximation processes must be used in such cases; the interested reader is invited to consult e.g. Braides (2002).

Consider a kinematically admissible field φ – an element of $SBV(\Omega)$ – satisfying (4.1.11). Define

$$d(x) := \operatorname{dist}(x, \overline{S(\varphi)}).$$

The volume of the area bounded by the s-level set of d is

$$\ell(s) := \left|\left\{x\in\mathbb{R}^2\,;\, d(x)\leq s\right\}\right|.$$

The distance function is 1-Lipschitz, *i.e.*, $|\nabla d(x)| = 1$ a.e., while, by the co-area formula for Lipschitz functions (see e.g. Ambrosio et al. (2000)),

$$\ell(s) = \int_0^s \mathcal{H}^1\left(\{x\,; d(x) = t\}\right)\, dt,$$

so that, in particular,

$$\ell'(s) = \mathcal{H}^1\left(\{x\,; d(x) = s\}\right). \tag{4.1.12}$$

Also, see Federer (1969)-3.2.39,

$$\lim_{s\to 0} \frac{\ell(s)}{2s} = \mathcal{H}^1(S(\varphi)).$$

We choose α_ε such that $\alpha_\varepsilon = o(\varepsilon)$, $\eta_\varepsilon = o(\alpha_\varepsilon)$, which is possible since $\eta_\varepsilon = o(\varepsilon)$, and define the functions

$$v_\varepsilon(x) := \begin{cases} 0 & \text{if } d(x) \le \alpha_\varepsilon \\ 1 - \exp\left(-\dfrac{d(x)-\alpha_\varepsilon}{2\varepsilon}\right) & \text{otherwise,} \end{cases} \tag{4.1.13}$$

and

$$\varphi_\varepsilon(x) := \begin{cases} \dfrac{d(x)}{\alpha_\varepsilon}\varphi(x) & \text{if } 0 \le d(x) \le \alpha_\varepsilon \\ \varphi(x) & \text{otherwise.} \end{cases}$$

Note that it is easily seen that $\varphi_\varepsilon \in W^{1,2}(\Omega)$. Further, $\varphi_\varepsilon \to \varphi$ in $L^2(\Omega)$, and $v_\varepsilon \to 1$ almost everywhere. Since $v_\varepsilon \le 1$,

$$\int_\Omega (v_\varepsilon^2 + \eta_\varepsilon)|\nabla\varphi_\varepsilon|^2 dx \le \int_{d(x)\le\alpha_\varepsilon} \eta_\varepsilon|\nabla\varphi_\varepsilon|^2 dx + \int_{d(x)\ge\alpha_\varepsilon} (1+\eta_\varepsilon)|\nabla\varphi|^2 dx.$$

Observe now that, for $d(x) \le \alpha_\varepsilon$, $\nabla\varphi_\varepsilon = d(x)/\alpha_\varepsilon\nabla\varphi + 1/\alpha_\varepsilon\varphi\nabla d$, so, in view of the 1-Lipschitz character of d and of the L^∞-bound on φ,

$$\begin{aligned}\int_\Omega (v_\varepsilon^2 + \eta_\varepsilon)|\nabla\varphi_\varepsilon|^2\, dx \le 2\left(\eta_\varepsilon \int_{d(x)\le\alpha_\varepsilon} |\nabla\varphi|^2\, dx + M^2\frac{\eta_\varepsilon}{\alpha_\varepsilon^2}\ell(\alpha_\varepsilon)\right)& \\ + \int_{d(x)\ge\alpha_\varepsilon} (1+\eta_\varepsilon)|\nabla\varphi|^2 dx.&\end{aligned}$$

Since $\int_\Omega |\nabla\varphi|^2\, dx < \infty$, the first term in the parenthesis on the right hand side above converges to 0 as $\varepsilon \to 0$. Recalling that $\ell(\alpha_\varepsilon)/\alpha_\varepsilon = O(1)$, while $\eta_\varepsilon/\alpha_\varepsilon = o(1)$ permits one to conclude that the limit of the second term in that parenthesis also converges to 0 with ε. We conclude that

$$\limsup_{\varepsilon\to 0} \int_\Omega \left(v_\varepsilon^2 + \eta_\varepsilon\right)|\nabla\varphi_\varepsilon|^2\, dx \le \int_\Omega |\nabla\varphi|^2\, dx. \tag{4.1.14}$$

Let us examine the surface energy term. Using once again the 1-Lipschitz character of d, together with the co-area formula, we get

$$\int_\Omega \left\{\varepsilon|\nabla v_\varepsilon|^2 + \frac{(1-v_\varepsilon)^2}{4\varepsilon}\right\} dx \leq \frac{\ell(\alpha_\varepsilon)}{4\varepsilon} + \int_{d(x)\geq\alpha_\varepsilon} \frac{1}{2\varepsilon}\exp(-\frac{d(x)-\alpha_\varepsilon}{\varepsilon})$$

$$\leq \frac{\ell(\alpha_\varepsilon)}{4\varepsilon} + \frac{1}{2\varepsilon}\int_{\alpha_\varepsilon}^{\infty} \exp(-\frac{s-\alpha_\varepsilon}{\varepsilon})\mathcal{H}^1\left(\{d(x)=s\}\right) ds. \quad (4.1.15)$$

Recalling (4.1.12),

$$\frac{1}{2\varepsilon}\int_{\alpha_\varepsilon}^{\infty} \exp(-\frac{s-\alpha_\varepsilon}{\varepsilon})\mathcal{H}^1\left(\{d(x)=s\}\right) ds = \frac{e^{\frac{\alpha_\varepsilon}{\varepsilon}}}{2\varepsilon}\int_{\alpha_\varepsilon}^{\infty} e^{-s/\varepsilon}\ell'(s)\, ds$$

$$= \frac{e^{\frac{\alpha_\varepsilon}{\varepsilon}}}{2}\int_{\alpha_\varepsilon/\varepsilon}^{\infty} e^{-t}\ell'(t\varepsilon)\, dt. \quad (4.1.16)$$

Since $\ell'(0) = \lim_{s\to 0}\ell(s)/s = 2\mathcal{H}^1(S(\varphi))$, $\alpha_\varepsilon = o(\varepsilon)$ and $\int_0^\infty e^{-t}\,dt = 1$, insertion of (4.1.16) into (4.1.15) and application of Lebesgue's dominated convergence theorem yields

$$\limsup_{\varepsilon\to 0}\int_\Omega \left\{\varepsilon|\nabla v_\varepsilon|^2 + \frac{(1-v_\varepsilon)^2}{4\varepsilon}\right\} dx \leq \mathcal{H}^1(S(\varphi)). \quad (4.1.17)$$

Collecting (4.1.14), (4.1.17) gives the upper Γ–limit inequality.

Remark 4.1.2 *The form of the field v_ε in (4.1.13) may seem somewhat ad-hoc. It is not. The choice of the profile for the field v_ε is derived from the solution of an "optimal profile" problem (see Alberti (2000)). Consider, in e.g. 2d, a point x on the crack and a line orthogonal to the crack and passing through x, parameterized by the variable s. Consider the restriction of the regularized surface energy to this line*

$$\mathcal{F}_{\varepsilon,x}(s) = k\int_0^\infty \left\{\frac{(1-v(s))^2}{4\varepsilon} + \varepsilon|v'(s)|^2\right\} ds.$$

Then the profile

$$v_\varepsilon(s) = 1 - \exp\left(-\frac{(s-\alpha_\varepsilon)}{2\varepsilon}\right)$$

corresponds to the minimizer of $\mathcal{F}_{\varepsilon,x}$ under the following boundary conditions:

$$v_\varepsilon(\alpha_\varepsilon) = 0; \ \lim_{s\to\infty} v_\varepsilon(s) = 1.$$

Indeed, it is also possible to construct the field v_ε for the upper Γ-limit along lines intersecting the crack set at 90^0 angles, using the solution to the optimal profile problem on each of those. Integration of the result along the crack set will also permit one to recover the upper Γ-limit.

The Γ–convergence result above can be extended to the restriction $\mathcal{F}_{\varepsilon,h}$ of $\mathcal{F}_\varepsilon$ to a linear finite element approximation, provided that the discretization parameter h is such that $h = o(\varepsilon)$ (see Bellettini and Coscia (1994), Bourdin (1999)). A closer look at the construction for the upper Γ–limit and at its adaptation to $\mathcal{F}_{\varepsilon,h}$ provides some useful insight into possible error estimates.

The construction of the sequence $(\varphi_{\varepsilon,h}, v_{\varepsilon,h})$ for the upper Γ–limit for $\mathcal{F}_{\varepsilon,h}$ can be obtained from that above. Let $\mathcal{T}_h$ be a conforming mesh of $\widetilde{\Omega} \setminus \partial\Omega_s$ and S_h be the set of all elements in $\mathcal{T}_h$ intersecting $S(\varphi)$. Let π_h be a linear finite element projection operator associated with $\mathcal{T}_h$, and consider

$$v_{\varepsilon,h}(x) := \begin{cases} 0 & \text{if } x \in S_h; \\ \pi_h\,(v_\varepsilon) & \text{otherwise,} \end{cases} \tag{4.1.18}$$

and

$$\varphi_{\varepsilon,h}(x) := \pi_h\,(\varphi_\varepsilon)\,. \tag{4.1.19}$$

Following a path similar that developed in the computation of the upper Γ-limit above, the first term $\ell(\alpha_\varepsilon)/4\varepsilon$ on the right hand-side of inequality (4.1.15) becomes $|S_h|/4\varepsilon \simeq \mathcal{H}^1(S(\varphi))h/4\varepsilon$, which converges to 0 only if $h = o(\varepsilon)$. The consideration of quadratic finite elements in lieu of linear ones would still induce an error on the surface energy of the order of h/ε, albeit with a different constant. This is why the proposed implementation only resorts to piecewise linear finite elements for φ and v.

In a different direction, this term links the anisotropy of the mesh to the quality of the approximation of the surface energy. In Negri (1999), M. Negri studied the effect of various types of structured meshes on the surface energy for the Mumford-Shah problem. In the numerical experiments, the isotropy of the surface term is ensured through the use of "almost" isotropic Delaunay meshes.

From the construction above, it is deduced that the relation $h = o(\varepsilon)$ only needs to be satisfied "close" to $S(\varphi)$. Of course, barring prior knowledge of $S(\varphi)$, uniformly homogeneous fine meshes are a must. However, *a posteriori* re-meshing the domain will then improve the accuracy of the energy estimate. However, *a priori* mesh adaption is not advisable because the local size of the mesh affects the quality of the approximation of the surface energy and can potentially create spurious local minimizers. So, *a posteriori* mesh refinement around the cracks shields the computations from artificial cracks that would correspond to local minima created by *a priori* mesh refinement! Note that some recent computations in Burke et al. (2010) seem to deal rather successfully with *a priori* mesh adaption.

The sequence for the upper Γ–limit is also admissible for the lower Γ–limit, so that, if φ if a minimizer for the total energy, the sequence $(\varphi_\varepsilon, v_\varepsilon)$ constructed above approximates a minimizing sequence for $\mathcal{F}_{\varepsilon,h}$ and this asymptotically in h, that is

$$\int_{\widetilde{\Omega}\setminus\partial\Omega_s} \left\{ \frac{(1-v_\varepsilon)^2}{4\varepsilon} + \varepsilon|\nabla v_\varepsilon|^2 \right\} dx = \left(1 + \frac{h}{4\varepsilon}\right) \mathcal{H}^1(S(\varphi)). \tag{4.1.20}$$

In practice, it is as if the fracture toughness had been amplified by a factor $1 + h/4\varepsilon$, which has to be accounted for when interpreting the results. The experiments in Section 4.3 highlight the effect of mesh isotropy on the results, and show how the fracture toughness is overestimated.

Quasi-static evolution The approximation scheme devised in Subsection 4.1 should now be reconciled with the evolutionary character of the weak discrete formulation. Irreversibility of the crack growth is enforced at the time-discrete level in the manner described below.

Consider a fixed ε and a fixed conforming mesh $\mathcal{T}_h$ of $\widetilde{\Omega} \setminus \partial\Omega_s$ with characteristic element size h. Introduce a small parameter $\eta > 0$, and at each step t_i, the set of vertices

$$K^i_{\varepsilon,h,\eta} := \left\{ s \in \mathcal{T}_h \,;\, v^i_{\varepsilon,h}(s) \le \eta \right\},\ i > 0;\ K^0_{\varepsilon,h,\eta} := \emptyset.$$

In the light of the Γ-convergence properties of $\mathcal{F}_{\varepsilon,h}$, the crack growth condition translates into a growth condition on the sets $K^i_{\varepsilon,h,\eta}$ and leads to the following fully spatially and temporally discrete evolution scheme:

`(Fde)` Find a sequence $\left(\varphi^{i+1}_{\varepsilon,h}, v^{i+1}_{\varepsilon,h}\right)_{i=0,\dots,n}$ of global minimizers for $\mathcal{F}_{\varepsilon,h}$ under the constraints

$$\varphi = g(t_{i+1}) \text{ on } \widetilde{\Omega} \setminus \Omega$$

and

$$v = 0 \text{ on } K^i_{\varepsilon,h,\eta}. \tag{4.1.21}$$

Recently, Giacomini conducted a rigorous analysis of a slightly different approach to the time evolution for $\mathcal{F}_\varepsilon$. In Giacomini (2005), crack growth is enforced through the monotonicity of v in time, *i.e.*, by successively minimizing $\mathcal{F}_\varepsilon$ among all (φ, v) such that $\varphi = g(t_{i+1})$ on $\widetilde{\Omega} \setminus \Omega$, and $v \le v^\varepsilon_i$ almost everywhere on Ω. In that setting, as both the time discretization parameter (Δ_n) and ε go to 0 (in a carefully ordered fashion), the discrete evolution converges to a continuous evolution satisfying the conclusions of Theorem 3.4.1.

In the forthcoming numerical experiments, crack growth is enforced as described in (4.1.21). Implementing monotonicity would not generate additional difficulties, but only slightly increase the computational cost as equality constraints would have to be replaced by inequality ("box") constraints.

Remark 4.1.3 *The Γ-convergence based approach to minimization is not so easily amenable to the treatment of local minimization. If $(\varphi, 1)$ is an* isolated *L^1-local minimizer for $\mathcal{F}$ (see (4.1.3)), then Theorem 2.1 in Kohn and Sternberg (1989) can be adapted to the current setting to prove the existence of a sequence of L^1-local minimizers $(\varphi_\varepsilon, v_\varepsilon)$ for $\mathcal{F}_\varepsilon$ converging to $(\varphi, 1)$ in L^1. Unfortunately, the isolation hypothesis is generically false: see for instance the 1d-traction experiment with a hard device in Paragraph 2.1.*

Even when the isolation hypothesis applies, the above-mentioned theorem grants the existence of a sequence of local minimizers for $\mathcal{F}_\varepsilon$ converging to a local minimizer of $\mathcal{F}$, but does not however guarantee that a converging sequence of local minimizers for $\mathcal{F}_\varepsilon$ converges to a local minimizer for $\mathcal{F}$.

The only positive results in this direction concern the numerically uninteresting 1d case Francfort et al. (2009).

4.2 Minimization algorithm

Recall that $\mathcal{F}_{\varepsilon,h}$ is the restriction of $\mathcal{F}_\varepsilon$ defined in (4.1.4) to a linear finite element approximation. Also note that, although $\mathcal{F}_\varepsilon$ is separately convex in its arguments φ and v, it is not convex in the pair (φ, v).

In the numerical experiments below, we fix the regularization parameter ε and generate a mesh with characteristic size h. We do not try to adapt the values on ε and h during the numerical minimization of $\mathcal{F}_{\varepsilon,h}$. Thus, the numerical implementation reduces to a sequence of minimizations for $\mathcal{F}_{\varepsilon,h}$, each corresponding to a separate time step. All presented experiments have been tested on meshes of various size and with different values of the parameter ε and/or of the time discretization length; the results seem impervious to such changes, at least for reasonably small choices of the parameters ε, h, Δ_n.

Because of the lack of convexity of $\mathcal{F}_{\varepsilon,h}$, the minimization scheme is purely heuristic. As per Section 4.1, we should choose a mesh size h which remains "small" compared to the regularization parameter, which in turn needs to be "small". In a 2d setting, this typically results in meshes with $(10)^5$ elements, while in 3d, meshes will consist of over a million elements.

Although the analysis of such large problems can be tackled thanks to the wider availability of massively parallel computers, there are, to our knowledge, no global minimization algorithms capable of handling them. At best, the algorithms will satisfy necessary optimality conditions for minimality.

The alternate minimization algorithm The first building block in the numerical implementation is an alternate minimization algorithm, leading to evolutions satisfying a first set of necessary conditions for optimality.

The functional $\mathcal{F}_\varepsilon$ – and therefore $\mathcal{F}_{\varepsilon,h}$ – is Gateaux-differentiable around any (φ, v). We compute the first order variation of $\mathcal{F}_{\varepsilon,h}$ around any kinematically admissible (φ, v) in the directions $(\tilde\varphi, 0)$ and $(0, \tilde v)$, where $\tilde\varphi$ and $\tilde v$ are admissible variations ($\tilde\varphi = 0$ on $\widetilde\Omega \setminus \Omega$ and $\tilde v = 0$ on $K^i_{\varepsilon,h,\eta}$) and obtain that the solution $(\varphi^{i+1}_{\varepsilon,h}, v^{i+1}_{\varepsilon,h})$ of the fully discrete evolution at time step t_{i+1} satisfies

$$\begin{cases} \displaystyle\int_\Omega \left((v^{i+1}_{\varepsilon,h})^2 + \eta_\varepsilon\right) \mathrm{D}W(\nabla\varphi^{i+1}_{\varepsilon,h}).\nabla\tilde\varphi\, dx = 0 \\ \displaystyle\int_\Omega \left(v^{i+1}_{\varepsilon,h}\tilde v\right) W(\nabla\varphi^{i+1}_{\varepsilon,h})\, dx + k\int_{\widetilde\Omega\setminus\partial\Omega_s} \left(\frac{v^{i+1}_{\varepsilon,h}\tilde v}{4\varepsilon} + \varepsilon\nabla v^{i+1}_{\varepsilon,h}\cdot\nabla\tilde v\right) dx \\ \displaystyle\qquad\qquad = k\int_{\widetilde\Omega\setminus\partial\Omega_s} \frac{\tilde v}{4\varepsilon}\, dx. \end{cases} \tag{4.2.1}$$

This leads to the following algorithm (δ is a fixed tolerance parameter):

Algorithm 1 *The alternate minimization algorithm:*

1: *let* $p = 0$ *and* $v^{(0)} := v^i_{\varepsilon,h}$.
2: ***repeat***
3: $\quad p \leftarrow p + 1$
4: $\quad$ *compute* $\varphi^{(p)} := \arg\min_\varphi \mathcal{F}_{\varepsilon,h}(\varphi, v^{(p-1)})$ *under the constraint* $\varphi^{(p)} = g(t_{i+1})$ *on* $\widetilde\Omega \setminus \Omega$.
5: $\quad$ *compute* $v^{(p)} := \arg\min_v \mathcal{F}_{\varepsilon,h}(\varphi^{(p)}, v)$ *under the constraint* $v^{(p)} = 0$ *on* $K^i_{\varepsilon,h,\eta}$
6: ***until*** $\|v^{(p)} - v^{(p-1)}\|_\infty \le \delta$
7: *set* $\varphi^{i+1}_{\varepsilon,h} := \varphi^{(p)}$ *and* $v^{i+1}_{\varepsilon,h} := v^{(p)}$

Since $\mathcal{F}_{\varepsilon,h}$ is separately convex in each of its arguments, the algorithm constructs at each time step a sequence with decreasing total energy; it is therefore unconditionally convergent in energy. A more detailed analysis conducted in Bourdin (2007) proves that, whenever the cracks are *a priori* known to propagate smoothly, the alternate minimization algorithm converges to the global minimizer of $\mathcal{F}_{\varepsilon,h}$ for fine enough time discretization steps. In cases where cracks propagate brutally, this algorithm can only be proved to converge to critical points of $\mathcal{F}_\varepsilon$, which may be a local (or global) minimizers, but also saddle points for $\mathcal{F}_\varepsilon$. As per Remark 4.1.3, local minimizers of $\mathcal{F}_\varepsilon$ can sometimes be proved to converge to local minimizers of $\mathcal{F}$, but similar results are lacking at present in the case of saddle points, except in 1d. The detection of saddle points require a detailed stability study. Because of the typical size of the problems, this is a difficult task which has yet to be implemented. Instead, we investigate additional necessary conditions for minimality and propose to devise compatible algorithms.

The backtracking algorithm, a tool for global minimization. When cracks propagate brutally, the alternate minimization algorithm, or any other descent-based algorithm for that matter, cannot be expected to converge to the global minimizer of $\mathcal{F}_{\varepsilon,h}$. Indeed, a numerical method that relies solely on (4.2.1) will lead to evolutions whose total energy $E(t)$ is not an absolutely continuous (or even continuous) function (see Figure 11 in Negri (2003) or Figure 3(b) in Bourdin et al. (2000)). This is incompatible with Theorem 3.4.1. So, since (4.2.1) is satisfied at each time step, those evolutions have to correspond to local minimizers or saddle points of the regularized energy. Such solutions – spurious from the standpoint of global minimization – can actually be eliminated by enforcing an additional optimality condition.

Consider a monotonically increasing load, as in Section 1.2.3, and suppose the elastic energy density W to be 2-homogeneous (adapting this argument to p-homogenous W is trivial). If $(\varphi^i_{\varepsilon,h}, v^i_{\varepsilon,h})$ is admissible for a time step t_i, then $\left(t_j/t_i\varphi^i_{\varepsilon,h}, v^i_{\varepsilon,h}\right)$ is admissible for all time steps t_j with $0 \le j \le i$, and

$$\mathcal{F}_{\varepsilon,h}\left(\frac{t_j}{t_i}\varphi^i_{\varepsilon,h}, v^i_{\varepsilon,h}\right) = \frac{t_j^2}{t_i^2}\mathcal{F}^b_{\varepsilon,h}(\varphi^i_{\varepsilon,h}, v^i_{\varepsilon,h}) + \mathcal{F}^s_{\varepsilon,h}(v^i_{\varepsilon,h}),$$

$\mathcal{F}^b_{\varepsilon,h}$ and $\mathcal{F}^s_{\varepsilon,h}$ denoting respectively the bulk and surface terms in $\mathcal{F}_{\varepsilon,h}$. But if the sequence $\{(\varphi^i_{\varepsilon,h}, v^i_{\varepsilon,h})\}$ is a solution of the fully discrete evolution, $(\varphi^j_{\varepsilon,h}, v^j_{\varepsilon,h})$ must minimize $\mathcal{F}_{\varepsilon,h}$ among all admissible pairs (φ, v), and in particular, for $0 \le j \le i \le n$,

$$\mathcal{F}^b_{\varepsilon,h}\left(\varphi^j_{\varepsilon,h}, v^j_{\varepsilon,h}\right) + \mathcal{F}^s_{\varepsilon,h}\left(v^j_{\varepsilon,h}\right) \le \frac{t_j^2}{t_i^2}\mathcal{F}^b_{\varepsilon,h}(\varphi^i_{\varepsilon,h}, v^i_{\varepsilon,h}) + \mathcal{F}^s_{\varepsilon,h}(v^i_{\varepsilon,h}). \quad (4.2.2)$$

In establishing (4.2.2), the *global* minimality of the evolution $\{(\varphi^i_{\varepsilon,h}, v^i_{\varepsilon,h})\}$ was used, so that (4.2.2) is a necessary condition for global minimality but it is neither necessary, nor sufficient for local minimality. Since $t_j \leq t_i$, the total energy, that is $\{\mathcal{F}_{\varepsilon,h}(\varphi^i_{\varepsilon,h}, v^i_{\varepsilon,h})\}$, associated with an evolution satisfying (4.2.2) is monotonically increasing.

Algorithmically, we check condition (4.2.2) against all previous time steps t_j, with j varying from 0 to i. If for some t_j, (4.2.2) is not satisfied, then $(\varphi^j_{\varepsilon,h}, v^j_{\varepsilon,h})$ cannot be the global minimizer for the time step t_j, and $(t_j/t_i\varphi^i_{\varepsilon,h}, v^i_{\varepsilon,h})$ provides an admissible field with a strictly smaller energy at time t_j. In this case, we backtrack to time step t_j, and restart the alternate minimization process, initializing the field v with $v^i_{\varepsilon,h}$. Because the alternate minimization algorithm constructs sequences with monotonically decreasing energy (at a given time step), repeated backtracking will converge to a solution such that (4.2.2) is satisfied for this particular choice of i and j.

4.3 The Tearing experiment

In order to illustrate the numerical method presented in the Sections above, we revisit the tearing experiment investigated in Section 2.2. Consider this time a rectangular domain $\Omega = (0, L) \times (-H, H)$ (with traction free boundary conditions at $x = L$). The analysis in Subsection 2.2 still applies and the field constructed there under assumption (2.2.1) is an admissible test field for this problem, provided of course that $0 \ll l(t) \leq L$.

However, when the domain has finite length, a crack splitting the whole domain is a minimizing competitor. Let φ_c represent that solution. Following the notation in Section 2.2, we set

$$\begin{cases} S(\varphi_c) = (0, L) \times \{0\} \\ u_c(t, x) = tH, \end{cases}$$

so that

$$E(\varphi_c) = kL.$$

A comparison of the energy of both types of evolutions demonstrates that, under assumption (2.2.1), the global minimizer for the tearing problem is such that $\mathbf{u}(x, y, t) = \text{sign}(y) u(t, x)\mathbf{e}_3$ and $S(\varphi) = [0, l(t)) \times \{0\}$, with

$$u(x,t) = \begin{cases} tH\left(1 - \dfrac{x}{l(t)}\right)^+ & \text{if } t \leq \dfrac{L}{2H}\sqrt{\dfrac{k}{\mu H}} \\ tH & \text{otherwise,} \end{cases} \tag{4.3.1}$$

where

$$l(t) = \begin{cases} tH\sqrt{\dfrac{\mu H}{k}} & \text{if } t \leq \dfrac{L}{2H}\sqrt{\dfrac{k}{\mu H}} \\ L & \text{otherwise.} \end{cases} \tag{4.3.2}$$

This corresponds to a crack that propagates at constant speed

$$\frac{\mathrm{d}l}{\mathrm{d}t} = H\sqrt{\frac{\mu H}{k}}$$

along the symmetry axis, until its length reaches $L/2$, and then jumps along the x-axis until the end point of that axis in the domain. Note that, during the smooth propagation phase, the bulk and surface energies of the sample are equal, and that, throughout the evolution, the total energy of the solution is

$$E(t) = \min\left(2tH\sqrt{\mu Hk}, kL\right). \tag{4.3.3}$$

Now, the anti-plane tearing problem is numerically solved by a method developed in Destuynder and Djaoua (1981), then compared to the crack evolution analytically obtained above.

We consider a domain with dimensions $H = 1$, $L = 5$. The material properties are $E = 1$, $\nu = .2$ (corresponding to $\mu \simeq .4167$), $k = 1.125\,(10)^{-2}$. Following Subsection 2.2, the analysis is restricted at first to symmetric solutions consisting of a single crack of length $l(t)$ propagating along the x–axis, starting from the left edge of the domain, with $l(0) = 0$. In order to estimate $l(t)$, we compute the equilibrium deformation $\varphi(1,l)$ corresponding to a unit load and a crack of length l, using finite element meshes consisting of approximately 70,000 nodes, automatically refined around the crack tip. For various choices of $l \in [0, L]$, we estimate the elastic energy $E_b(1,l)$ associated with $\varphi(1,l)$, as well as the energy release rate $G(1,l) = -\partial E_b/\partial l(1,l)$, using classical formulae for the derivative of E_b with respect to the domain shape. Figures 4.3.1, 4.3.2 respectively represent the evolution of $E_b(1,l)$ and $G(1,l)$.

From now onward, we refer to the analytical solution as the "1d solution" in all figures, as well as in the text.

A quick analysis of the numerical results shows that $G(1,l)$ is strictly decreasing (and therefore that E_b is strictly convex) for $0 \leq l < l_c^*$, with $l_c^* \simeq 4.19$. For $l_c^* \leq l \leq 5$, G is an increasing function of l. Recalling Remark 1.2.3 in Section 1, we deduce that the crack will first propagate smoothly, following Griffith's criterion. When it reaches the length l_c^* , it will then jump brutally to the right edge of the domain because not doing so would violate the constraint that $G \leq k$. It could be argued that such

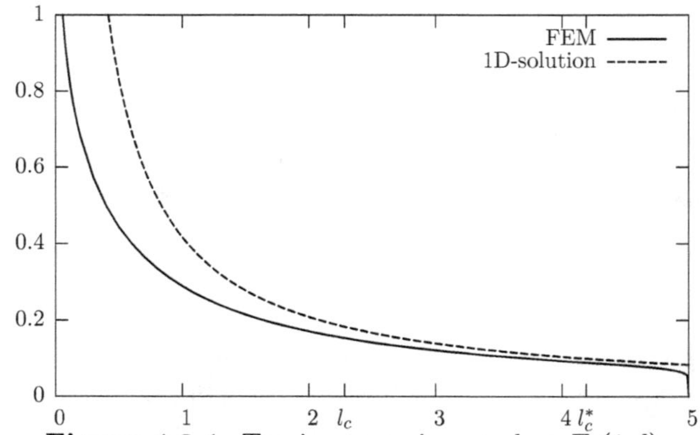

Figure 4.3.1. Tearing experiment: $l \mapsto E_b(1, l)$.

an evolution satisfies (necessary conditions for) `(Ulm)`. We will comment further on this evolution in the discussion of Figure 4.3.6.

The numerical values of $E_b(1, l)$ lead to an estimate of the position of the crack tip as a function of the load. Let $\varphi(t, l)$ be the equilibrium deformation associated with the load t, and $E_b(t, l) := t^2 E_b(1, l)$ the associated bulk

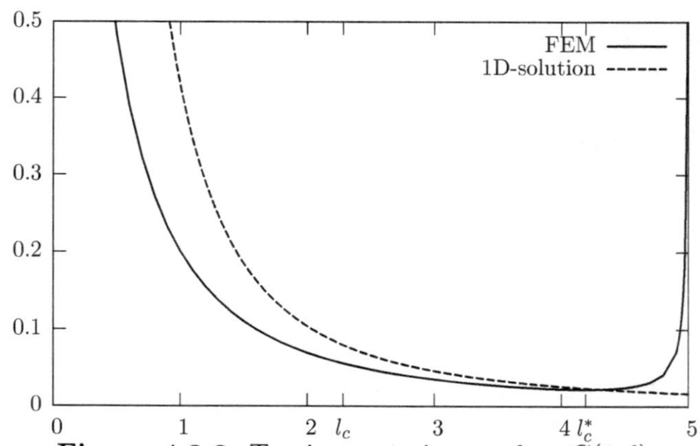

Figure 4.3.2. Tearing experiment: $l \mapsto G(1, l)$.

energy. If the crack keeps on propagating smoothly, then

$$-t^2 \frac{\partial E_b}{\partial l}(1, l) = k. \tag{4.3.4}$$

That relation is used to compute the load t for which the crack length is l, and thereafter $l(t)$.

Once again, a crack splitting the whole domain along the x–axis is a minimizing competitor. Consider t_c and $l_c := l(t_c)$ such that $E_b(t_c, l_c) + kl_c = kL$. For $t > t_c$, splitting the domain is energetically preferable. The value of t_c can be estimated from the computations of $E_b(1, l)$. Using the finite element computations described above, we get $t_c \simeq .47$. The critical length l_c is such that

$$E_b(1, l_c) = -(L - l_c)\frac{\partial E_b}{\partial l}(1, l_c).$$

Numerically, we obtain $l_c \simeq 2.28$. That value is strictly less than the length l_c^* for which the constraint $G \leq k$ can no longer be met, as expected when global energy minimization presides. Indeed, the energetic landscape is explored in its entirety through global minimization, allowing the crack to decrease its energy at l_c, rather than waiting for G to reach k at l_c^*.[2]

Notice the sudden jump introduced in Griffith's evolution – that satisfying (4.3.4) – at t_c. Classically, such a jump would not be allowed to take place and Griffith's evolution would cease to hold at l_c^*.

Figure 4.3.3 represents the numerically computed globally minimizing evolution of the bulk, surface, and total energies (thin lines), together with the analytically computed energies of the 1d solution – see (4.3.1), (4.3.2) – obtained in Section 2.2 and above (thick lines).

The computed evolution has the crack propagating smoothly for $0 \leq t < t_c$, until it reaches the critical length l_c, then cutting brutally through the domain. For small loads, the one-dimensional analysis overestimates the crack length; note that as $l \to 0$, $G(1, l) \to \infty$, and that the accuracy of our finite element computations cannot be guaranteed. When t, and therefore l, become large enough, the values of $dE_b(t)/dt$ and $dE_s(t)/dt$ become very close to those obtained in Section 2.2. Numerically we obtain $dE_b(t)/dt \simeq 7.43\ (10)^{-2}$ and $dE_s(t)/dt \simeq 6.83\ (10)^{-2}$ while the 1d result is $dE_b(t)/dt = dE_s(t)/dt = H\sqrt{k\mu H} \simeq 7.22\ (10)^{-2}$.

Next, a numerical experiment that uses the algorithms developed in this section is conducted. So as to favor symmetric solution, we use a

[2] As an aside, note that the critical length l_c does not depend upon the fracture toughness k!

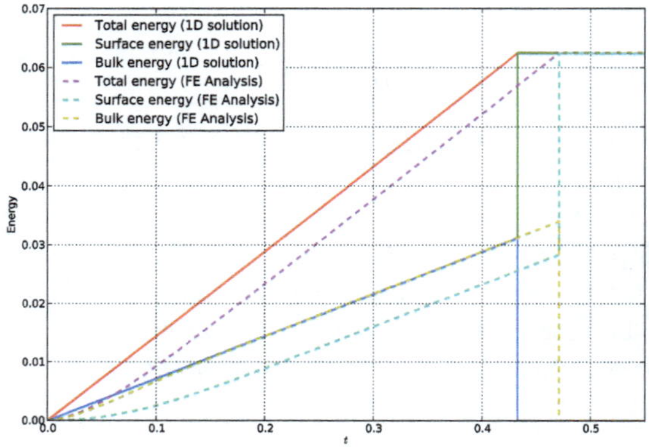

Figure 4.3.3. Evolution of the bulk surface and total energies following (`Ugm`), as a function of the load t. They are computed using a classical finite element analysis and compared to the 1d solution.

structured mesh obtained by a splitting of each square in a structured grid into two right triangles. It consists of 154,450 nodes and 307,298 elements. The mesh size is $(10)^{-2}$; the regularization parameters are $\varepsilon = (10)^{-2}$ and $\eta_\varepsilon = (10)^{-9}$. We consider 100 equi-distributed time steps between 0 and 1. Recalling (4.1.20), the effective fracture toughness in the computations is $k_{\text{eff}} = (1 + h/4\varepsilon)\, k = .0125$.

Figure 4.3.4 represents the computed bulk, surface and total energy, as well as their values obtained via the proposed algorithm, as a function of t. Once again, the backtracking algorithm leads to an evolution with a monotonically increasing and continuous total energy.

Figure 4.3.5 represents the v field, representing the crack for $t = .49$ and $t = .5$. The values $v = 0$ are coded in red and $v = 1$ in blue.

The agreement with the classical solution is remarkable. The bulk energies are within 1% of each others, and the surface energies within 10%. For long enough cracks, the surface and bulk energies grow at a constant rate, and $dE_b(t)/dt \simeq 6.95\,(10)^{-2}$ and $dE_s(t)/dt \simeq 7.03\,(10)^{-2}$. The critical load upon which the crack propagates brutally is $.49 \leq t_c \leq .5$ (vs. a estimated value of .47), and the critical length is $l_c := l(.49) \simeq 2.46$ which, again, is in agreement with the finite element analysis presented above ($l_c \simeq 2.28$). The final surface energy is $6.38\,(10)^{-2}$, which is consistent with the estimate we

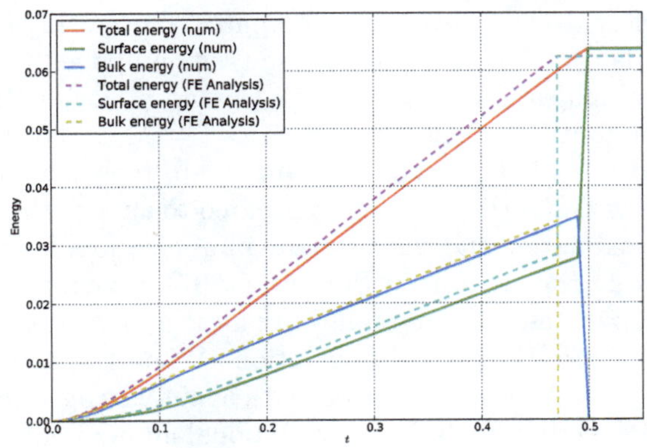

Figure 4.3.4. Evolution of the bulk surface and total energies following (`Ugm`), as a function of the load t. Comparison of values obtained through the variational approximation with backtracking and through finite element analysis.

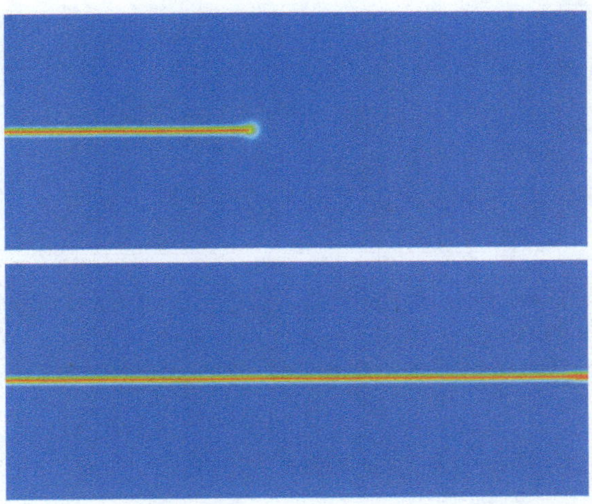

Figure 4.3.5. Position of the crack set in the tearing experiment for $t = .49$ (top) and $t = .50$ (bottom).

gave in Section 4.1 ($k(1 + h/4\varepsilon)L = 6.25\,(10)^{-2}$).

As noted before, the first evolution computed above using finite element analysis – that is that following Griffith until it jumps at $l_c^* \simeq 4.19$ – can be argued to be one satisfying (necessary conditions for) `(Ulm)`. It propagates smoothly until it reaches $l_c^* \simeq 4.19$ at $t = t_c^* \simeq .75$, then brutally to the right end-side of the domain. Figure 4.3.6, represent the bulk, surface and total energies of this solution, compared to an experiment using the variational approximation and the alternate minimization, *but without backtracking.* Following the analysis in Bourdin (2007), we expect that, as long as the crack propagates smoothly following local minimizers, the alternate minimization will provide the right evolution. When the crack propagates brutally, nothing can be said. However, once again, the agreement between our experiments is striking. Using the variational approximation, we obtain $t_c^* \simeq .82$ (instead of .75 using the finite element analysis). The estimate for the critical length is $l_c^* \simeq 4.08$ (vs. 4.19 for the finite element computations).

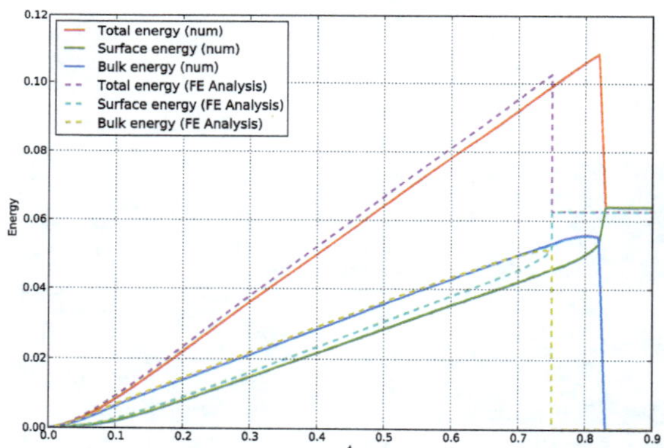

Figure 4.3.6. Evolution of the bulk surface and total energies following `(Ulm)`, as a function of the load t, computed using a classical finite element analysis. Comparison to the variational approximation without backtracking.

The symmetry assumption about the x-axis was instrumental in deriving the theoretical results in Subsection 2.2; it was also imposed as a meshing restriction in the previous computation. In its absence, a bona fide theoretical prediction is difficult to make, but an educated guess may provide insight into the possible crack path. We thus introduce a third class of solutions: a crack propagating along the symmetry axis with length $l(t)$ until some critical t_c at which it brutally bifurcates, reaching one of the sides of the domain. The crack for $t \geq t_c$ is assumed L- shaped, *i.e.*, of the form $(0, l(t_c)) \times \{0\} \cup \{l(t_c)\} \times (0, -H)$ or its mirror image with respect to the x–axis. It then remains to minimize in t_c. Appealing to (4.3.3), (4.3.2) and comparing the energy associated with the straight crack, *i.e.*, $2tH\sqrt{\mu H k}$, to that associated with the bifurcated crack, *i.e.*, $k(tH\sqrt{\mu H k} + H)$, yields

$$t_c = \sqrt{\frac{k}{\mu H}},$$

and

$$l(t_c) = H.$$

The total energy of this branch of solution as a function of the loading parameter t is

$$E(\varphi) = \min\left(2tH\sqrt{\mu H k}, 2kH\right).$$

If $L > 2H$, this asymmetric solution has a lower energy than its symmetric counterpart as soon as $t \geq \sqrt{k/\mu H}$.

We propose a second set of experiments that use a non-symmetric Delaunay-Voronoi mesh. The mesh size is still $h = (10)^{-2}$, and the other parameters are those of the previous experiment.

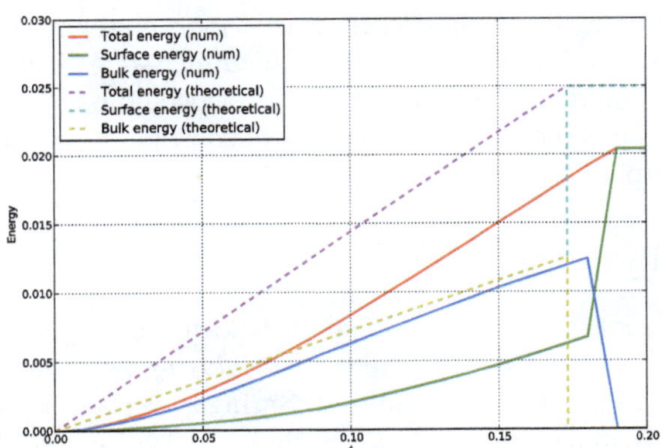

Figure 4.3.7. Evolution of the bulk surface and total energies as a function of the load t. Numerical and expected values ($t_c \simeq .17$).

The energy plot Figure 4.3.7 shows that the evolution is qualitatively as expected, *i.e.*, smooth propagation of the crack tip, then brutal propagation.

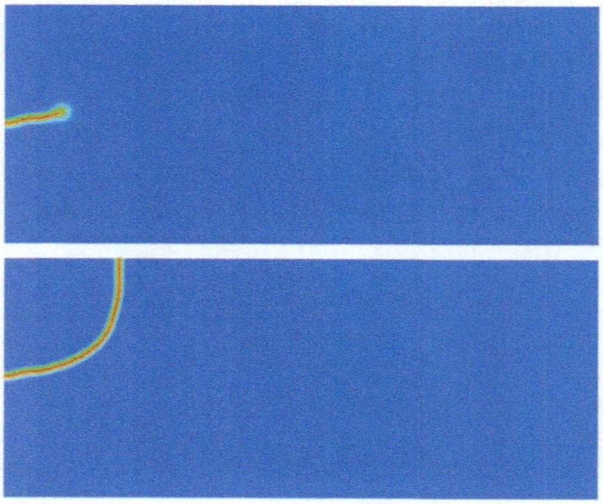

Figure 4.3.8. Position of the crack set in the tearing experiment for $t = .18$ (top) and $t = .19$ (bottom).

Once again, the position of the crack tip lags behind its theoretical position and the comparison between the numerical and theoretical energies is difficult.

Figure 4.3.8 shows the crack tip just before (top) and after (bottom) brutal propagation. The evolution is clearly not globally minimizing: connecting the tip of the crack for $t = .18$ to the upper edge of the domain at a near 90° angle would cost less surface energy. It would be unwise at present to view the perhaps more realistic numerical solution as an outcome of the true minimization.

Bibliography

G. Alberti. Variational models for phase transitions, an approach via Γ–convergence. In G. Buttazzo, editor, *Calculus of Variations and Partial Differential Equations*, pages 95–114. Springer–Verlag, 2000.

L. Ambrosio. Existence theory for a new class of variational problems. *Arch. Ration. Mech. An.*, 111:291–322, 1990.

L. Ambrosio. On the lower semicontinuity of quasiconvex integrals in $SBV(\Omega, \mathbf{R}^k)$. *Nonlinear Anal.-Theor.*, 23(3):405–425, 1994.

L. Ambrosio and V.M. Tortorelli. Approximation of functionals depending on jumps by elliptic functionals via Γ-convergence. *Commun. Pur. Appl. Math.*, 43(8):999–1036, 1990.

L. Ambrosio and V.M. Tortorelli. On the approximation of free discontinuity problems. *Boll. Un. Mat. Ital. B (7)*, 6(1):105–123, 1992.

L. Ambrosio, N. Fusco, and D. Pallara. *Functions of Bounded Variation and Free Discontinuity Problems.* Oxford University Press, 2000.

J. M. Ball and F. Murat. $W^{1,p}$-quasiconvexity and variational problems for multiple integrals. *J. Funct. Anal.*, 58(3):225–253, 1984.

G. Bellettini and A. Coscia. Discrete approximation of a free discontinuity problem. *Numer. Func. Anal. opt.*, 15(3-4):201–224, 1994.

B. Bourdin. *Une méthode variationnelle en mécanique de la rupture. Théorie et applications numériques.* Thèse de doctorat, Université Paris-Nord, 1998.

B. Bourdin. Image segmentation with a finite element method. *ESAIM Math. Model. Numer. Anal.*, 33(2):229–244, 1999.

B. Bourdin. Numerical implementation of the variational formulation for quasi-static brittle fracture. *Interfaces Free Bound.*, 9(3):411–430, 2007.

B. Bourdin and A. Chambolle. Implementation of an adaptive finite-element approximation of the Mumford-Shah functional. *Numer. Math.*, 85(4): 609–646, 2000.

B. Bourdin, G. A. Francfort, and J.-J. Marigo. Numerical experiments in revisited brittle fracture. *J. Mech. Phys. Solids*, 48:797–826, 2000.

B. Bourdin, G. A. Francfort, and J.-J. Marigo. *The variational approach to fracture.* Springer, New York, 2008. Reprinted from J. Elasticity **91** (2008), no. 1-3, With a foreword by Roger Fosdick.

A. Braides. *Γ-convergence for Beginners*, volume 22 of *Oxford Lecture Series in Mathematics and its Applications.* Oxford University Press, 2002.

D. Bucur and N. Varchon. Boundary variation for a Neumann problem. *Ann. Scuola Norm.-Sci.*, 29(4):807–821, 2000.

S. Burke, C. Ortner, and E. Süli. An adaptive finite element approximation of a variational model of brittle fracture. *SIAM J. Numer. Anal.*, 48(3): 980–1012, 2010.

A. Chambolle. A density result in two-dimensional linearized elasticity, and applications. *Arch. Ration. Mech. An.*, 167(3):211–233, 2003.

A. Chambolle. An approximation result for special functions with bounded variations. *J. Math. Pure Appl.*, 83:929–954, 2004.

A. Chambolle. Addendum to "an approximation result for special functions with bounded deformation" [j. math. pures appl. (9) 83 (7) (2004) 929–954]: the n-dimensional case. *J. Math. Pure Appl.*, 84:137–145, 2005.

A. Chambolle and F. Doveri. Continuity of Neumann linear elliptic problems on varying two-dimensional bounded open sets. *Commun. Part. Diff. Eq.*, 22(5-6):811–840, 1997.

A. Chambolle, A. Giacomini, and M. Ponsiglione. Crack initiation in brittle materials. *Arch. Ration. Mech. Anal.*, 188(2):309–349, 2008.

B. Dacorogna. *Direct Methods in the Calculus of Variations.* Springer Verlag, Berlin, Heidelberg, 1989.

G. Dal Maso. *An introduction to* Γ*-convergence.* Birkhäuser, Boston, 1993.

G. Dal Maso and G. Lazzaroni. Quasistatic crack growth in finite elasticity with non-interpenetration. *Ann. Inst. H. Poincaré Anal. Non Linéaire*, 27(1):257–290, 2010.

G. Dal Maso and R. Toader. A model for the quasi-static growth of brittle fractures: Existence and approximation results. *Arch. Ration. Mech. An.*, 162:101–135, 2002.

G. Dal Maso, G. A. Francfort, and R. Toader. Quasistatic crack growth in nonlinear elasticity. *Arch. Ration. Mech. An.*, 176(2):165–225, 2005.

G. Dal Maso, A. Giacomini, and M. Ponsiglione. A variational model for quasistatic crack growth in nonlinear elasticity: some qualitative properties of the solutions. *Boll. Unione Mat. Ital. (9)*, 2(2):371–390, 2009.

E. De Giorgi, M. Carriero, and A. Leaci. Existence theorem for a minimum problem with free discontinuity set. *Arch. Ration. Mech. An.*, 108:195–218, 1989.

P. Destuynder and M. Djaoua. Sur une interprétation mathématique de l'intégrale de rice en théorie de la rupture fragile. *Math. Met. Appl. Sc*, 3:70–87, 1981.

L.C. Evans and R.F. Gariepy. *Measure theory and fine properties of functions.* CRC Press, Boca Raton, FL, 1992.

H. Federer. *Geometric measure theory.* Springer-Verlag, New York, 1969.

I. Fonseca and N. Fusco. Regularity results for anisotropic image segmentation models. *Ann. Scuola Norm. Sup. Pisa Cl. Sci. (4)*, 24(3):463–499, 1997.

G.A. Francfort and C. Larsen. Existence and convergence for quasi-static evolution in brittle fracture. *Commun. Pur. Appl. Math.*, 56(10):1465–1500, 2003.

G.A. Francfort, N. Le, and S. Serfaty. Critical points of ambrosio-tortorelli converge to critical points of mumford-shah in the one-dimensional dirichlet case. *ESAIM Control Optim. Calc. Var.*, 15(3):576–598, 2009.

A. Giacomini. Ambrosio-Tortorelli approximation of quasi-static evolution of brittle fractures. *Calc. Var. Partial Dif.*, 22(2):129–172, 2005.

A. Giacomini and M. Ponsiglione. A discontinuous finite element approximation of quasi-static growth of brittle fractures. *Numer. Func. Anal. Opt.*, 24(7-8):813–850, 2003.

A. Giacomini and M. Ponsiglione. Discontinuous finite element approximation of quasistatic crack growth in nonlinear elasticity. *Math. Mod. Meth. Appl. S.*, 16(1):77–118, 2006.

A.A. Griffith. The phenomena of rupture and flow in solids. *Philos. T. Roy. Soc. A*, CCXXI-A:163–198, 1920.

M.E. Gurtin. *Configurational forces as basic concepts of continuum physics*, volume 137 of *Applied Mathematical Sciences.* Springer-Verlag, New York, 2000.

H. Hahn. Über Annäherung an Lebesgue'sche integrale durch Riemann'sche summen. *Sitzungsber. Math. Phys. Kl. K. Akad. Wiss. Wien*, 123:713–743, 1914.

R.V. Kohn and P. Sternberg. Local minimizers and singular perturbations. *Proc. Roy. Soc. Edin. A*, 111(A):69–84, 1989.

F. Leonetti and F. Siepe. Maximum principle for vector valued minimizers. *J. Convex Anal.*, 12(2):267–278, 2005.

D. Mumford and J. Shah. Optimal approximations by piecewise smooth functions and associated variational problems. *Commun. Pur. Appl. Math.*, XLII:577–685, 1989.

F. Murat. The Neumann sieve. In *Nonlinear variational problems (Isola d'Elba, 1983)*, volume 127 of *Res. Notes in Math.*, pages 24–32. Pitman, Boston, MA, 1985.

M. Negri. The anisotropy introduced by the mesh in the finite element approximation of the Mumford-Shah functional. *Numer. Func. Anal. opt.*, 20(9-10):957–982, 1999.

M. Negri. A finite element approximation of the Griffith model in fracture mechanics. *Numer. Math.*, 95:653–687, 2003.

M. Negri and M. Paolini. Numerical minimization of the Mumford-Shah functional. *Calcolo*, 38(2):67–84, 2001.

C. A. Rogers. *Hausdorff measures.* Cambridge University Press, London, 1970.

Multiphase Flow Modeling via Hamilton's Principle

Sergey Gavrilyuk [*‡]

[*] Aix-Marseille University, CNRS UMR 6595, IUSTI, Marseille, France
[‡] CISM, International Center for Mechanical Sciences, Udine, Italy

Abstract We present here a variational approach to derivation of multiphase flow models. Two basic ingredients of this method are as follows. First, a conservative part of the model is derived based on the Hamilton principle of stationary action. Second, phenomenological dissipative terms are added which are compatible with the entropy inequality. The variational technique is shown up, and mathematical models (classical and non-classical) describing fluid-fluid and fluid-solid mixtures and interfaces are derived.

1 Introduction

For any vectors $\mathbf{a}, \mathbf{b}$ we shall use the notations $\mathbf{a} \cdot \mathbf{b}$ for their *scalar product* (the line vector (covector) is multiplied by the column vector) and $\mathbf{a} \otimes \mathbf{b}$ for their *tensor product* (the column vector is "multiplied" by the line vector). Linear transformations of vectors into vectors (second order tensors) are denoted by uppercase letters. The divergence of a linear transformation A is the covector div$A$ such that, for any constant vector $\mathbf{a}$

$$\mathrm{div}(A) \cdot \mathbf{a} = \mathrm{div}(A\mathbf{a}).$$

In particular, for any vector field $\mathbf{v}$

$$\mathrm{div}(A\mathbf{v}) = divA \cdot \mathbf{v} + tr\left(A\frac{\partial \mathbf{v}}{\partial \mathbf{x}}\right)$$

$$\mathrm{div}(\mathbf{u} \otimes \mathbf{v}) = \mathbf{v}\mathrm{div}\mathbf{u} + \frac{\partial \mathbf{v}}{\partial \mathbf{x}}\mathbf{u}$$

In the last formula we have been identified the line vectors and the column vectors. The identical transformation is denoted by I, and the gradient operator by ∇. The elements of a linear transformation A in a given base

$\mathbf{e}_j$, $j = 1, 2, 3$ are denoted be A^i_j : $A\mathbf{e}_j = A^i_j \mathbf{e}_i$ where i means lines and j columns. The repeated Latin indices imply summation. In particular,

$$(\mathbf{a} \otimes \mathbf{b})^i_j = a^i b_j.$$

If $f(A)$ is a scalar function of A, the matrix of $\dfrac{\partial f}{\partial A}$ is defined by the formula

$$\left(\frac{\partial f}{\partial A}\right)^j_i = \frac{\partial f}{\partial A^i_j}.$$

The definition of the divergence of a linear transformation used in France is opposite to the one used in USA where the divergence of A is defined as follows

$$\operatorname{div}(A) \cdot \mathbf{a} = \operatorname{div}(A^T \mathbf{a})$$

for any constant vector $\mathbf{a}$. Once an appropriate definition has been selected, it should remain unchanged during calculations.

2 Variations

The results of this Section are well known and can be partially found, for example, in the book by Berdichevsky (2009).

Let $\mathcal{D}(t)$ be a volume of the physical space occupied by a fluid at time t. A particle is labeled by its initial position $\mathbf{X}$ in the reference space $\mathcal{D}(t_0)$, the motion of the continuum is defined as a diffeomorphism from $\mathcal{D}(t_0)$ into $\mathcal{D}(t)$:

$$\mathbf{x} = \varphi_t(\mathbf{X}).$$

The velocity $\mathbf{u}$ and the deformation gradient F are defined by

$$\mathbf{v} = \frac{\partial \varphi_t(\mathbf{X})}{\partial t}, \quad F = \frac{\partial \mathbf{x}}{\partial \mathbf{X}} = \frac{\partial \varphi_t(\mathbf{X})}{\partial \mathbf{X}},$$

Let

$$\mathbf{x} = \mathbf{\Phi}_t(\mathbf{X}, \varepsilon),$$

be a one-parameter *family of virtual motions* of the medium such that

$$\mathbf{\Phi}_t(\mathbf{X},\varepsilon)|_{\varepsilon=0} = \varphi_t(\mathbf{X}).$$

Here ε is a scalar defined in the vicinity of zero. We define the *virtual displacement* $\delta\mathbf{x}\,(t, \mathbf{X})$ associated with the virtual motion :

$$\delta\mathbf{x}\,(t, \mathbf{X}) = \left.\frac{\partial \mathbf{\Phi}_t(\mathbf{X},\varepsilon)}{\partial \varepsilon}\right|_{\varepsilon=0}. \tag{1}$$

Let us consider any quantity f defined in the Eulerian coordinates (or in the Lagrangian coordinates)

$$(t, \mathbf{x}) \longrightarrow f(t, \mathbf{x})$$

or

$$(t, \mathbf{X}) \rightarrow f(t, \mathbf{X})$$

We abuse the notation by using the same letter f for completely different functions, because the last one is obtained from the first one by replacing $\mathbf{x} = \boldsymbol{\varphi}_t(\mathbf{X})$. To avoid a possible confusion, we will often indicate the arguments of these functions. The corresponding virtual families of f in the Eulerian coordinates will be denoted by $\hat{f}(t, \mathbf{x}, \varepsilon)$ and $\tilde{f}(t, \mathbf{X}, \varepsilon)$. The relation between them is as follows :

$$\tilde{f}(t, \mathbf{X}, \varepsilon) = \hat{f}(t, \Phi_t(\mathbf{X}, \varepsilon), \varepsilon) \tag{2}$$

Let us define the Eulerian variations of f

$$\hat{\delta} f(t, \mathbf{x}) = \left.\frac{\partial \hat{f}}{\partial \varepsilon}\right|_{\varepsilon=0}$$

and the Lagrangian variation :

$$\tilde{\delta} f(t, \mathbf{X}) = \left.\frac{\partial \tilde{f}}{\partial \varepsilon}\right|_{\varepsilon=0}$$

Differentiating (2) with respect to ε we obtain the following relation between two types of variations :

$$\tilde{\delta} f = \hat{\delta} f + \frac{\partial f}{\partial \mathbf{x}} \delta \mathbf{x} = \hat{\delta} f + \nabla f \cdot \delta \mathbf{x} \tag{3}$$

Variation of density
The density $\rho\,(t, \mathbf{X})$ is determined by

$$\rho \det F = \rho_0(\mathbf{X}),$$

where $\rho_0(\mathbf{X})$ is the reference density.
Lemma

$$\tilde{\delta} F = \frac{\partial \delta \mathbf{x}}{\partial \mathbf{X}}, \quad \tilde{\delta}\,(\det F) = \det F \ \mathrm{tr}\left(F^{-1} \tilde{\delta} F\right) = \det F \ \mathrm{div}\,(\delta \mathbf{x})$$

Here the variation $\delta \mathbf{x}$ defined by (1) is considered as a function of the Eulerian coordinates. In the following, all operations from the vector analysis (div, $\nabla, \dots$) will be applied only to functions of the Eulerian coordinates.

Proof

By definition,

$$\tilde{F}(t, \mathbf{X}, \varepsilon) = \frac{\partial \mathbf{\Phi}_t(\mathbf{X}, \varepsilon)}{\partial \mathbf{X}}.$$

Differentiating it ones with respect to ε we obtain the first formula. The proof of the second formula we present below is due to P. Casal. Let $\mathbf{w}_1$, $\mathbf{w}_2, \dots, \mathbf{w}_n$ be n constant vectors. We denote by $\det(\mathbf{w}_1, \mathbf{w}_2, \dots, \mathbf{w}_n)$ the determinant of the matrix having columns $\mathbf{w}_1, \mathbf{w}_2, \dots, \mathbf{w}_n$. For any linear transformation A its determinant and its trace satisfy the following relations:

$$\det(A\mathbf{w}_1, A\mathbf{w}_2, \dots, A\mathbf{w}_n) = \det A \det(\mathbf{w}_1, \mathbf{w}_2, \dots, \mathbf{w}_n),$$

$$tr(A) \det(\mathbf{w}_1, \mathbf{w}_2, \dots, \mathbf{w}_n)$$

$$= \det(A\mathbf{w}_1, \mathbf{w}_2, \dots, \mathbf{w}_n) + \dots + \det(\mathbf{w}_1, \mathbf{w}_2, \dots, A\mathbf{w}_n).$$

Hence

$$d(\det A) \det(\mathbf{w}_1, \mathbf{w}_2, \dots, \mathbf{w}_n) = \det(dA\mathbf{w}_1, A\mathbf{w}_2, \dots, A\mathbf{w}_n)$$

$$+ \dots + \det(A\mathbf{w}_1, A\mathbf{w}_2, \dots, dA\mathbf{w}_n)$$

$$= \det(dAA^{-1}A\mathbf{w}_1, A\mathbf{w}_2, \dots, A\mathbf{w}_n) + \dots$$

$$+ \det(A\mathbf{w}_1, A\mathbf{w}_2, \dots, dAA^{-1}A\mathbf{w}_n)$$

$$= tr(dAA^{-1}) \det(A\mathbf{w}_1, A\mathbf{w}_2, \dots, A\mathbf{w}_n)$$

$$= tr(dAA^{-1}) \det(A) \det(\mathbf{w}_1, \mathbf{w}_2, \dots, \mathbf{w}_n)$$

$$= \det(A)\, tr(A^{-1}dA) \det(\mathbf{w}_1, \mathbf{w}_2, \dots, \mathbf{w}_n).$$

Hence,

$$d(detA) = \det A \ \mathrm{tr}\left(A^{-1}dA\right).$$

It is now sufficient to take $d = \tilde{\delta}$ and to apply this result to $\det(F)$. The Lemma has been proved.

In particular, the Lemma and (3) imply :

$$\tilde{\delta}\rho = -\rho \mathrm{div}(\delta \mathbf{x}), \qquad \hat{\delta}\rho = -\mathrm{div}(\rho \delta \mathbf{x}).$$

Variation of entropy

The entropy per unit mass η is conserved along trajectories, so it is a function only of the Lagrangian coordinates : $\eta = \eta_0(\mathbf{X})$. Hence, its Lagrangian and Eulerian variations are :

$$\tilde{\delta}\eta = 0, \quad \hat{\delta}\eta = -\nabla\eta \cdot \delta\mathbf{x}.$$

Variation of velocity

The variation of the velocity is direct :

$$\tilde{\delta}\mathbf{v} = \frac{\partial \delta\mathbf{x}}{\partial t}, \quad \hat{\delta}\mathbf{v} = \frac{D\delta\mathbf{x}}{Dt} - \frac{\partial \mathbf{v}}{\partial \mathbf{x}}\delta\mathbf{x},$$

where the material derivative $\frac{D}{Dt}$ is defined as

$$\frac{D}{Dt} = \frac{\partial}{\partial t} + \mathbf{v} \cdot \nabla.$$

3 Hamilton's principle

Let us consider the Hamilton action :

$$a = \int_{t_0}^{t_1} L dt,$$

where $[t_0, t_1]$ is a time interval and L is the corresponding Lagrangian. The Hamilton principle is : for any virtual motion satisfying

$$\mathbf{\Phi}_{t_0}(\mathbf{X},\varepsilon) = \varphi_{t_0}(\mathbf{X}), \quad \mathbf{\Phi}_{t_1}(\mathbf{X},\varepsilon) = \varphi_{t_1}(\mathbf{X}), \tag{4}$$

$$\mathbf{\Phi}_t(\mathbf{X},\varepsilon) = \varphi_t(\mathbf{X}), \quad \mathbf{X} \in \partial\mathcal{D}(t_0), \tag{5}$$

and imposed constraints, the variation of the Hamilton action is stationary :

$$\delta a = 0.$$

Let us remark that (4), (5) imply

$$\delta\mathbf{x}|_{t=t_0} = \delta\mathbf{x}|_{t=t_1} = 0, \quad \delta\mathbf{x}|_{\partial\mathcal{D}(t)} = 0.$$

We will recall first the Lagrangians for classical models of continuum mechanics and the corresponding constraints.

Lagrangian of an incompressible fluid

$$L = \int_{\mathcal{D}(t)} \frac{\rho\,|\mathbf{v}|^2}{2} dD$$

The density ρ and the velocity field $\mathbf{v}$ verify the constraints :

$$\frac{D\rho}{Dt} = 0, \quad div\,(\mathbf{v}) = 0.$$

Lagrangian of a compressible fluid

$$L = \int_{\mathcal{D}(t)} \rho \left(\frac{|\mathbf{v}|^2}{2} - e\,(\rho, \eta) \right) dD$$

The density ρ, the velocity field $\mathbf{v}$ and the entropy η verify the constraints :

$$\frac{D\rho}{Dt} + \rho div\,(\mathbf{v}) = 0, \quad \frac{D\eta}{Dt} = 0.$$

The specific internal energy $e\,(\rho, \eta)$ is a given function satisfying the Gibbs identity :

$$\theta d\eta = de + pd\left(\frac{1}{\rho}\right),$$

where θ is the temperature and p is the pressure.

Lagrangian of an elastic body

$$L = \int_{\mathcal{D}(t)} \rho \left(\frac{|\mathbf{v}|^2}{2} - e\,(G, \eta) \right) dD.$$

Here

$$G = \left(FF^T\right)^{-1} = \left(F^T\right)^{-1} F^{-1}$$

is the Finger tensor. In the case of isotropic solids the energy is a function of the invariants of G. This Lagrangian is suitable for the Eulerian description of solids. Another possibility commonly used in the literature is to consider the specific energy in the form $e\,(C, \eta)$ where $C = F^T F$ is the right Cauchy-Green deformation tensor.

4 Governing equations of classical continua

4.1 Incompressible fluids

Let

$$a = \int_{t_0}^{t_1} L dt = \int_{t_0}^{t_1} \int_{\mathcal{D}(t)} \frac{\rho\,|\mathbf{v}|^2}{2} dD$$

The constraints

$$\frac{D\rho}{Dt} = 0, \quad div\,(\mathbf{v}) = 0$$

can be written in the form

$$\rho = \rho_0(\mathbf{X}), \quad \det F = 1.$$

Hence

$$\hat{\delta}\rho = -\nabla\rho \cdot \delta\mathbf{x}, \quad div\,(\delta\mathbf{x}) = 0.$$

To satisfy the last constraint we take

$$\delta\mathbf{x} = \mathrm{rot}\,(\mathbf{A})$$

where $\mathbf{A}$ is a smooth vector field vanishing at the boundary $\mathcal{D}(t)$:

$$\mathbf{A}|_{\partial\mathcal{D}(t)} = 0.$$

The variation of the Hamilton action is

$$\delta a = \left.\frac{d}{d\varepsilon}\right|_{\varepsilon=0} \int_{t_0}^{t_1}\int_{\mathcal{D}(t)} \frac{\hat{\rho}\,|\hat{\mathbf{v}}|^2}{2} dD = \int_{t_0}^{t_1}\int_{\mathcal{D}(t)} \hat{\delta}\left(\frac{\rho\,|\mathbf{v}|^2}{2}\right) dD$$

$$= \int_{t_0}^{t_1}\int_{\mathcal{D}(t)} \left(\hat{\delta}\rho\frac{|\mathbf{v}|^2}{2} + \rho\mathbf{v}\cdot\hat{\delta}\mathbf{v}\right) dD$$

$$= \int_{t_0}^{t_1}\int_{\mathcal{D}(t)} \left(-\frac{|\mathbf{v}|^2}{2}\nabla\rho\cdot\delta\mathbf{x} + \rho\mathbf{v}\cdot\left(\frac{D\delta\mathbf{x}}{Dt} - \frac{\partial\mathbf{v}}{\partial\mathbf{x}}\delta\mathbf{x}\right)\right) dD$$

$$= \int_{t_0}^{t_1}\int_{\mathcal{D}(t)} \left(-\nabla\left(\frac{\rho\,|\mathbf{v}|^2}{2}\right)\cdot\delta\mathbf{x} + \rho\mathbf{v}\cdot\frac{D\delta\mathbf{x}}{Dt}\right) dD$$

$$= \int_{t_0}^{t_1}\int_{\mathcal{D}(t)} \left(-\nabla\left(\frac{\rho\,|\mathbf{v}|^2}{2}\right)\cdot\delta\mathbf{x} + \rho\mathbf{v}\cdot\left(\frac{\partial\delta\mathbf{x}}{\partial t} + \frac{\partial\delta\mathbf{x}}{\partial\mathbf{x}}\mathbf{v}\right)\right) dD$$

$$= \int_{t_0}^{t_1}\int_{\mathcal{D}(t)} \left(\frac{\partial\,(\rho\mathbf{v}\cdot\delta\mathbf{x})}{\partial t} + div\left((\rho\mathbf{v}\otimes\mathbf{v})\cdot\delta\mathbf{x} - \frac{\rho\,|\mathbf{v}|^2}{2}\delta\mathbf{x}\right)\right) dD$$

$$-\int_{t_0}^{t_1}\int_{\mathcal{D}(t)} \left(\frac{\partial\rho\mathbf{v}}{\partial t} + div\,(\rho\mathbf{v}\otimes\mathbf{v})\right)\cdot\delta\mathbf{x}dD$$

Using the Gauss-Ostrogradski formula and taking into account the fact that $\delta\mathbf{x}$ vanishes at $\partial\mathcal{D}(t)$ and $\delta\mathbf{x}|_{t=t_0} = \delta\mathbf{x}|_{t=t_1} = 0$, we finally get

$$\delta a = -\int_{t_0}^{t_1}\int_{\mathcal{D}(t)} \left(\frac{\partial\rho\mathbf{v}}{\partial t} + div\,(\rho\mathbf{v}\otimes\mathbf{v})\right)\cdot\delta\mathbf{x}dD$$

Let us remark that for any function f

$$\int_{t_0}^{t_1}\int_{\mathcal{D}(t)} \nabla f \cdot rot\,(\mathbf{A})\,dD = \int_{t_0}^{t_1}\int_{\mathcal{D}(t)} div\,(f\;rot\,(\mathbf{A}))\,dD$$

$$= \int_{t_0}^{t_1}\int_{\partial\mathcal{D}(t)} f\;rot\,(\mathbf{A})\cdot\mathbf{n}dD = 0$$

if

$$rot\,(\mathbf{A})\cdot\mathbf{n} = \mathbf{0}.$$

Here $\mathbf{n}$ is the external unit normal vector to $\partial\mathcal{D}(t)$. But the last condition is verified if the field $\mathbf{A}$ is orthogonal to $\partial\mathcal{D}(t)$:

$$\mathbf{A}\cdot\mathbf{s} = \mathbf{0}$$

for any vector $\mathbf{s}$ tangent to $\partial\mathcal{D}(t)$. But

$$\mathbf{A}|_{\partial\mathcal{D}(t)} = 0.$$

Hence

$$\delta a = -\int_{t_0}^{t_1}\int_{\mathcal{D}(t)} \left(\frac{\partial\rho\mathbf{v}}{\partial t} + div\,(\rho\mathbf{v}\otimes\mathbf{v})\right)\cdot\delta\mathbf{x}dD$$

$$= -\int_{t_0}^{t_1}\int_{\mathcal{D}(t)} \left(\frac{\partial\rho\mathbf{v}}{\partial t} + div\,(\rho\mathbf{v}\otimes\mathbf{v})\right)\cdot rot\,(\mathbf{A})\,dD.$$

Hence, there exists a scalar function p called pressure such that

$$\frac{\partial\rho\mathbf{v}}{\partial t} + div\,(\rho\mathbf{v}\otimes\mathbf{v}) = -\nabla p.$$

So, we have obtained the Euler equations of incompressible non-homogeneous fluids :

$$\frac{\partial\rho\mathbf{v}}{\partial t} + div\,(\rho\mathbf{v}\otimes\mathbf{v}+p\mathbf{I}) = 0,$$

$$\frac{D\rho}{Dt} = 0, \quad div\,(\mathbf{v}) = 0.$$

When the mass forces $\mathbf{f}$ are added, the equations become

$$\frac{\partial\rho\mathbf{v}}{\partial t} + div\,(\rho\mathbf{v}\otimes\mathbf{v}+p\mathbf{I}) = \rho\mathbf{f}, \tag{6}$$

$$\frac{D\rho}{Dt} = 0, \quad div\,(\mathbf{v}) = 0.$$

In the case of potential forces ($\mathbf{f} = -\nabla\mathbf{\Psi}\,(\mathbf{x})$) the equations admit the energy conservation law :

$$\frac{\partial}{\partial t}\left(\rho\left(\frac{|\mathbf{v}|^2}{2} + \mathbf{\Phi}\right)\right) + div\left(\rho\mathbf{v}\left(\frac{|\mathbf{v}|^2}{2} + \mathbf{\Phi}\right) + p\mathbf{v}\right) = 0.$$

4.2 Compressible fluids

Let us consider the Hamilton action for compressible fluids :

$$a = \int_{t_0}^{t_1} L dt = \int_{t_0}^{t_1} \int_{\mathcal{D}(t)} \left(\frac{\rho\,|\mathbf{v}|^2}{2} - \rho e(\rho, \eta) \right) dD$$

The constraints

$$\frac{D\rho}{Dt} + \rho div\,(\mathbf{v}) = 0, \quad \frac{D\eta}{Dt} = 0,$$

can by rewritten in the Lagrangian coordinates in the form

$$\rho \det F = \rho_0\,(\mathbf{X})\,, \quad \eta = \eta_0(\mathbf{X}).$$

Hence

$$\hat{\delta}\rho = -div(\rho\delta\mathbf{x}), \quad \hat{\delta}\eta = -\nabla\eta \cdot \delta\mathbf{x}.$$

The variation of the Hamilton action is :

$$\delta a = \frac{d}{d\varepsilon}\bigg|_{\varepsilon=0} \int_{t_0}^{t_1} \int_{\mathcal{D}(t)} \left(\frac{\hat{\rho}\,|\hat{\mathbf{v}}|^2}{2} - \hat{\rho} e(\hat{\rho}, \hat{\eta}) \right) dD$$

$$= \int_{t_0}^{t_1} \int_{\mathcal{D}(t)} \hat{\delta} \left(\frac{\rho\,|\mathbf{v}|^2}{2} - \rho e(\rho, \eta) \right) dD$$

$$= \int_{t_0}^{t_1} \int_{\mathcal{D}(t)} \left(\hat{\delta}\rho \left(\frac{|\mathbf{v}|^2}{2} - e - \frac{p}{\rho} \right) + \rho\mathbf{v}\cdot\hat{\delta}\mathbf{v} - \frac{\partial e(\rho,\eta)}{\partial\eta}\hat{\delta}\eta \right) dD$$

$$= - \int_{t_0}^{t_1} \int_{\mathcal{D}(t)} div(\rho\delta\mathbf{x}) \left(\frac{|\mathbf{v}|^2}{2} - \left(e + \frac{p}{\rho} \right) \right) dD$$

$$+ \int_{t_0}^{t_1} \int_{\mathcal{D}(t)} \left(\rho\mathbf{v}\cdot \left(\frac{D\delta\mathbf{x}}{Dt} - \frac{\partial\mathbf{v}}{\partial\mathbf{x}}\delta\mathbf{x} \right) + \frac{\partial e(\rho,\eta)}{\partial\eta} \nabla\eta\cdot\delta\mathbf{x} \right) dD$$

$$= \int_{t_0}^{t_1} \int_{\mathcal{D}(t)} \rho\delta\mathbf{x}\cdot\nabla \left(\frac{|\mathbf{v}|^2}{2} - \left(e + \frac{p}{\rho} \right) \right) dD$$

$$+ \int_{t_0}^{t_1} \int_{\mathcal{D}(t)} \left(\rho\mathbf{v}\cdot \left(\frac{D\delta\mathbf{x}}{Dt} - \frac{\partial\mathbf{v}}{\partial\mathbf{x}}\delta\mathbf{x} \right) + \frac{\partial e(\rho,\eta)}{\partial\eta} \nabla\eta\cdot\delta\mathbf{x} \right) dD$$

$$= - \int_{t_0}^{t_1} \int_{\mathcal{D}(t)} \rho\delta\mathbf{x}\cdot\nabla \left(e + \frac{p}{\rho} \right) dD$$

$$+\int_{t_0}^{t_1}\int_{\mathcal{D}(t)}\left(\rho\mathbf{v}\cdot\left(\frac{\partial\delta\mathbf{x}}{\partial t}+\frac{\partial\delta\mathbf{x}}{\partial\mathbf{x}}\mathbf{v}\right)+\frac{\partial e(\rho,\eta)}{\partial\eta}\nabla\eta\cdot\delta\mathbf{x}\right)dD$$

$$=\int_{t_0}^{t_1}\int_{\mathcal{D}(t)}\left(-\rho\delta\mathbf{x}\cdot\frac{\nabla p}{\rho}+\rho\mathbf{v}\cdot\left(\frac{\partial\delta\mathbf{x}}{\partial t}+\frac{\partial\delta\mathbf{x}}{\partial\mathbf{x}}\mathbf{v}\right)\right)dD$$

$$=\int_{t_0}^{t_1}\int_{\mathcal{D}(t)}-\left(\frac{\partial\rho\mathbf{v}}{\partial t}+div\left(\rho\mathbf{v}\otimes\mathbf{v}\right)+\nabla p\right)\cdot\delta\mathbf{x}dD=0.$$

In this derivation, we have used the Gauss-Ostrogradski formula and have taken into account the fact that $\delta\mathbf{x}$ vanishes at $\partial\mathcal{D}(t)$ and $\delta\mathbf{x}|_{t=t_0}=\delta\mathbf{x}|_{t=t_1}=0$. So, we have obtained the Euler equations of compressible fluids :

$$\frac{\partial\rho\mathbf{v}}{\partial t}+div\left(\rho\mathbf{v}\otimes\mathbf{v}+p\mathbf{I}\right)=0,$$

$$\frac{D\rho}{Dt}+\rho div\left(\mathbf{v}\right)=0,$$

$$\frac{D\eta}{Dt}=0.$$

These equations admit the energy conservation law :

$$\frac{\partial}{\partial t}\left(\rho\left(\frac{|\mathbf{v}|^2}{2}+e(\rho,\eta)\right)\right)+div\left(\rho\mathbf{v}\left(\frac{|\mathbf{v}|^2}{2}+e(\rho,\eta)\right)+p\mathbf{v}\right)=0.$$

4.3 Isotropic elastic bodies

Let

$$L=\int_{\mathcal{D}(t)}\rho\left(\frac{|\mathbf{v}|^2}{2}-e\left(G,\eta\right)\right)dD$$

Here

$$G=\left(FF^T\right)^{-1}=F^{-T}F^{-1}$$

is the Finger tensor. The tensor G is more convenient for the Eulerian formulation of governing equations of isotropic elastic materials. Let us recall that the material is *isotropic* if for any orthogonal transformation O

$$e\left(OGO^T,\eta\right)=e\left(G,\eta\right).$$

Hence, the internal energy depends only on the invariants of G. In particular, we choose such invariants in the form :

$$J_i=tr\left(G^i\right),\ i=1,2,3.$$

Let $\mathbf{X} = (X^\alpha)$ be the Lagrangian coordinates, $\alpha = 1,2,3$, $\mathbf{x} = (x^i)$ be the Eulerian coordinates, $i = 1,2,3$. We introduce the *curvilinear cobasis*

$$\mathbf{e}^\alpha = \nabla X^\alpha, \quad \alpha = 1,2,3$$

which is dual to the natural *curvilinear basis*

$$\mathbf{e}_\alpha = \frac{\partial \mathbf{x}}{\partial X^\alpha}, \quad \alpha = 1,2,3.$$

In particular, the vector $\mathbf{e}^\alpha$ is the α - th column of F^{-T} :

$$F^{-T} = \begin{pmatrix} \frac{\partial X^1}{\partial x^1} & \frac{\partial X^2}{\partial x^1} & \frac{\partial X^3}{\partial x^1} \\ \frac{\partial X^1}{\partial x^2} & \frac{\partial X^2}{\partial x^2} & \frac{\partial X^3}{\partial x^2} \\ \frac{\partial X^1}{\partial x^3} & \frac{\partial X^2}{\partial x^3} & \frac{\partial X^3}{\partial x3} \end{pmatrix} = \begin{pmatrix} \mathbf{e}^1, & \mathbf{e}^2, & \mathbf{e}^3 \end{pmatrix}$$

$$= \begin{pmatrix} e_1^1 & e_1^2 & e_1^3 \\ e_2^1 & e_2^2 & e_2^3 \\ e_3^1 & e_3^2 & e_3^3 \end{pmatrix}, \quad e_j^\alpha = (\mathbf{e}^\alpha)_j = \frac{\partial X^\alpha}{\partial x^j}.$$

The scalar product of these vectors satisfies

$$\mathbf{e}^\beta \cdot \mathbf{e}_\alpha = \delta_\alpha^\beta$$

where δ_α^β are the Kronecker symbols. With these definitions

$$G = \sum_{\alpha=1}^{3} \mathbf{e}^\alpha \otimes \mathbf{e}^\alpha, \quad G^{-1} = \sum_{\alpha=1}^{3} \mathbf{e}_\alpha \otimes \mathbf{e}_\alpha. \tag{7}$$

Since

$$\frac{DX^\alpha}{Dt} = 0$$

we have by taking the gradient of that equation :

$$\frac{\partial \mathbf{e}^\alpha}{\partial t} + \nabla(\mathbf{v} \cdot \mathbf{e}^\alpha) = 0, \quad \text{rot}\mathbf{e}^\alpha = 0. \tag{8}$$

In particular, equations (7) and (8) imply that

$$\frac{DG}{Dt} + G\frac{\partial \mathbf{v}}{\partial \mathbf{x}} + \left(\frac{\partial \mathbf{v}}{\partial \mathbf{x}}\right)^T G = 0.$$

Remark 1. It is interesting to note that the left hand side of this equation can be considered as an *objective derivative.* Indeed, let $O(t)$ be a time dependent orthogonal transformation : $O(t)O(t)^T = I$. Consider new variables denoted with primes :

$$\mathbf{x}' = O(t)\mathbf{x},\ \mathbf{v}' = \frac{dO(t)}{dt}\mathbf{x}+O(t)\mathbf{v}.$$

The aim is to show that the equation for

$$G' = O(t)GO^T(t)$$

will be the same as for G. We get :

$$\frac{DG}{Dt} = \frac{\partial G}{\partial t} + \frac{\partial G}{\partial \mathbf{x}}\mathbf{v}$$

$$= \frac{\partial\left(O^T(t)G'O(t)\right)}{\partial t'} + \frac{\partial\left(O^T(t)G'O(t)\right)}{\partial \mathbf{x}'}\frac{\partial \mathbf{x}'}{\partial t}$$

$$+\frac{\partial\left(O^T(t)G'O(t)\right)}{\partial \mathbf{x}'}\frac{\partial \mathbf{x}'}{\partial \mathbf{x}}\left(O^T(t)\mathbf{v}' - O^T(t)\frac{dO(t)}{dt}O^T(t)\mathbf{x}'\right)$$

$$= \frac{\partial\left(O^T(t)G'O(t)\right)}{\partial t'} + \frac{\partial\left(O^T(t)G'O(t)\right)}{\partial \mathbf{x}'}\frac{dO(t)}{dt}O^T(t)\mathbf{x}'$$

$$+\frac{\partial\left(O^T(t)G'O(t)\right)}{\partial \mathbf{x}'}\left(\mathbf{v}' - \frac{dO(t)}{dt}O^T(t)\mathbf{x}'\right)$$

$$= \frac{\partial\left(O^T(t)G'O(t)\right)}{\partial t'} + \frac{\partial\left(O^T(t)G'O(t)\right)}{\partial \mathbf{x}'}\mathbf{v}'$$

$$= \frac{dO^T(t)}{dt}G'O(t) + O^T(t)G'\frac{dO(t)}{dt} + O^T(t)\frac{D'G'}{Dt'}O(t).$$

Hence

$$\frac{DG}{Dt} + G\frac{\partial \mathbf{v}}{\partial \mathbf{x}} + \left(\frac{\partial \mathbf{v}}{\partial \mathbf{x}}\right)^T G =$$

$$= \frac{dO^T(t)}{dt}G'O(t) + O^T(t)G'\frac{dO(t)}{dt} + O^T(t)\frac{D'G'}{Dt'}O(t)$$

$$+O^T(t)G'O(t)\left(O^T(t)\frac{\partial \mathbf{v}'}{\partial \mathbf{x}'}O(t) - O^T(t)\frac{dO(t)}{dt}\right)$$

$$+\left(O^T(t)\frac{\partial \mathbf{v}'}{\partial \mathbf{x}'}O(t) - O^T(t)\frac{dO(t)}{dt}\right)O^T(t)G'O(t)$$

$$= O^T(t)\left(\frac{D'G'}{Dt'} + G'\frac{\partial \mathbf{v}'}{\partial \mathbf{x}'} + \left(\frac{\partial \mathbf{v}'}{\partial \mathbf{x}'}\right)^T G'\right)O(t).$$

Finally,

$$\frac{D'G'}{Dt'} + G'\frac{\partial \mathbf{v}'}{\partial \mathbf{x}'} + \left(\frac{\partial \mathbf{v}'}{\partial \mathbf{x}'}\right)^T G'$$

$$= O(t)\left(\frac{DG}{Dt} + G\frac{\partial \mathbf{v}}{\partial \mathbf{x}} + \left(\frac{\partial \mathbf{v}}{\partial \mathbf{x}}\right)^T G\right) O^T(t).$$

Remark 2. The differential constraint $\mathrm{rot}\mathbf{e}^\alpha = 0$ in (8) is compatible with the evolution equation for $\mathbf{e}^\alpha$: if $\mathrm{rot}\mathbf{e}^\alpha$ vanishes at time $t = 0$, it vanishes at all $t > 0$. This constraint is automatically satisfied in the one-dimensional case. In the multi-dimensional case equation (8) can also be replaced by

$$\frac{\partial \mathbf{e}^\alpha}{\partial t} + \nabla\left(\mathbf{v}\cdot\mathbf{e}^\alpha\right) = -\mathrm{rot}\mathbf{e}^\alpha \wedge \mathbf{v}. \tag{9}$$

We will obtain now the expression for the stress tensor. Let e be the specific internal energy :

$$e = e\left(G, \eta\right)$$

where η is the specific entropy. Let us consider the variation of the internal energy $\mathcal{E}_i$ at fixed value of η :

$$\delta\mathcal{E}_i = \delta\int_{\mathcal{D}(t)} \rho e\left(G,\eta\right) dD$$

We have

$$\delta\mathcal{E}_i = \delta\int_{\mathcal{D}(t)} \rho e\left(G,\eta\right) dD = \int_{\mathcal{D}(0)} \rho_0 \tilde{\delta} e\left(G,\eta\right) dD_0$$

$$= \int_{\mathcal{D}(0)} \rho_0 tr\left(\frac{\partial e}{\partial G}\tilde{\delta}G\right) dD_0.$$

The matrix $\partial e/\partial G$ is symmetric. Since

$$\tilde{\delta}G = \left(\tilde{\delta}F^{-1}\right)^T F^{-1} + \left(F^{-1}\right)^T \tilde{\delta}F^{-1}$$

$$= -\left(F^{-1}\tilde{\delta}FF^{-1}\right)^T F^{-1} - \left(F^{-1}\right)^T F^{-1}\tilde{\delta}FF^{-1}$$

$$= -\left(\frac{\partial \delta\mathbf{x}}{\partial \mathbf{x}}\right)^T G - G\frac{\partial \delta\mathbf{x}}{\partial \mathbf{x}}$$

we obtain

$$\delta\mathcal{E}_i = -\int_{\mathcal{D}(0)} 2\rho_0 tr\left(\frac{\partial e}{\partial G}G\frac{\partial \delta\mathbf{x}}{\partial \mathbf{x}}\right) dD_0$$

$$= \int_{\mathcal{D}(t)} tr\left(\sigma \frac{\partial \delta \mathbf{x}}{\partial \mathbf{x}}\right) dD.$$

The tensor

$$\sigma = -2\rho \frac{\partial e}{\partial G} G$$

is called the Murnaghan stress tensor (Godunov, 1978, Godunov and Romenskii, 2003). It is symmetric, if e is isotropic. Indeed, in the isotropic case, e depends only on the invariants of G :

$$J_i = tr\left(G^i\right), \quad i = 1, 2, 3$$

In particular, the determinant of G (denoted by $|G|$) can be expressed in terms of J_i in the form

$$|G| = \frac{2J_3 - 3J_1 J_2 + J_1^3}{6}$$

It can be proved that

$$\frac{\partial J_i}{\partial G} = iG^{i-1}, i = 1, 2, 3.$$

In particular (see also the Lemma)

$$\frac{\partial |G|}{\partial G} = |G| G^{-1}.$$

Hence

$$\sigma = -2\rho \frac{\partial e}{\partial G} G = -2\rho \left(\frac{\partial e}{\partial J_1} I + 2\frac{\partial e}{\partial J_2} G + 3\frac{\partial e}{\partial J_3} G^2 \right) G = -2\rho G \frac{\partial e}{\partial G} = \sigma^T$$

Here the density ρ is expressed by

$$\rho = \rho_0 |G|^{1/2}$$

EOS formulation in separable form It is natural to present the energy in a separable form : the energy is the sum of a "hydrodynamic" part and an "elastic" part :

$$e = e^h(\rho, \eta) + e^e(g)$$

where

$$g = \frac{G}{|G|^{1/3}}.$$

The hydrodynamic part of the energy $e^h(\rho,\eta)$ can be taken in the form of stiffened gas equation of state :

$$e^h = \frac{p+\gamma p_\infty}{\rho(\gamma-1)}, \quad p+p_\infty = A\exp\left(\frac{\eta-\eta_0}{c_v}\right)\rho^\gamma, \quad \eta_0 = const, \quad A = const \tag{10}$$

Here γ is the polytropic exponent, p_∞ is a constant, c_v is the specific heat at constant volume.

The elastic part of the internal energy e^e depends only on g. The tensor g has a unit determinant, so it is unaffected by the volume change. This idea to take the arguments of the internal energy in this form was first proposed by Gouin and Debieve (1986) (see also Plohr and Plohr, 2005), but for the dependence of the energy on the right Cauchy-Green tensor. The simplest example of the elastic energy is

$$e^e(g,\eta) = \frac{\mu_s}{4\rho_0} tr\left((g-I)^2\right) = \frac{\mu_s}{4\rho_0}\left(\frac{J_2}{|G|^{2/3}} - \frac{2J_1}{|G|^{1/3}} + 3\right) \tag{11}$$

where μ_s is the shear modulus. The stress tensor will be then

$$\sigma = -2\rho\frac{\partial e}{\partial G}G = -pI + S, \qquad tr(S) = 0, \tag{12}$$

where the deviatoric part is

$$S = -\mu_s\frac{\rho}{\rho_0}\left(\frac{1}{|G|^{2/3}}\left(G^2 - \frac{J_2}{3}I\right) - \frac{1}{|G|^{1/3}}\left(G - \frac{J_1}{3}I\right)\right)$$

and the thermodynamic pressure is

$$p = \rho^2\frac{\partial e^h}{\partial \rho}.$$

The elastic part of the energy has no influence on the pressure, it is determined only by the hydrodynamic part. In the case of small displacements, we obtain the classical Hooke law.

Governing equations The governing equations can be written in the form :

$$\frac{\partial \mathbf{e}^\alpha}{\partial t} + \nabla(\mathbf{v}\cdot\mathbf{e}^\alpha) = 0, \quad \mathrm{rot}\mathbf{e}^\alpha = 0,$$

$$\frac{\partial \rho}{\partial t} + div(\rho\mathbf{v}) = 0,$$

$$\frac{\partial \rho\mathbf{v}}{\partial t} + div(\rho\mathbf{v}\otimes\mathbf{v} - \sigma) = 0,$$

$$\frac{\partial\left(\rho\left(e + \frac{\mathbf{v}\cdot\mathbf{v}}{2}\right)\right)}{\partial t} + div\left(\rho\mathbf{v}\left(e + \frac{\mathbf{v}\cdot\mathbf{v}}{2}\right) - \sigma\mathbf{v}\right) = 0.$$

For the energy in the form (10), 11), the equations are hyperbolic at least in the one-dimensional case. The proof will be given in a more general situation involving solid-fluid mixtures.

5 Multiphase flow modeling: general definitions

Consider a multiphase mixture of two compressible components. Each a-th component ($a = 1, 2$) has its own phase average characteristics: the velocity $\mathbf{v}_a$; the density ρ_a ; entropy η_a, the internal energy per unit mass $e_a(\rho_a, \eta_a)$, the temperature θ_a etc. The apparent densities $\bar{\rho}_a$ are defined by : $\bar{\rho}_a = \alpha_a\rho_a$. The mixture density is : $\rho = \alpha_1\rho_1 + \alpha_2\rho_2$. Here α_a is the volume fraction of the a-th component, $\alpha_1 + \alpha_2 = 1$.

The definition of the volume fraction is as follows. At each point $(t, \mathbf{x})$ consider a representative mixture volume $\Omega : \Omega = \Omega_1 \cup \Omega_2$ where Ω_a is the volume occupied by the $\alpha - th$ phase. Consider the characteristic function of the $a - th$ phase :

$$\chi_a(t, \mathbf{x}') = \begin{cases} 1, \text{ if } (t, \mathbf{x}') \in \Omega_a \\ 0, \text{ if } (t, \mathbf{x}') \notin \Omega_a \end{cases}$$

Then

$$\alpha_a(t, \mathbf{x}) = \langle\chi_a\rangle = \frac{\int_\Omega \chi_a(t, \mathbf{x}')\, d\omega'}{Vol(\Omega)} = \frac{Vol(\Omega_a)}{Vol(\Omega)}$$

The phase average characteristics are defined in standard way :

$$\rho_a(t, \mathbf{x}) = \frac{\int_{\Omega_a} \rho_a'(t, \mathbf{x}')\, d\omega'}{Vol(\Omega_a)}$$

$$\mathbf{v}_a(t,\mathbf{x}) = \frac{\int\limits_{\Omega_a} \rho'_a\,(t,\mathbf{x}')\,\mathbf{v}'_a\,(t,\mathbf{x}')\,d\omega'}{\int\limits_{\Omega_a} \rho'_a\,(t,\mathbf{x}')\,d\omega'}$$

etc. Other types of average can be introduced: time average, space-time average etc. (see, for example, Drew and Passman (1998), Ishii and Hibiki (2006)). The usual technique in obtaining the governing equations of multiphase flows is to multiply the system of balance laws by the characteristic function and average it. In particular, the mass conservation laws become :

$$\frac{\partial\,(\alpha_a \rho_a)}{\partial t} + div\ (\alpha_a \rho_a \mathbf{v}_a) = 0.$$

However, applying the same averaging procedure to the conservation laws of momentum and energy, we immediately arrive to the closure problem : the balance equations contain more unknowns than equations. A possible remedy to that is to formulate the Lagrangian of the system directly in terms of average variables. And then to proceed in usual way by writing the corresponding Euler-Lagrange equations. The principal difficulty in applying such an approach is : how to determine this Lagrangian? At least two approaches can be used. The first is a phenomenological formulation of such a Lagrangian. The validation of such an approach should be established by comparing to experiments. The second possibility is an asymptotic modeling where the development with respect to a small parameter allows us to obtain an approximate Lagrangian from the exact one.

Also, it is not sufficient just to obtain the Euler-Lagrange equations : dissipation terms consistent with the second law of thermodynamics should be introduced to describe real physics.

6 Equilibrium one-velocity model

Quite often one can suppose that the average velocities of phases are equal : $\mathbf{v}_1 = \mathbf{v}_2 = \mathbf{v}$. We introduce the volume kinetic energy of the system :

$$T = \sum_{a=1}^{2} \alpha_a \rho_a \frac{|\mathbf{v}_a|^2}{2} = \left(\sum_{a=1}^{2} \alpha_a \rho_a\right) \frac{|\mathbf{v}|^2}{2} = \rho \frac{|\mathbf{v}|^2}{2},$$

and the potential energy

$$W = \sum_{a=1}^{2} \alpha_a \rho_a e_a = \rho \sum_{a=1}^{2} Y_a e_a, \qquad Y_a = \frac{\alpha_a \rho_a}{\rho}, \qquad Y_1 + Y_2 = 1.$$

The variable Y_a is called *mass fraction*. The energies $e_a(\rho_a, \eta_a)$ verify the Gibbs identity:

$$\theta_a d\eta_a = de_a + p_a d(\frac{1}{\rho_a})$$

where p_a are the pressures. The Lagrangian of the mixture is :

$$L = \int_{\mathcal{D}(t)} \rho \left(\frac{|\mathbf{v}|^2}{2} - e \right) dD$$

where we assume the additivity of the mixture energy defined as :

$$e = Y_1 e_1 \left(\frac{Y_1 \rho}{\alpha_1}, \eta_1 \right) + Y_2 e_2 \left(\frac{Y_2 \rho}{\alpha_2}, \eta_2 \right).$$

The variation of the Hamilton action can be found under the usual constraints :

$$\frac{\partial \rho}{\partial t} + div\ (\rho \mathbf{v}) = 0,$$

$$\frac{DY_1}{Dt} = 0,$$

$$\frac{D\eta_a}{Dt} = 0, \qquad a = 1, 2.$$

The novelty is that a new independent variable appears, the volume fraction, which is not related to the virtual displacements. Hence, an additional variation should be taken. The variations of the density, mass fraction and entropies are given by :

$$\hat{\delta}\rho = -div(\rho \delta \mathbf{x}), \quad \hat{\delta}\eta_a = -\nabla \eta_a \cdot \delta \mathbf{x}, \qquad \hat{\delta} Y_a = -\nabla Y_a \cdot \delta \mathbf{x}.$$

The corresponding variation of the Lagrangian can be found in the same way as for classical compressible fluids :

$$\delta a = \int_{t_0}^{t_1} \int_{\mathcal{D}(t)} - \left(\frac{\partial \rho \mathbf{v}}{\partial t} + div\ (\rho \mathbf{v} \otimes \mathbf{v}) + \nabla p \right) \cdot \delta \mathbf{x} dD = 0$$

where

$$p = \alpha_1 p_1 + \alpha_2 p_2$$

is the *mixture pressure*. It gives the momentum equation :

$$\frac{\partial \rho \mathbf{v}}{\partial t} + div\ (\rho \mathbf{v} \otimes \mathbf{v}) + \nabla p = 0.$$

Doing so, we have supposed that the volume fraction is fixed. Now, it is necessary to find the variation of the Lagrangian with respect to α_1 :

$$\delta L = \int_{\mathcal{D}(t)} \rho\left(-Y_1\frac{\partial e_1}{\partial \rho_1}\left(-\frac{Y_1\rho}{\alpha_1^2}\right) - Y_2\frac{\partial e_2}{\partial \rho_2}\left(\frac{Y_2\rho}{\alpha_2^2}\right)\right)\delta\alpha_1 dD$$

$$= \int_{\mathcal{D}(t)} (p_1 - p_2)\,\delta\alpha_1 dD = 0.$$

Since it is valid for any $\delta\alpha_1$, it follows from here that

$$p_1 - p_2 = 0.$$

The equilibrium of velocities implies also the pressure equilibrium : $p = p_1 = p_2$. This is an algebraic equation for the volume fraction. The energy equation is a consequence of the mass conservation laws, the momentum equation and the entropy equations :

$$\left(\rho\left(e + \frac{|\mathbf{v}|^2}{2}\right)\right)_t + div\left(\rho\mathbf{v}\left(e + \frac{|\mathbf{v}|^2}{2}\right) + p\mathbf{v}\right) = 0. \tag{13}$$

6.1 What is the sound speed in multiphase flow models?

The sound speeds of pure phases are determined as :

$$c_a^2 = \left.\frac{\partial p_a(\rho_a, \eta_a)}{\partial \rho_a}\right|_{\eta_a = const}, \qquad a = 1, 2.$$

They correspond to the phase velocity of propagation of linear perturbations of the corresponding Euler equations linearized at rest. The corresponding sound speed c_w in a two-fluid mixture is called *Wood sound speed* and is defined as

$$\frac{1}{\rho c_w^2} = \frac{\alpha_1}{\rho_1 c_1^2} + \frac{\alpha_2}{\rho_2 c_2^2}. \tag{14}$$

Consider first an euristic method of derivation of this formula. The definition of the specific mixture volume is :

$$\tau = Y_1\tau_1 + Y_2\tau_2,$$

where

$$\tau = \frac{1}{\rho}, \qquad \tau_a = \frac{1}{\rho_a}, \qquad a = 1, 2.$$

Let us differentiate it with respect to the mixture pressure $p = p_1 = p_2$:

$$\frac{d\tau}{dp} = -\frac{1}{\rho^2 c_w^2} = Y_1 \frac{d\tau_1}{dp_1} + Y_2 \frac{d\tau_2}{dp_2} = -\frac{Y_1}{\rho_1^2 c_1^2} - \frac{Y_2}{\rho_2^2 c_2^2}.$$

It implies immediately (14). Another way, more rigorous one, is to calculate the characteristic speeds of the governing system. For simplicity, we consider only one-dimensional case ($\mathbf{v} = (u), \mathbf{x} = (x)$). The equations are :

$$\frac{D\tau}{Dt} - u_x = 0,$$

$$\frac{Du}{Dt} + \frac{p_x}{\rho} = 0,$$

$$\frac{DY_1}{Dt} = 0,$$

$$\frac{D\eta_1}{Dt} = 0,$$

$$\frac{D\eta_2}{Dt} = 0.$$

The mean specific volume is given by :

$$\tau = \tau(p, \eta_1, \eta_2, Y_1) = Y_1 \tau_1(p, \eta_1) + Y_2 \tau_2(p, \eta_2).$$

Obviously,

$$\frac{D\tau}{Dt} = Y_1 \frac{D\tau_1(p, \eta_1)}{Dt} + Y_2 \frac{D\tau_1(p, \eta_1)}{Dt} =$$

$$-\left(\frac{Y_1}{\rho_1^2 c_1^2} + \frac{Y_2}{\rho_2^2 c_2^2}\right)\frac{Dp}{Dt} = -\left(\frac{\alpha_1 \tau}{\rho_1 c_1^2} + \frac{\alpha_2 \tau}{\rho_2 c_2^2}\right)\frac{Dp}{Dt}.$$

Hence, the system is

$$\frac{Dp}{Dt} + \rho c_w^2 u_x = 0, \tag{15}$$

$$\frac{Du}{Dt} + \frac{p_x}{\rho} = 0,$$

$$\frac{DY_1}{Dt} = 0,$$

$$\frac{D\eta_1}{Dt} = 0,$$

$$\frac{D\eta_2}{Dt} = 0.$$

System (15) can be rewritten in terms of unknowns $\mathbf{U} = (p, u, Y_1, \eta_1, \eta_2)^T$:

$$\mathbf{U}_t + A(\mathbf{U})\,\mathbf{U}_x = 0,$$

where

$$A(\mathbf{U}) = \begin{pmatrix} u & \rho c_w^2 & 0 & 0 & 0 \\ \frac{1}{\rho} & u & 0 & 0 & 0 \\ 0 & 0 & u & 0 & 0 \\ 0 & 0 & 0 & u & 0 \\ 0 & 0 & 0 & 0 & u \end{pmatrix}$$

The characteristic values of the matrix A are

$$\lambda_{1,2} = u \pm c_w, \qquad \lambda_{3,4,5} = u.$$

On can see that the system is hyperbolic (obviously, the eigenvalue $\lambda = u$ of multiplicity three has exactly three eigenvectors).

The Wood sound speed c_w has very interesting physical properties. Indeed, let us fix the phase densities and entropies and plot the Wood sound speed as a function of the volume fraction.

$$c_w = \frac{1}{\sqrt{(\alpha_1\rho_1 + \alpha_2\rho_2)\left(\frac{\alpha_1}{\rho_1 c_1^2} + \frac{\alpha_2}{\rho_2 c_2^2}\right)}}.$$

Its minimal value is much lower than the sound speeds in either medium. For example, for the "air-water" mixture with $c_1 = 330m/s$, $\rho_1 = 1,24$ kg/m^3, $c_2 = 1500m/s$, $\rho_2 = 1000\ kg/m^3$ the minimal sound speed is about $23m/s$. This very surprising fact has also been validated experimentally (Micaelli, 1982).

The model (15) we have derived is usually called "5 equations model" (see, for example, Kapila *et al.* (2001) for the derivation of this model from a more general two-velocity model). As we have seen, it is hyperbolic. However, it is not in conservative form. The last renders questionable a formal determination of the Rankine-Hugoniot relations for this model. A possible solution to this problem can be found in Saurel *et al.* (2007).

7 Nonequilibrium one-velocity model

The introduction of dissipation is always a phenomenological procedure which however should satisfy the entropy inequality. A natural relaxation equation can be written to replace the algebraic relation $p_1 - p_2 = 0$. Introducing the dissipation function

$$\mathcal{D} = \frac{1}{2\mu}\left(\frac{D\alpha_1}{Dt}\right)^2$$

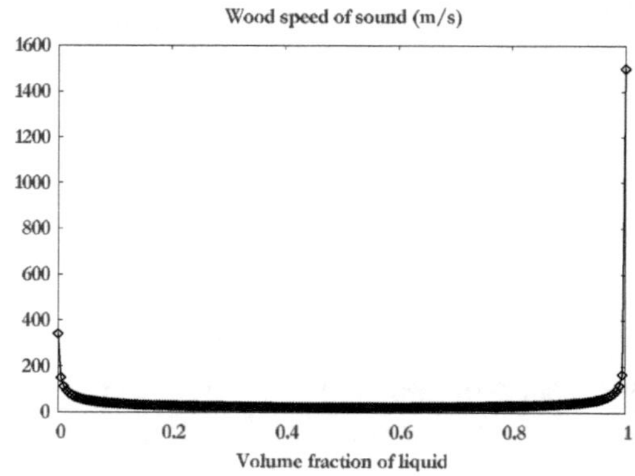

Figure 1. The minimal value of the Wood sound speed is much lower than the sound speeds in either components.

one can replace the equilibrium condition by the following one :

$$p_1 - p_2 = \frac{\partial \mathcal{D}}{\partial \left(\frac{D\alpha_1}{Dt}\right)} = \frac{1}{\mu}\frac{D\alpha_1}{Dt}.$$

Here μ is some constant. The physical sense of this equation is clear : the volume fraction changes due to the pressure difference. This equation is the simplest one. In more general case it may also contain inertia terms (see Gavrilyuk and Saurel (2002) and Saurel, Gavrilyuk and Renaud (2003) for details). The equations for entropies should satisfy the entropy inequality in the form :

$$\frac{\partial (\rho\eta)}{\partial t} + \operatorname{div}(\rho\eta\mathbf{v}) \geq 0 \qquad (16)$$

where η is the mixture entropy defined by

$$\rho\eta = \alpha_1\rho_1\eta_1 + \alpha_2\rho_2\eta_2.$$

We take the equations for the entropies in the form :

$$\alpha_a\rho_a\theta_a\frac{D\eta_a}{Dt} = f_a, \qquad a = 1, 2$$

where f_a should be chosen to satisfy the entropy inequality (16). The production terms f_a should also be compatible with the energy equation (13). Developing the energy equation one can obtain

$$\frac{De}{Dt} + p\frac{D}{Dt}\left(\frac{1}{\rho}\right) = 0.$$

Or

$$\alpha_1\rho_1\theta_1\frac{D\eta_1}{Dt}+\alpha_2\rho_2\theta_2\frac{D\eta_2}{Dt}=\mu\left(p_1-p_2\right)^2\geq 0. \tag{17}$$

To satisfy (17) one can choose

$$\alpha_a\rho_a\theta_a\frac{D\eta_a}{Dt}=f_a=\left(p_a-p_I\right)\frac{D\alpha_a}{Dt},\qquad a=1,2. \tag{18}$$

where the interface pressure p_I can be taken as :

$$p_I=\beta_2 p_1+\beta_1 p_2,\qquad \beta_1+\beta_2=1,\qquad \beta_a\geq 0. \tag{19}$$

Here β_i are some constants. In particular, the interface pressure p_I can be taken as the pressure at the contact discontinuity obtained as the solution of the linearized Riemann problem with initial pressures p_1, p_2 (Saurel, Gavrilyuk and Renaud, 2003). For that case

$$\beta_2=\frac{Z_2}{Z_1+Z_2},\qquad \beta_1=\frac{Z_1}{Z_1+Z_2},$$

where Z_a are acoustical impedances of pure phases. With such a choice, the entropies of each phase increase :

$$\alpha_1\rho_1\frac{D\eta_1}{Dt}=\frac{1}{\theta_1}\left(p_1-p_I\right)\frac{D\alpha_1}{Dt}=\frac{\mu\beta_1}{\theta_1}\left(p_1-p_2\right)^2\geq 0,$$

$$\alpha_2\rho_2\frac{D\eta_2}{Dt}=\frac{1}{\theta_2}\left(p_2-p_I\right)\frac{D\alpha_2}{Dt}=\frac{\mu\beta_2}{\theta_2}\left(p_1-p_2\right)^2\geq 0.$$

Summing these equations we obtain the entropy inequality (16). Finally, the corresponding 1D non-equilibrium model is :

$$\frac{\partial\rho}{\partial t}+\frac{\partial\left(\rho u\right)}{\partial x}=0, \tag{20}$$

$$\frac{Du}{Dt}+\frac{1}{\rho}\frac{\partial p}{\partial x}=0,$$

$$p=\alpha_1 p_1\left(\frac{Y_1\rho}{\alpha_1},\eta_1\right)+\alpha_2 p_2\left(\frac{Y_2\rho}{\alpha_2},\eta_2\right),$$

$$\frac{D\alpha_1}{Dt}=\mu\left(p_1-p_2\right),$$

$$\frac{DY_1}{Dt}=0,$$

$$\frac{D\eta_1}{Dt} = \frac{\mu\beta_1}{\alpha_1\rho_1\theta_1}(p_1 - p_2)^2,$$
$$\frac{D\eta_2}{Dt} = \frac{\mu\beta_2}{\alpha_2\rho_2\theta_2}(p_1 - p_2)^2.$$

The equations are also hyperbolic, however the characteristic speeds are now : $\lambda_{1,2} = u \pm c_f$, $\lambda_k = u$, $k = 3, 4, 5, 6$. Here c_f is the "frozen" sound speed :

$$c_f^2 = Y_1 c_1^2 + Y_2 c_2^2.$$

One can prove that

$$c_f^2 > c_w^2. \tag{21}$$

When $\mu \to \infty$, we recover in this asymptotic limit the equilibrium model (15). The condition (21) is known as Whitham's stability condition (or Whitham's subcharacteristic condition) (Whitham, 1974). The Wood sound speed and the "frozen" sound speed are the velocities of propagation of long and short waves, respectively, for (20).

8 Bubbly fluids

In applications (for example, shock wave propagation in bubbly fluids, high velocity impacts etc.) it is necessary to take into account inertia effects. In particular, in such media, the internal energy of a mixture is a function not only of the flow parameters but also their derivatives. Consider a particular case of fluids containing gas bubbles ("bubbly fluids"). For simplicity, we will suppose that the liquid phase is incompressible : only gas bubbles are compressible.

We will use the index "1" for a liquid and the index "2" for a gas. The kinetic energy of an incompressible fluid of density $\rho_{10} = const$ due to single bubble oscillations is (Iordansky (1960), Kogarko (1961), Wijngaarden (1968)) :

$$2\pi R^3 \rho_{10} \left(\frac{dR}{dt}\right)^2$$

where R is the bubble radius. If N_2 is the number of bubbles per unit volume of the mixture satisfying the conservation law

$$\frac{\partial N_2}{\partial t} + div\ (N_2 \mathbf{v}) = 0,$$

the corresponding energy per unit volume of N_2 moving bubbles (having the same mean velocity as a surrounding fluid) will be :

$$2\pi R^3 N_2 \rho_{10} \left(\frac{DR}{Dt}\right)^2, \qquad \frac{D}{Dt} = \frac{\partial}{\partial t} + \mathbf{v}\cdot\nabla,$$

The bubble volume fraction is :

$$\alpha_2 = \frac{4}{3}\pi R^3 N_2.$$

The Lagrangian of the bubbly fluid can be written in the form :

$$L = \int_{\mathcal{D}(t)} \left(\frac{\rho\left|\mathbf{v}\right|^2}{2} + \frac{3}{2}\alpha_2\rho_{10}\left(\frac{DR}{Dt}\right)^2 - \rho Y_2 e_2\left(\frac{Y_2\rho}{\alpha_2}, \eta_2\right)\right) dD \tag{22}$$

We consider the case where the mass concentration of bubbles $Y_2 = \frac{\alpha_2\rho_2}{\rho}$, the number of bubbles per unit mass $n_2 = \frac{N_2}{\rho}$ and the gas entropy η_2 are constant. The bubble radius and the gas density can be expressed through the average density ρ as :

$$\frac{4}{3}\pi R^3 = \left(\frac{1}{\rho} - \frac{Y_1}{\rho_{10}}\right)/n_2, \quad \rho_2 = Y_2\left(\frac{1}{\rho} - \frac{Y_1}{\rho_{10}}\right)^{-1}. \tag{23}$$

We rewrite (22) in the following generic form :

$$L = \int_{\mathcal{D}(t)} \left(\frac{\rho\left|\mathbf{v}\right|^2}{2} - W\left(\rho, \frac{D\rho}{Dt}\right)\right) dD \tag{24}$$

In our particular case

$$W\left(\rho, \frac{D\rho}{Dt}\right) = \rho\left(Y_2 e_2\left(\rho_2, \eta_2\right) - 2\pi R^3 n_2\rho_{10}\left(\frac{DR}{Dt}\right)^2\right)$$

The usual constraint should be respected :

$$\frac{\partial\rho}{\partial t} + div\ (\rho\mathbf{v}) = 0.$$

Let us define the variation of such a generic Lagrangian (24). Obviously,

$$\delta a$$

$$= \int_{t_0}^{t_1}\int_{\mathcal{D}(t)} \left(\hat{\delta}\rho\left(\frac{\left|\mathbf{v}\right|^2}{2} - \frac{\delta W}{\delta\rho}\right) + \rho\mathbf{v}\cdot\hat{\delta}\mathbf{v} - \frac{\partial W}{\partial\left(\frac{D\rho}{Dt}\right)}\hat{\delta}\left(\frac{D\rho}{Dt}\right)\right) dDdt$$

$$= \int_{t_0}^{t_1}\int_{\mathcal{D}(t)} \left(\left(\rho\mathbf{v} - \frac{\partial W}{\partial\left(\frac{D\rho}{Dt}\right)}\nabla\rho\right)\cdot\hat{\delta}\mathbf{v} + \hat{\delta}\rho\left(\frac{\left|\mathbf{v}\right|^2}{2} - \frac{\delta W}{\delta\rho}\right)\right) dDdt$$

$$+\int_{t_0}^{t_1}\int_{\mathcal{D}(t)}\left(-\frac{\partial}{\partial t}\left(\frac{\partial W}{\partial\left(\frac{D\rho}{Dt}\right)}\hat{\delta}\rho\right)-\operatorname{div}\left(\frac{\partial W}{\partial\left(\frac{D\rho}{Dt}\right)}\mathbf{v}\hat{\delta}\rho\right)\right)dDdt$$

$$=\int_{t_0}^{t_1}\int_{\mathcal{D}(t)}\left(\left(\frac{|\mathbf{v}|^2}{2}-\frac{\delta W}{\delta\rho}\right)\hat{\delta}\rho+\left(\rho\mathbf{v}-\frac{\partial W}{\partial\left(\frac{D\rho}{Dt}\right)}\nabla\rho\right)\cdot\hat{\delta}\mathbf{v}\right)dDdt$$

$$+\int_{t_0}^{t_1}\int_{\mathcal{D}(t)}\left(-\frac{\partial}{\partial t}\left(\frac{\partial W}{\partial\left(\frac{D\rho}{Dt}\right)}\hat{\delta}\rho\right)-\operatorname{div}\left(\frac{\partial W}{\partial\left(\frac{D\rho}{Dt}\right)}\mathbf{v}\hat{\delta}\rho\right)\right)dDdt$$

$$=\int_{t_0}^{t_1}\int_{\mathcal{D}(t)}\left(\left(\frac{|\mathbf{v}|^2}{2}-\frac{\delta W}{\delta\rho}\right)\hat{\delta}\rho+\left(\rho\mathbf{v}-\frac{\partial W}{\partial\left(\frac{D\rho}{Dt}\right)}\nabla\rho\right)\cdot\hat{\delta}\mathbf{v}\right)dDdt.$$

since the conservative terms are vanishing at the boundary $\partial\left([t_0,t_1]\times\mathcal{D}(t)\right)$. Here we have introduced the classical notation for the variational derivative of W:

$$\frac{\delta W}{\delta\rho}=\frac{\partial W}{\partial\rho}-\frac{\partial}{\partial t}\left(\frac{\partial W}{\partial\left(\frac{\partial\rho}{\partial t}\right)}\right)-\operatorname{div}\left(\frac{\partial W}{\partial(\nabla\rho)}\mathbf{v}\right)$$

$$=\frac{\partial W}{\partial\rho}-\frac{\partial}{\partial t}\left(\frac{\partial W}{\partial\left(\frac{D\rho}{Dt}\right)}\right)-\operatorname{div}\left(\frac{\partial W}{\partial\left(\frac{D\rho}{Dt}\right)}\mathbf{v}\right)$$

$$=\frac{\partial W}{\partial\rho}-\rho\frac{D}{Dt}\left(\frac{1}{\rho}\frac{\partial W}{\partial\left(\frac{D\rho}{Dt}\right)}\right)$$

Integrating by parts we obtain

$$\delta a=\int_{t_0}^{t_1}\int_{\mathcal{D}(t)}\delta\mathcal{L}dDdt$$

where

$$\delta\mathcal{L}=\mathbf{l}\cdot\delta\mathbf{x},\quad \mathbf{l}=-\rho\frac{D\mathbf{v}}{Dt}-\nabla\left(\rho\frac{\delta W}{\delta\rho}-W\right)$$

$$-\nabla W+\frac{\delta W}{\delta\rho}\nabla\rho+\rho\frac{D}{Dt}\left(\frac{1}{\rho}\frac{\partial W}{\partial\left(\frac{D\rho}{Dt}\right)}\nabla\rho\right)+\frac{\partial W}{\partial\left(\frac{D\rho}{Dt}\right)}\left(\frac{\partial\mathbf{v}}{\partial\mathbf{x}}\right)^T\nabla\rho$$

Let us show that the following term vanishes :

$$-\nabla W + \frac{\delta W}{\delta \rho}\nabla \rho + \rho \frac{D}{Dt}\left(\frac{1}{\rho}\frac{\partial W}{\partial\left(\frac{D\rho}{Dt}\right)}\nabla \rho\right) + \frac{\partial W}{\partial\left(\frac{D\rho}{Dt}\right)}\left(\frac{\partial \mathbf{v}}{\partial \mathbf{x}}\right)^T \nabla \rho = 0.$$

Indeed,

$$-\nabla W + \frac{\delta W}{\delta \rho}\nabla \rho + \rho \frac{D}{Dt}\left(\frac{1}{\rho}\frac{\partial W}{\partial\left(\frac{D\rho}{Dt}\right)}\nabla \rho\right) + \frac{\partial W}{\partial\left(\frac{D\rho}{Dt}\right)}\left(\frac{\partial \mathbf{v}}{\partial \mathbf{x}}\right)^T \nabla \rho$$

$$= -\frac{\partial W}{\partial \rho}\nabla \rho - \frac{\partial W}{\partial\left(\frac{D\rho}{Dt}\right)}\nabla\left(\frac{D\rho}{Dt}\right)$$

$$+\left(\frac{\partial W}{\partial \rho} - \rho\frac{D}{Dt}\left(\frac{1}{\rho}\frac{\partial W}{\partial\left(\frac{D\rho}{Dt}\right)}\right)\right)\nabla \rho + \rho\frac{D}{Dt}\left(\frac{1}{\rho}\frac{\partial W}{\partial\left(\frac{D\rho}{Dt}\right)}\right)\nabla \rho$$

$$+\frac{\partial W}{\partial\left(\frac{D\rho}{Dt}\right)}\frac{D\nabla \rho}{Dt} + \frac{\partial W}{\partial\left(\frac{D\rho}{Dt}\right)}\left(\frac{\partial \mathbf{v}}{\partial \mathbf{x}}\right)^T \nabla \rho$$

$$= -\frac{\partial W}{\partial\left(\frac{D\rho}{Dt}\right)}\nabla\left(\frac{D\rho}{Dt}\right) + \frac{\partial W}{\partial\left(\frac{D\rho}{Dt}\right)}\frac{D\nabla \rho}{Dt} + \frac{\partial W}{\partial\left(\frac{D\rho}{Dt}\right)}\left(\frac{\partial \mathbf{v}}{\partial \mathbf{x}}\right)^T \nabla \rho = 0.$$

Finally, the momentum equation is :

$$\rho\frac{D\mathbf{v}}{Dt} + \nabla\left(\rho\frac{\delta W}{\delta \rho} - W\right) = 0.$$

The governing equations can also be rewritten in conservative form :

$$\frac{\partial \rho}{\partial t} + div\left(\rho \mathbf{v}\right) = 0, \tag{25}$$

$$\frac{\partial \rho \mathbf{v}}{\partial t} + \mathrm{div}\left(\rho \mathbf{v} \otimes \mathbf{v}\right) + \nabla p = 0. \tag{26}$$

where the pressure is defined by

$$p = \rho\frac{\delta W}{\delta \rho} - W. \tag{27}$$

It admits also the energy conservation law :

$$\frac{\partial}{\partial t}\left(\frac{\rho\left|\mathbf{v}\right|^2}{2}+E\right)+\operatorname{div}\left(\mathbf{v}\left(\frac{\rho\left|\mathbf{v}\right|^2}{2}+E+p\right)\right)=0 \tag{28}$$

where

$$E=W-\frac{D\rho}{Dt}\frac{\partial W}{\partial\left(\frac{D\rho}{Dt}\right)}.$$

If W depends only on ρ, then $W=E$ and equations (25), (26) coincide with the classical Euler equations of barotropic fluids.

Let us calculate the pressure equation (27) in explicit form for the bubbly fluids. We introduce the specific potential w :

$$W=\rho w,$$

$$w=Y_2e_2\left(\rho_2,\eta_2\right)-2\pi R^3n_2\rho_{10}\left(\frac{DR}{Dt}\right)^2.$$

Obviously,

$$p=\rho^2\left(\frac{\partial w}{\partial\rho}-\frac{D}{Dt}\left(\frac{\partial w}{\partial\left(\frac{D\rho}{Dt}\right)}\right)\right).$$

Consider now $w\left(\rho,\frac{D\rho}{Dt}\right)$ as a function of R and $\frac{DR}{Dt}$. Then

$$p=-\frac{1}{4\pi R^2n_2}\left(\frac{\partial w}{\partial R}-\frac{D}{Dt}\left(\frac{\partial w}{\partial\left(\frac{DR}{Dt}\right)}\right)\right)$$

$$=-\frac{1}{4\pi R^2n_2}\left(-p_2\frac{3\left(\frac{4\pi}{3}R^3n_2\right)^2}{\frac{4\pi}{3}R^4n_2}-6\pi R^2n_2\rho_{10}\left(\frac{DR}{Dt}\right)^2\right)$$

$$-\frac{1}{4\pi R^2n_2}\frac{D}{Dt}\left(4\pi R^3n_2\rho_{10}\left(\frac{DR}{Dt}\right)\right)$$

$$=-\frac{1}{4\pi R^2n_2}\left(-p_2\frac{3\left(\frac{4\pi}{3}R^3n_2\right)^2}{\frac{4\pi}{3}R^4n_2}-6\pi R^2n_2\rho_{10}\left(\frac{DR}{Dt}\right)^2\right.$$

$$\left.+4\pi R^3n_2\rho_{10}\frac{D^2R}{Dt^2}+12\pi R^2n_2\rho_{10}\left(\frac{DR}{Dt}\right)^2\right)$$

$$=p_2+\rho_{10}\left(\frac{3}{2}\left(\frac{DR}{Dt}\right)^2+R\frac{D^2R}{Dt^2}\right).$$

The equation

$$p = p_2 + \rho_{10}\left(\frac{3}{2}\left(\frac{DR}{Dt}\right)^2 + R\frac{D^2R}{Dt^2}\right) \tag{29}$$

is known as the Rayleigh-Lamb equation describing oscillations of a single bubble in an infinite fluid. Model (25), (26) and (29) is called in the literature Iordansky-Kogarko-Wijngaarden model (Iordansky (1960), Kogarko (1961), van Wijngaarden (1968)).

8.1 Dispersive shallow water flows as an analogue of bubbly flows

Much work has been done in the derivation of relatively simple mathematical models of long non-linear water waves. One of popular models is the Green-Naghdi model obtained in the one-dimensional case by Su and Gardner (1969) and in the multi-dimensional case by Green, Laws and Naghdi (1974, 1976) within the context of a homogeneous one-layer fluid. In the literature, this model is usually called Green-Naghdi model (GN model or GN system). A derivation of the GN model based on the variational formulation of the Euler equations was done by Miles and Salmon (1985) (see also Salmon (1988, 1998)). A mathematical justification of the GN model was done by Makarenko (1986) and Alvarez-Samaniego and Lannes (2008). Solitary wave solutions of the GN model were obtained by Su and Gardner (1969). The linear stability of solitary waves has been proved by Li (2001). A criterium of stability of shear flows for the GN model was proposed by Gavrilyuk and Teshukov (2004). A wide class of multi-dimensional solutions and approximate solutions of nonlinear multi-dimensional GN model has been found by Gavrilyuk and Teshukov (2001) and Teshukov and Gavrilyuk (2006). Unsteady undular bores were described by El, Grimshaw and Smyth (2006).

It is interesting to note that such a model can be viewed as an analogue of "bubbly fluids" flow model. The Green-Naghdi equations describing shallow water flows over a plane bottom are :

$$\frac{\partial h}{\partial t} + div\,(h\mathbf{v}) = 0,$$

$$\frac{\partial h\mathbf{v}}{\partial t} + \mathrm{div}\left(h\mathbf{v}\otimes\mathbf{v} + \left(\frac{gh^2}{2} + \frac{h^2}{3}\frac{D^2h}{Dt^2}\right)I_2\right) = 0.$$

Here h is the position of free surface, $\mathbf{v}$ is the average with respect the vertical direction $2D$ velocity field, g is the gravity acceleration, and I_2 is the two-dimensional identity transformation. The Green-Naghdi model is formally described with the same type of Lagrangian :

$$L = \int_{\mathcal{D}(t)} \left(\frac{h\left|\mathbf{v}\right|^2}{2} - W\left(h, \frac{Dh}{Dt}\right) \right) dD$$

where the function W is :

$$W(h, \frac{Dh}{Dt}) = \frac{gh^2}{2} - \frac{h}{6}\left(\frac{Dh}{Dt}\right)^2.$$

Obviously,

$$p = h\frac{\delta W}{\delta h} - W = \frac{gh^2}{2} + \frac{h^2}{3}\frac{D^2 h}{Dt^2}.$$

Such an analogy between "bubbly fluids" and dispersive shallow water flows is very useful. Indeed, the parameters of waves propagating in bubbly fluids can not be observed, they can only be measured, while the position of the free surface (playing the role of average density) can be at least easily observed.

9 Two-velocity one-pressure model

Consider the case where the averaging velocities of phases are not equal. The Lagrangian of the mixture is :

$$L = \int_{\mathcal{D}} \left(\sum_{a=1}^{2} \alpha_a \rho_a \frac{\left|\mathbf{v}_a\right|^2}{2} - \sum_{a=1}^{2} \alpha_a \rho_a e_a \right) dD$$

Here $\mathcal{D}$ is not a material volume, this notion has no sense in the case of two velocity continua. We will use the Eulerian variations to define the governing equations. The imposed constraints are

$$(\alpha_a \rho_a)_t + \text{div}\,(\alpha_a \rho_a \mathbf{v}_a) = 0,$$

$$(\alpha_a \rho_a \eta_a)_t + \text{div}\,(\alpha_a \rho_a \mathbf{v}_a \eta_a) = 0.$$

Or, in terms of apparent densities $\bar{\rho}_a$:

$$(\bar{\rho}_a)_t + \text{div}\,(\bar{\rho}_a \mathbf{v}_a) = 0,$$

$$(\bar{\rho}_a \eta_a)_t + \text{div}\,(\bar{\rho}_a \mathbf{v}_a \eta_a) = 0.$$

At least three types of independent variations should be considered. The first one is the variation with respect to the volume fraction giving us the equilibrium condition :

$$p_1 = p_2 = p, \qquad p = \alpha_1 p_1 + \alpha_2 p_2.$$

The second and the third variations in terms of virtual displacements $\delta\mathbf{x}_a$ (which can be introduced in the same way as in the case of one-velocity continua) give us usual expressions of variations of the density, entropy and velocity :

$$\hat{\delta}\bar{\rho}_a = -div(\bar{\rho}_a\delta\mathbf{x}_a), \quad \hat{\delta}\eta_a = -\nabla\eta_a\cdot\delta\mathbf{x}_a, \qquad \hat{\delta}\mathbf{v}_a = \frac{D_a\delta\mathbf{x}_a}{Dt} - \frac{\partial\mathbf{v}_a}{\partial\mathbf{x}}\delta\mathbf{x}_a,$$

$$\frac{D_a}{Dt} = \frac{\partial}{\partial t} + \mathbf{v}_a\cdot\nabla$$

The Euler-Lagrange equations correspond to the classical momentum equations for each phase :

$$\bar{\rho}_a\frac{D_a\mathbf{v}_a}{Dt} + \alpha_a\nabla p_a = 0.$$

Taking into account the equilibrium condition they can also be rewritten in non-conservative form :

$$\frac{\partial\alpha_1\rho_1\mathbf{v}_1}{\partial t} + div\left(\alpha_1\rho_1\mathbf{v}_1\otimes\mathbf{v}_1 + \alpha_1 p_1 I\right) = p\nabla\alpha_1,$$

$$\frac{\partial\alpha_2\rho_2\mathbf{v}_2}{\partial t} + div\left(\alpha_2\rho_2\mathbf{v}_2\otimes\mathbf{v}_2 + \alpha_2 p_2 I\right) = p\nabla\alpha_2.$$

Summing them we obtain the equation of the total momentum :

$$\left(\sum_{a=1}^{2}\alpha_a\rho_a\mathbf{v}_a\right)_t + \mathrm{div}\left(\sum_{a=1}^{2}\alpha_a\rho_a\mathbf{v}_a\otimes\mathbf{v}_a + pI\right) = 0.$$

The total energy is :

$$\left(\sum_{a=1}^{2}\alpha_a\rho_a\left(e_a + \frac{|\mathbf{v}_a|^2}{2}\right)\right)_t$$

$$+\mathrm{div}\left(\sum_{a=1}^{2}\alpha_a\rho_a\mathbf{v}_a\left(e_a + \frac{|\mathbf{v}_a|^2}{2}\right) + \alpha_a p_a\mathbf{v}_a\right) = 0.$$

9.1 Analysis of the two-velocity one-pressure model

For simplicity, consider the one-dimensional isentropic case ($\eta_a = const$). The equations are :

$$(\alpha_1\rho_1)_t + (\alpha_1\rho_1 u_1)_x = 0, \tag{30}$$

$$(\alpha_2\rho_2)_t + (\alpha_2\rho_2 u_2)_x = 0,$$

$$\frac{D_1 u_1}{Dt} + \frac{p_x}{\rho_1} = 0,$$

$$\frac{D_2 u_2}{Dt} + \frac{p_x}{\rho_2} = 0.$$

The pressures are then functions of densities:

$$p_1\left(\frac{\bar\rho_1}{\alpha_1}\right) = p_2\left(\frac{\bar\rho_2}{\alpha_2}\right) = p. \tag{31}$$

Using the expression for the differential of the volume fraction obtained from the equilibrium condition (31)

$$d\alpha_1 = \frac{\frac{c_1^2}{\alpha_1}d\bar\rho_1 - \frac{c_2^2}{\alpha_2}d\bar\rho_2}{\frac{c_1^2\bar\rho_1}{\alpha_1^2} + \frac{c_2^2\bar\rho_2}{\alpha_2^2}} = \frac{\frac{c_1^2}{\alpha_1}d\bar\rho_1 - \frac{c_2^2}{\alpha_2}d\bar\rho_2}{\frac{c_1^2\rho_1}{\alpha_1} + \frac{c_2^2\rho_2}{\alpha_2}},$$

and rewriting the equations in variables $(\bar\rho_1, \bar\rho_2, u_1, u_2)$, one can obtain the following characteristic polynomial for (30) :

$$(\lambda - u_1)^2(\lambda - u_2)^2 - \frac{\alpha_2\rho c_w^2}{\rho_2}(\lambda - u_1)^2 - \frac{\alpha_1\rho c_w^2}{\rho_1}(\lambda - u_2)^2 = 0.$$

Transforming it to the form

$$1 = \frac{\alpha_2\rho c_w^2}{\rho_2}\frac{1}{(\lambda - u_2)^2} + \frac{\alpha_1\rho c_w^2}{\rho_1}\frac{1}{(\lambda - u_1)^2}$$

and looking for the minimal value of the right-hand side, one obtain the following criterion of hyperbolicity (existence of four real eigenvalues):

$$(u_2 - u_1)^2 > \rho c_w^2\left(\left(\frac{\alpha_1}{\rho_1}\right)^{1/3} + \left(\frac{\alpha_2}{\rho_2}\right)^{1/3}\right)^3.$$

However, this model is not hyperbolic for small relative velocity. One can prove that the corresponding volume energy

$$E = \alpha_1\rho_1 e_1 + \alpha_2\rho_2 e_2 = \bar\rho_1 e_1 + \bar\rho_2 e_2$$

has the following differential :

$$dE = h_1 d\bar\rho_1 + h_2 d\bar\rho_2.$$

where $h_i, i = 1, 2$ are the phase enthalpies :

$$dh_i = \frac{dp_i}{\rho_i} = \frac{dp}{\rho_i}.$$

Also,

$$\frac{\partial^2 E}{\partial \bar{\rho}_1^2} = \frac{\rho c_w^2}{\rho_1^2},$$

$$\frac{\partial^2 E}{\partial \bar{\rho}_2^2} = \frac{\partial h_2}{\partial \bar{\rho}_2} = \frac{c_2^2}{\rho_2}\left(\frac{1}{\alpha_2} + \frac{\bar{\rho}_2}{\alpha_2^2}\frac{\partial \alpha_1}{\partial \bar{\rho}_2}\right) = \frac{\rho c_w^2}{\rho_2^2},$$

$$\frac{\partial^2 E}{\partial \bar{\rho}_1 \partial \bar{\rho}_2} = \frac{\rho c_w^2}{\rho_1 \rho_2}.$$

In particular this implies that the corresponding Hessian matrix is degenerate:

$$\det\begin{pmatrix} \frac{\partial^2 E}{\partial \bar{\rho}_1^2} & \frac{\partial^2 E}{\partial \bar{\rho}_1 \partial \bar{\rho}_2} \\ \frac{\partial^2 E}{\partial \bar{\rho}_1 \partial \bar{\rho}_2} & \frac{\partial^2 E}{\partial \bar{\rho}_2^2} \end{pmatrix} = 0.$$

In mechanics, the non-convexity of the internal energy shows up non-stability of the mechanical system. In our case, the degeneracy of the internal energy is responsible for the non-hyperbolicity of the governing equations and, as a consequence, the ill-posedness of the corresponding Cauchy problem.

10 Two-velocity two-pressure model

The following non-equilibrium model obtained by adding relaxation terms in the pressure equilibrium model is called Baer-Nunziato model (1986) (or BN model) :

$$\frac{\partial \alpha_1 \rho_1}{\partial t} + div\ (\alpha_1 \rho_1 \mathbf{v}_1) = 0, \tag{32}$$

$$\frac{\partial \alpha_2 \rho_2}{\partial t} + div\ (\alpha_2 \rho_2 \mathbf{v}_2) = 0,$$

$$\frac{\partial \alpha_1 \rho_1 \mathbf{v}_1}{\partial t} + div\,(\alpha_1 \rho_1 \mathbf{v}_1 \otimes \mathbf{v}_1 + \alpha_1 p_1 I) = p_I \nabla \alpha_1 + \lambda\,(\mathbf{v}_2 - \mathbf{v}_1),$$

$$\frac{\partial \alpha_2 \rho_2 \mathbf{v}_2}{\partial t} + div\,(\alpha_2 \rho_2 \mathbf{v}_2 \otimes \mathbf{v}_2 + \alpha_2 p_2 I) = p_I \nabla \alpha_2 - \lambda\,(\mathbf{v}_2 - \mathbf{v}_1),$$

$$\frac{\partial}{\partial t}\left(\alpha_1 \rho_1 \left(\frac{|\mathbf{v}_1|^2}{2} + e_1\right)\right) + div\left(\alpha_1 \rho_1 \mathbf{v}_1 \left(\frac{|\mathbf{v}_1|^2}{2} + e_1\right) + \alpha_1 p_1 \mathbf{v}_1\right) =$$

$$= -p_I \frac{\partial \alpha_1}{\partial t} + \lambda \mathbf{v}_1 \cdot (\mathbf{v}_2 - \mathbf{v}_1),$$

$$\frac{\partial}{\partial t}\left(\alpha_2 \rho_2 \left(\frac{|\mathbf{v}_2|^2}{2} + e_2\right)\right) + div\left(\alpha_2 \rho_2 \mathbf{v}_2 \left(\frac{|\mathbf{v}_2|^2}{2} + e_2\right) + \alpha_2 p_2 \mathbf{v}_2\right) =$$

$$= -p_I \frac{\partial \alpha_2}{\partial t} - \lambda \mathbf{v}_1 \cdot (\mathbf{v}_2 - \mathbf{v}_1),$$

$$\frac{D_I \alpha_1}{Dt} = \mu(p_1 - p_2), \qquad \frac{D_I}{Dt} = \frac{\partial}{\partial t} + \mathbf{v}_I \cdot \nabla.$$

Here p_I is the "interface" pressure, $\mathbf{v}_I$ is the "interface" velocity, and λ is the friction coefficient. In the classical BN model the following duality is supposed $p_I = p_2$ and $\mathbf{v}_I = \mathbf{v}_1$, or $p_I = p_1$ and $\mathbf{v}_I = \mathbf{v}_2$. The right-hand side in the momentum equations is the sum of "nozzling terms" $p_I \nabla \alpha_a$ (the word "nozzling" is to underline the analogy between these equations and those describing gas flow in the duct of variable cross section) and Stokes type terms $\lambda(\mathbf{v}_2 - \mathbf{v}_1)$. Other choice of "interface" variables is possible (see, for example, Saurel, Gavrilyuk and Renaud, 2003). Such a system of equations satisfies the entropy inequality and has only real eigenvalues.

11 Equilibrium diffuse interface model of solid-fluid interactions

Pioneering works by Karni (1994), Abgrall (1996) and Saurel and Abgrall (1999) have shown the attractivity the diffuse interface approach for modelling interfaces between ideal fluids having different thermodynamic characteristics. In such an approach the interface between different materials is considered as a diffusion zone : it is a mixture of two components where the volume fraction of the phases varies continuously from zero to one. To assure the non-degeneracy of the governing equations, one also supposes that even in the bulk (where, formally, we have only pure phases) a negligible quantity of the other phase is present. Hence, the interface can be viewed as an effective fluid mixture. We will generalize this approach to the case of solid-fluid interfaces by using Hamilton's principle (Gavrilyuk *et al.* (2008), Favrie *et al.* (2009)).

Consider the following Lagrangian for a solid-fluid mixture :

$$L = \int_{\Omega_t} \rho \left(\frac{\mathbf{v} \cdot \mathbf{v}}{2} - e\right) d\Omega$$

where the average density and the average volume energy are defined as

$$\rho = \alpha_s \rho_s + \alpha_g \rho_g, \qquad \rho e = \alpha_s \rho_s e_s + \alpha_g \rho_g e_g.$$

The indices "s" and "g" mean "solid" and "gas", respectively. The mass conservation laws can be rewritten in terms of the mass fractions Y_s and Y_g :

$$\frac{DY_s}{Dt} = 0, \qquad \frac{DY_g}{Dt} = 0, \qquad Y_s = \frac{\alpha_s \rho_s}{\rho}, \qquad Y_g = \frac{\alpha_g \rho_g}{\rho}, \qquad Y_s + Y_g = 1.$$

Also, the entropies of each phase are conserved :

$$\frac{D\eta_s}{Dt} = 0, \qquad \frac{D\eta_g}{Dt} = 0.$$

The energy of the solid phase e_s is a function of the entropy η_s and the mixture deformation tensor

$$G = \sum_{\beta=1}^{3} \mathbf{E}^\beta \otimes \mathbf{E}^\beta$$

where

$$\mathbf{E}^\beta = \nabla X^\beta$$

and $\mathbf{X} = (X^\beta)$ are the Lagrangian coordinates of a mixture particle. The mixture density is related to G by the formula :

$$\rho(t, \mathbf{X}) = \rho_0(\mathbf{X}) \left|G\right|^{1/2}$$

We use here capital letters $\mathbf{E}^\beta$ to distinguish the mixture case from the pure solid one. The gas energy e_g is a function of the gas density ρ_g and the gas entropy η_s.

To apply the Hamilton principle we find the variations of ρ, $\mathbf{v}$, Y_s, Y_g, η_g, η_s as functions of virtual displacements $\delta\mathbf{x}$ at fixed Lagrangian coordinates :

$$\tilde{\delta}\rho = -\rho \text{div}\left(\delta\mathbf{x}\right), \qquad \tilde{\delta}\mathbf{v} = \frac{\partial \delta\mathbf{x}}{\partial t}, \qquad \tilde{\delta}\eta_i = 0, \qquad \tilde{\delta}Y_i = 0. \tag{33}$$

$$\tilde{\delta}\mathbf{E}^\beta = -\left(\frac{\partial \delta\mathbf{x}}{\partial \mathbf{x}}\right)^T \mathbf{E}^\beta.$$

As usually, the symbol $\tilde{\delta}$ means the variation at fixed Lagrangian coordinates. The last formula giving the variation of $\mathbf{E}^\beta$ can be found in the following way. There exist potentials X^β, $\beta = 1, 2, 3$ (the mixture Lagrangian coordinates) such that

$$\frac{DX^\beta}{Dt} = 0.$$

It follows from the definition of X^β that its Lagrangian variation is zero :

$$\tilde{\delta} X^\beta = 0.$$

Since

$$\tilde{\delta} F^{-1} = -F^{-1}\tilde{\delta} F F^{-1} = -F^{-1}\frac{\partial \delta \mathbf{x}}{\partial \mathbf{X}} F^{-1} = -F^{-1}\frac{\partial \delta \mathbf{x}}{\partial \mathbf{x}}$$

we get from the definition of $\mathbf{E}^\beta$:

$$\tilde{\delta}\mathbf{E}^\beta = \tilde{\delta}\nabla X^\beta = \tilde{\delta}\left(F^{-T}\nabla_{\mathbf{X}} X^\beta\right) = -\left(F^{-1}\frac{\partial \delta \mathbf{x}}{\partial \mathbf{x}}\right)^T \nabla_{\mathbf{X}} X^\beta$$

$$= -\left(\frac{\partial \delta \mathbf{x}}{\partial \mathbf{x}}\right)^T F^{-T}\nabla_{\mathbf{X}} X^\beta = -\left(\frac{\partial \delta \mathbf{x}}{\partial \mathbf{x}}\right)^T \mathbf{E}^\beta.$$

In particular, the variation of $G = \sum_{\beta=1}^{3} \mathbf{E}^\beta \otimes \mathbf{E}^\beta$ is given by :

$$\tilde{\delta} G = \tilde{\delta}\left(\sum_{\beta=1}^{3} \mathbf{E}^\beta \otimes \mathbf{E}^\beta\right) = -\left(\frac{\partial \delta \mathbf{x}}{\partial \mathbf{x}}\right)^T G - G\left(\frac{\partial \delta \mathbf{x}}{\partial \mathbf{x}}\right). \tag{34}$$

The energy of the solid is taken in separable form

$$e_s = e_s^h(\rho_s, \eta_s) + e_s^e(g), \qquad g = \frac{G}{|G|^{1/3}}.$$

Its variation is :

$$\tilde{\delta} e_s = \tilde{\delta} e_s^h\left(\frac{Y_s \rho}{\alpha_s}, \eta_s\right) + \tilde{\delta} e_s^e(g)$$

$$= \frac{\partial e_s^h}{\partial \rho_s}\tilde{\delta}\left(\frac{Y_s\rho}{\alpha_s}\right) + tr\left(\frac{\partial e_s^e}{\partial G}\tilde{\delta} G\right) + \theta_s \tilde{\delta}\eta_s$$

$$= \frac{\rho Y_s}{\alpha_s}\frac{\partial e_s^h}{\partial \rho_s}\left(\frac{\tilde{\delta}\rho}{\rho} - \frac{\tilde{\delta}\alpha_s}{\alpha_s}\right) + tr\left(\frac{\partial e_s^e}{\partial G}\tilde{\delta} G\right)$$

$$= \frac{p_s}{\rho_s}\left(\frac{\tilde{\delta}\rho}{\rho} - \frac{\tilde{\delta}\alpha_s}{\alpha_s}\right) + tr\left(\frac{\partial e_s^e}{\partial G}\tilde{\delta} G\right)$$

$$= \frac{p_s}{\rho_s}\left(\frac{\tilde{\delta}\rho}{\rho} - \frac{\tilde{\delta}\alpha_s}{\alpha_s}\right) - \frac{1}{2\rho_s} tr\left(S_s G^{-1}\tilde{\delta} G\right)$$

$$= \frac{p_s}{\rho_s}\left(\frac{\tilde{\delta}\rho}{\rho} - \frac{\tilde{\delta}\alpha_s}{\alpha_s}\right) + \frac{1}{\rho_s} tr\left(S_s \frac{\partial \delta \mathbf{x}}{\partial \mathbf{x}}\right)$$

$$= \frac{p_s}{\rho_s}\left(\frac{\tilde{\delta}\rho}{\rho} - \frac{\tilde{\delta}\alpha_s}{\alpha_s}\right) + \frac{1}{\rho_s} tr\left(S_s \frac{\partial \delta \mathbf{x}}{\partial \mathbf{x}}\right)$$

$$= \frac{p_s}{\rho_s}\left(-\mathrm{div}\,(\delta \mathbf{x}) - \frac{\tilde{\delta}\alpha_s}{\alpha_s}\right) + \frac{1}{\rho_s} tr\left(S_s \frac{\partial \delta \mathbf{x}}{\partial \mathbf{x}}\right)$$

$$= \frac{p_s}{\rho_s}\left(-\frac{\tilde{\delta}\alpha_s}{\alpha_s}\right) + \frac{1}{\rho_s} tr\left((-p_s I + S_s)\frac{\partial \delta \mathbf{x}}{\partial \mathbf{x}}\right)$$

$$= -\frac{1}{\rho_s} tr\left(\sigma_s \frac{\partial \delta \mathbf{x}}{\partial \mathbf{x}}\right) - \frac{p_s}{\alpha_s \rho_s}\tilde{\delta}\alpha_s.$$

Here

$$S_s = -2\rho_s \frac{\partial e_s^e}{\partial G} G$$

is the deviatoric part of the stress tensor

$$\sigma_s = -p_s I + S_s, \qquad tr(S_s) = 0.$$

Finally,

$$\tilde{\delta} e_s = -\frac{1}{\rho_s} tr\left(\sigma_s \frac{\partial \delta \mathbf{x}}{\partial \mathbf{x}}\right) - \frac{p_s}{\alpha_s \rho_s}\tilde{\delta}\alpha_s \tag{35}$$

Analogous considerations give a simpler variation for the gas phase :

$$\tilde{\delta} e_g = -\frac{1}{\rho_g} tr\left(\sigma_g \frac{\partial \delta \mathbf{x}}{\partial \mathbf{x}}\right) - \frac{p_g}{\alpha_g \rho_g}\tilde{\delta}\alpha_s = \frac{p_g}{\rho_g}\mathrm{div}\delta\mathbf{x} - \frac{p_g}{\alpha_g \rho_g}\tilde{\delta}\alpha_g \tag{36}$$

Now, we are ready to take the variation of the Hamilton action. We use formulas (33), (35), (36), the mass conservation law $\rho d\Omega = \rho_0(\mathbf{X})\, d\Omega_0$ to present the variation of a in the form :

$$0 = \delta a = \delta \int_{t_1}^{t_2} \int_{\Omega_t} \rho \left(\frac{\mathbf{v}\cdot\mathbf{v}}{2} - e\right) d\Omega dt$$

$$= \delta \int_{t_1}^{t_2} \int_{\Omega_0} \rho_0(\mathbf{X}) \left(\frac{\mathbf{v}\cdot\mathbf{v}}{2} - e\right) d\Omega_0 dt$$

$$= \int_{t_1}^{t_2} \int_{\Omega_0} \rho_0 (\mathbf{X}) \left(\tilde{\delta} \mathbf{v} \cdot \mathbf{v} - \tilde{\delta} e \right) d\Omega_0 dt$$

$$= \int_{t_1}^{t_2} \int_{\Omega_0} \rho_0 (\mathbf{X}) \left(\frac{\partial \delta \mathbf{x}}{\partial t} \cdot \mathbf{v} + Y_s \left(\frac{1}{\rho_s} tr \left(\sigma_s \frac{\partial \delta \mathbf{x}}{\partial \mathbf{x}} \right) + \frac{p_s}{\alpha_s \rho_s} \tilde{\delta} \alpha_s \right) \right) d\Omega_0 dt$$

$$- \int_{t_1}^{t_2} \int_{\Omega_0} \rho_0 (\mathbf{X}) \left(Y_g \left(\frac{p_g}{\rho_g} \mathrm{div} \delta \mathbf{x} - \frac{p_g}{\alpha_g \rho_g} \tilde{\delta} \alpha_g \right) \right) d\Omega_0 dt$$

$$= \int_{t_1}^{t_2} \int_{\Omega_0} \rho_0 (\mathbf{X}) \left(\frac{\partial \delta \mathbf{x}}{\partial t} \cdot \mathbf{v} \right) d\Omega_0 dt$$

$$+ \int_{t_1}^{t_2} \int_{\Omega_0} \frac{\rho_0 (\mathbf{X})}{\rho} \left(tr \left((\alpha_s \sigma_s - \alpha_g p_g) \frac{\partial \delta \mathbf{x}}{\partial \mathbf{x}} \right) + p_s \tilde{\delta} \alpha_s + p_g \tilde{\delta} \alpha_g \right) d\Omega_0 dt$$

Since $\tilde{\delta} \alpha_s + \tilde{\delta} \alpha_g = 0$, it implies the equilibrium condition :

$$p_s - p_g = 0. \tag{37}$$

Here

$$p_s = -\frac{1}{3} tr (\sigma_s), \qquad p_g = -\frac{1}{3} tr (\sigma_g).$$

Turning back to the Eulerian coordinates we transform the last variation to the following one :

$$0 = \delta a = \int_{t_1}^{t_2} \int_{\Omega_t} \left(\frac{D \delta \mathbf{x}}{Dt} \cdot \rho \mathbf{v} + tr \left((\alpha_s \sigma_s - \alpha_g p_g) \frac{\partial \delta \mathbf{x}}{\partial \mathbf{x}} \right) \right) d\Omega_t dt$$

Integrating by parts and taking into account the fact that $\delta \mathbf{x}$ is vanishing at the boundary $\Omega_t \times [t_1, t_2\,]$ we obtain

$$0 = \delta a = - \int_{t_1}^{t_2} \int_{\Omega_t} \left(\frac{\partial \rho \mathbf{v}}{\partial t} + \mathrm{div} \left(\rho \mathbf{v} \otimes \mathbf{v} - (\alpha_s \sigma_s + \alpha_g \sigma_g) \right) \right) \cdot \delta \mathbf{x} d\Omega dt$$

for any $\delta \mathbf{x}$ vanishing at the boundary $\partial \left(\Omega_t \times [t_1, t_2\,] \right)$. It implies the momentum equation :

$$\frac{\partial \rho \mathbf{v}}{\partial t} + \mathrm{div} \left(\rho \mathbf{v} \otimes \mathbf{v} - (\alpha_s \sigma_s + \alpha_g \sigma_g) \right) = 0. \tag{38}$$

The entropy equations

$$\frac{D\eta_s}{Dt} = 0, \qquad \frac{D\eta_g}{Dt} = 0$$

complemented by the mass and momentum balance laws imply the mixture energy conservation law:

$$\frac{\partial}{\partial t}\left(\rho\left(e+\frac{\mathbf{v}\cdot\mathbf{v}}{2}\right)\right) + \operatorname{div}\left(\rho\mathbf{v}\left(e+\frac{\mathbf{v}\cdot\mathbf{v}}{2}\right) - \left(\alpha_s\sigma_s + \alpha_g\sigma_g\right)\mathbf{v}\right) = 0. \quad (39)$$

Finally, the equilibrium solid-gas governing equations are

$$\frac{\partial \mathbf{E}^{\beta}}{\partial t} + \nabla\left(\mathbf{E}^{\beta}\cdot\mathbf{v}\right) = -\mathrm{rot}\mathbf{E}^{\beta}\wedge\mathbf{v}, \quad (40a)$$

$$\mathrm{rot}\mathbf{E}^{\beta} = 0, \quad (40b)$$

$$\frac{\partial\left(\alpha_g\rho_g\right)}{\partial t} + \operatorname{div}\left(\alpha_g\rho_g\mathbf{v}\right) = 0, \quad (40c)$$

$$\frac{\partial \rho\mathbf{v}}{\partial t} + \operatorname{div}\left(\rho\mathbf{v}\otimes\mathbf{v} - \left(\alpha_s\sigma_s + \alpha_g\sigma_g\right)\right) = 0, \quad (40d)$$

$$\frac{D\eta_s}{Dt} = 0, \qquad \frac{D\eta_g}{Dt} = 0, \quad (40e)$$

$$p_s = p_g. \quad (40f)$$

Equations (40) admit the energy conservation law (39).

12 Non-equilibrium diffuse interface model of solid-fluid interactions

For numerical reasons it is preferable to use a non-equilibrium model where the algebraic equation for the volume fraction $p_s - p_g = 0$ is replaced by a differential equation. This procedure will be done in the same way as for fluid-fluid mixtures. The relaxation equation will be :

$$p_s - p_g = \frac{1}{\mu}\frac{D\alpha_s}{Dt}$$

Now, we postulate the mixture energy equation :

$$\frac{\partial}{\partial t}\left(\rho\left(e+\frac{\mathbf{v}\cdot\mathbf{v}}{2}\right)\right) + \operatorname{div}\left(\rho\mathbf{v}\left(e+\frac{\mathbf{v}\cdot\mathbf{v}}{2}\right) - \left(\alpha_s\sigma_s + \alpha_g\sigma_g\right)\mathbf{v}\right) = 0. \quad (41)$$

As in the case of fluid-fluid mixtures, we can transform the energy equation (41) to the following one :

$$\alpha_s\rho_s\theta_s\frac{D\eta_s}{Dt} + \alpha_g\rho_g\theta_g\frac{D\eta_g}{Dt} = \left(p_s - p_g\right)\frac{D\alpha_s}{Dt}. \quad (42)$$

Indeed,

$$0=\frac{\partial}{\partial t}\left(\rho\left(e+\frac{\mathbf{v}\cdot\mathbf{v}}{2}\right)\right)+\operatorname{div}\left(\rho\mathbf{v}\left(e+\frac{\mathbf{v}\cdot\mathbf{v}}{2}\right)-\left(\alpha_s\sigma_s+\alpha_g\sigma_g\right)\mathbf{v}\right)$$

$$=\rho\frac{De}{Dt}-tr\left(\left(\alpha_s\sigma_s+\alpha_g\sigma_g\right)\frac{\partial\mathbf{v}}{\partial\mathbf{x}}\right)$$

$$=\rho\left(Y_s\frac{\partial e_s^h}{\partial\rho_s}\frac{D\rho_s}{Dt}+Y_s tr\left(\frac{\partial e_s^e}{\partial G}\frac{DG}{Dt}\right)+Y_s\frac{\partial e_s}{\partial\eta_s}\frac{D\eta_s}{Dt}\right.$$

$$\left.+Y_g\frac{\partial e_g}{\partial\rho_g}\frac{D\rho_g}{Dt}+Y_g\frac{\partial e_g}{\partial\eta_g}\frac{D\eta_g}{Dt}\right)-tr\left(\left(\alpha_s\sigma_s-\alpha_g p_g I\right)\frac{\partial\mathbf{v}}{\partial\mathbf{x}}\right)$$

$$=\alpha_s p_s\left(-\frac{1}{\alpha_s}\frac{D\alpha_s}{Dt}-\operatorname{div}\mathbf{v}\right)+tr\left(\alpha_s S_s\frac{\partial\mathbf{v}}{\partial\mathbf{x}}\right)+\rho Y_s\frac{\partial e_s}{\partial\eta_s}\frac{D\eta_s}{Dt}$$

$$+\alpha_g p_g\left(-\frac{1}{\alpha_g}\frac{D\alpha_g}{Dt}-\operatorname{div}\mathbf{v}\right)+\rho Y_g\frac{\partial e_g}{\partial\eta_g}\frac{D\eta_g}{Dt}-tr\left(\left(\alpha_s\sigma_s-\alpha_g p_g I\right)\frac{\partial\mathbf{v}}{\partial\mathbf{x}}\right)$$

$$=\rho\left(Y_s\theta_s\frac{D\eta_s}{Dt}+Y_g\theta_g\frac{D\eta_g}{Dt}\right)-\left(p_s-p_g\right)\frac{D\alpha_s}{Dt}.$$

As in the case of fluid mixtures, the same equations (18) for the entropy production can be used :

$$\alpha_s\rho_s\theta_s\frac{D\eta_s}{Dt}=\left(p_s-p_I\right)\frac{D\alpha_s}{Dt},\qquad \alpha_g\rho_g\theta_g\frac{D\eta_g}{Dt}=\left(p_g-p_I\right)\frac{D\alpha_s}{Dt}.$$

Obviously, they are compatible with (42). Finally, the non-equilibrium model is :

$$\frac{\partial\left(\mathbf{E}^\beta\right)}{\partial t}+\nabla\left(\mathbf{E}^\beta\cdot\mathbf{v}\right)=-\operatorname{rot}\mathbf{E}^\beta\wedge\mathbf{v},\tag{43a}$$

$$\frac{\partial\left(\alpha_g\rho_g\right)}{\partial t}+\operatorname{div}\left(\alpha_g\rho_g\mathbf{v}\right)=0,\tag{43b}$$

$$\frac{\partial\rho\mathbf{v}}{\partial t}+\operatorname{div}\left(\rho\mathbf{v}\otimes\mathbf{v}-\left(\alpha_s\sigma_s+\alpha_g\sigma_g\right)\right)=0,\tag{43c}$$

$$\alpha_s\rho_s\theta_s\frac{D\eta_s}{Dt}=\left(p_s-p_I\right)\frac{D\alpha_s}{Dt},\qquad \alpha_g\rho_g\theta_g\frac{D\eta_g}{Dt}=-\left(p_g-p_I\right)\frac{D\alpha_s}{Dt},\tag{43d}$$

$$\frac{D\alpha_s}{Dt}=\mu\left(p_s-p_g\right),\qquad \mu>0.\tag{43e}$$

The geometric equations (43a) and the entropy equations (43d) can also be rewritten in different forms which are more convenient for numerical computations :

$$\frac{\partial \mathbf{E}^{\beta}}{\partial t} + \frac{\partial \mathbf{E}^{\beta}}{\partial \mathbf{x}}\mathbf{v} + \left(\frac{\partial \mathbf{v}}{\partial \mathbf{x}}\right)^{T} \mathbf{E}^{\beta} = 0, \tag{44a}$$

$$\frac{\partial}{\partial t}\left(\alpha_s \rho_s e_s\right) + \operatorname{div}\left(\alpha_s \rho_s e_s \mathbf{v}\right) - \alpha_s tr\left(\sigma_s \frac{\partial \mathbf{v}}{\partial \mathbf{x}}\right) = -p_I \mu \left(p_s - p_g\right), \tag{44b}$$

$$\frac{\partial}{\partial t}\left(\alpha_g \rho_g e_g\right) + \operatorname{div}\left(\alpha_g \rho_g e_g \mathbf{v}\right) - \alpha_g tr\left(\sigma_g \frac{\partial \mathbf{v}}{\partial \mathbf{x}}\right) = -p_I \mu \left(p_g - p_s\right). \tag{44c}$$

12.1 Hyperbolicity of the non-equilibrium model

Let us show that the non-equilibrium model (43) is hyperbolic. We present here the proof of hyperbolicity in the one-dimensional case. We denote $\mathbf{v} = (u, v, w)^T$. Let $\mathbf{E}^{\beta} = (A^{(\beta)}, B^{(\beta)}, C^{(\beta)})$. Relation $\mathrm{rot}\mathbf{E}^{\beta} = 0$ implies that only components $A^{(\beta)}$ vary, the other components are constant. We can always suppose that initially the vectors $\mathbf{E}^{\beta}$coincide with the Cartesian basis. Hence, $B^{(2)} = C^{(3)} = 1$, $B^{(1)} = B^{(3)} = C^{(1)} = C^{(2)} = 0$. The matrix G becomes :

$$G = \begin{pmatrix} \left(A^{(1)}\right)^2 + \left(A^{(2)}\right)^2 + \left(A^{(3)}\right)^2 & A^{(2)} & A^{(3)} \\ A^{(2)} & 1 & 0 \\ A^{(3)} & 0 & 1 \end{pmatrix}.$$

Obviously,

$$|G| = \left(A^{(1)}\right)^2,$$

$$J_1 = tr(G) = \left(A^{(1)}\right)^2 + \left(A^{(2)}\right)^2 + \left(A^{(3)}\right)^2 + 2,$$

$$J_2 = tr(G^2) = \left(\left(A^{(1)}\right)^2 + \left(A^{(2)}\right)^2 + \left(A^{(3)}\right)^2\right)^2$$

$$+2\left(\left(A^{(2)}\right)^2 + \left(A^{(3)}\right)^2\right) + 2.$$

We remark that the invariants J_1 and J_2 depend only on $A^{(1)}$ and $Z =$

$\sqrt{\left(A^{(2)}\right)^2 + \left(A^{(3)}\right)^2}$. The one-dimensional governing equations become :

$$\frac{\partial A^{(1)}}{\partial t} + \frac{\partial \left(A^{(1)}u\right)}{\partial x} = 0, \tag{45a}$$

$$\frac{\partial A^{(2)}}{\partial t} + \frac{\partial \left(A^{(2)}u + v\right)}{\partial x} = 0, \tag{45b}$$

$$\frac{\partial A^{(3)}}{\partial t} + \frac{\partial \left(A^{(3)}u + w\right)}{\partial x} = 0, \tag{45c}$$

$$\frac{\partial \rho}{\partial t} + \frac{\partial (\rho u)}{\partial x} = 0, \tag{45d}$$

$$\frac{\partial \rho Y_g}{\partial t} + \frac{\partial (\rho Y_g u)}{\partial x} = 0, \tag{45e}$$

$$\frac{\partial \rho u}{\partial t} + \frac{\partial \left(\rho u^2 - \alpha_s \sigma_{s11} - \alpha_g \sigma_{g11}\right)}{\partial x} = 0, \tag{45f}$$

$$\frac{\partial \rho v}{\partial t} + \frac{\partial \left(\rho u v - \alpha_s \sigma_{s12} - \alpha_g \sigma_{g12}\right)}{\partial x} = 0, \tag{45g}$$

$$\frac{\partial \rho w}{\partial t} + \frac{\partial \left(\rho u w - \alpha_s \sigma_{s13} - \alpha_g \sigma_{g13}\right)}{\partial x} = 0, \tag{45h}$$

$$\frac{\partial \rho \alpha_g}{\partial t} + \frac{\partial (\rho \alpha_g u)}{\partial x} = \rho \mu \left(p_g - p_s\right), \qquad \mu > 0, \tag{45i}$$

$$\frac{\partial (\rho \eta_s)}{\partial t} + \operatorname{div}(\rho \eta_s \mathbf{v}) = \frac{(p_s - p_I)}{Y_s \theta_s} \frac{D\alpha_s}{Dt}, \tag{45j}$$

$$\frac{\partial (\rho \eta_g)}{\partial t} + \operatorname{div}(\rho \eta_g \mathbf{v}) = \frac{(p_g - p_I)}{Y_g \theta_g} \frac{D\alpha_g}{Dt}. \tag{45k}$$

System (45) admits the energy conservation law :

$$\frac{\partial}{\partial t}\left(\rho\left(e + \frac{u^2 + v^2 + w^2}{2}\right)\right) + \frac{\partial}{\partial x}\left(\rho u \left(e + \frac{u^2 + v^2 + w^2}{2}\right) - \sigma_{11} u - \sigma_{12} v - \sigma_{13} w\right) = 0, \tag{46}$$

where

$$\sigma_{ij} = \alpha_s \sigma_{sij} + \alpha_g \sigma_{gij},$$

$$\rho e = \rho \left(Y_g e_g \left(\frac{Y_g \rho}{\alpha_g}, \frac{\rho \eta_g}{\rho}\right) + Y_s e_s^h \left(\frac{Y_s \rho}{1 - \alpha_g}, \frac{\rho \eta_s}{\rho}\right) + Y_s e_s^e \left(\frac{G}{|G|^{1/3}}\right)\right) \tag{47}$$

Let us recall a general result proved by Godunov, Friedrichs and Lax : if a system of conservation laws

$$\mathbf{V}_t + \mathbf{G}(\mathbf{V})_x = \mathbf{0} \tag{48}$$

admits an additional conservation law

$$\varphi(\mathbf{V})_t + \psi(\mathbf{V})_x = 0$$

where $\varphi(\mathbf{V})$ is a convex function of $\mathbf{V}$, the system (48) is hyperbolic (see Godunov and Romenskii (2003) for detail). In our case this function φ is the total volume energy defined from (46) :

$$E = \rho\left(e + \frac{u^2+v^2+w^2}{2}\right)$$

and $\mathbf{V} = \left(A^{(1)},\ A^{(2)},\ A^{(3)},\ \rho, \rho Y_g, \rho u, \rho v, \rho w, \rho\alpha_g, \rho\eta_s, \rho\eta_g\right)^T$. Let us remark that the equations for the volume and mass fractions, and the entropies evolve along contact characteristics. Hence, for hyperbolicity it is sufficient to check the convexity of the energy with respect to a lower number of variables $\hat{\mathbf{V}} = \left(A^{(1)},\ A^{(2)},\ A^{(3)}, \rho, \rho u, \rho v, \rho w\right)^T$. The total energy is in the form :

$$E = \mathcal{E}^k(\rho, \rho u, \rho v, \rho w) + \mathcal{E}^h(\rho, \alpha_g, Y_g, \eta_g, \eta_s) + \mathcal{E}_s^e\left(A^{(1)}, A^{(2)}, A^{(3)}\right)$$

where the kinetic, hydrodynamic and elastic parts of the volume energy are defined as :

$$\mathcal{E}^k(\rho, \rho u, \rho v, \rho w) = \frac{(\rho u)^2 + (\rho v)^2 + (\rho w)^2}{2\rho},$$

$$\mathcal{E}^h = \mathcal{E}^h(\rho, \alpha_g, Y_g, \eta_g, \eta_s) = \rho Y_g e_g^h\left(\rho\frac{Y_g}{\alpha_g}, \eta_g\right) + \rho Y_s e_s^h\left(\rho\frac{1-Y_g}{1-\alpha_g}, \eta_s\right),$$

$$\mathcal{E}_s^e\left(A^{(1)}, A^{(2)}, A^{(3)}\right) = \rho Y_s e_s^e(g) = Y_s\rho_0 |G|^{1/2} e_s^e(g).$$

The energy E is a convex function of $\hat{\mathbf{V}}$, if

$$\frac{\mathcal{E}^k + \mathcal{E}^h}{\rho} = \frac{u^2+v^2+w^2}{2} + Y_g\varepsilon_g^h(\tau_g, \eta_g) + Y_g\varepsilon_s^h(\tau_s, \eta_s)$$

is convex with respect to $(\tau = 1/\rho, u, v, w)$, and

$$\mathcal{E}_s^e = Y_s\rho_0 |G|^{1/2} e_s^e\left(\frac{G}{|G|^{1/3}}\right)$$

is convex with respect to $A^{(\beta)}$. The function $\left(\mathcal{E}^k + \mathcal{E}^h\right)/\rho$ is convex with respect $(\tau = 1/\rho, u, v, w)$ if the hydrodynamic energies of pure phases e_g^h and e_s^h are convex with respect to τ_g and τ_s, respectively. This is the case, for example, of the stiffened gas equation of state we use for applications. We prove the convexity of $\mathcal{E}_s^e$ in a special case where

$$e_s^e(g) = \frac{\mu_s}{4\rho_{s0}} tr\left((g-I)^2\right) = \frac{\mu_s}{4\rho_{s0}}\left(\frac{J_2}{|G|^{2/3}} - \frac{2J_1}{|G|^{1/3}} + 3\right). \tag{49}$$

Here ρ_{s0} is the solid reference density. Let us show the convexity of $\mathcal{E}_s^e$ in the case (49). This energy is in the form :

$$\mathcal{E}_s^e = Y_s \frac{\rho_0}{\rho_{s0}} \frac{\mu_s}{4} |G|^{1/2} tr\left(\left(\frac{G}{|G|^{1/3}} - I\right)^2\right)$$

$$= Y_s \frac{\rho_0}{\rho_{s0}} \frac{\mu_s}{4} A^{(1)} \left(\frac{J_2}{\left(A^{(1)}\right)^{4/3}} - \frac{2J_1}{\left(A^{(1)}\right)^{2/3}} + 3\right)$$

$$= Y_s \frac{\rho_0}{\rho_{s0}} \frac{\mu_s}{4} \left(\frac{\left(\left(A^{(1)}\right)^2 + Z^2\right)^2 + 2Z^2 + 2}{\left(A^{(1)}\right)^{1/3}}\right.$$

$$\left. - 2\left(A^{(1)}\right)^{1/3}\left(\left(A^{(1)}\right)^2 + Z^2 + 2\right) + 3\right)$$

where

$$Z^2 = \left(A^{(2)}\right)^2 + \left(A^{(3)}\right)^2.$$

If the energy $\mathcal{E}_s^e$ is convex with respect to $A^{(1)}$ and Z , and $\partial \mathcal{E}_s^e / \partial Z > 0$, it will be convex with respect to $A^{(1)}$, $A^{(2)}$ and $A^{(3)}$. Obviously,

$$\frac{\partial \mathcal{E}_s^e}{\partial Z} = Y_s \frac{\rho_0}{\rho_{s0}} \frac{\mu_s}{\left(A^{(1)}\right)^{1/3}} \left(Z^3 + Z\left(\left(A^{(1)}\right)^2 - \left(A^{(1)}\right)^{2/3} + 1\right)\right) > 0$$

since for any positive x we have $x^3 - x + 1 > 0$. Finally, to prove the convexity of $\mathcal{E}_s^e$ with respect to variables $\left(A^{(1)}, Z\right)$ we have to prove that the corresponding 2×2 Hessian matrix is positive definite in all domain of parameters $Z > 0$, $A^{(1)} > 0$:

$$\begin{pmatrix} \frac{\partial^2 \mathcal{E}_s^e}{\partial \left(A^{(1)}\right)^2} & \frac{\partial^2 \mathcal{E}_s^e}{\partial A^{(1)} \partial Z} \\ \frac{\partial^2 \mathcal{E}_s^e}{\partial A^{(1)} \partial Z} & \frac{\partial^2 \mathcal{E}_s^e}{\partial Z^2} \end{pmatrix} > 0$$

The proof can be done by direct calculations. The symmetrization method assures the hyperbolicity but does not give the characteristic eigenvalues in explicit form.

12.2 Applications

We made a comparison with experimental results on cylindrical gel samples (cylindrical drops of jelly-like materials) impacting a rigid surface (see Figure 2). Jelly-like materials remain elastic even for extreme deformations (between 100 % and 200 %). This regime was recently studied experimentally by Luu Lia-Hua and Forterre (2009). A cylindrical gel sample (carbopol) impacting a rigid hydrophobic surface, first spreads until some limit size and then a full elastic recoil is observed which may be followed by a complete rebound. We reproduced this quasi-reversible behavior numerically. The separable form of the energy equation was also used with $\mu_s = 85Pa$. The comparison during the spreading phase at a given time instant is shown in Figure 3.

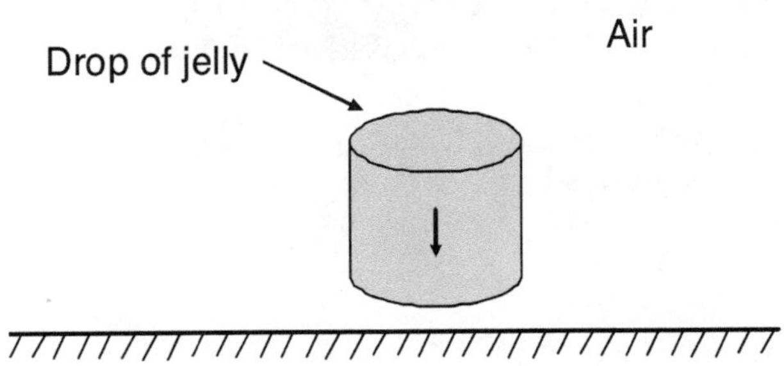

Figure 2. A cylindrical drop of jelly is impacting a rigid surface.

Figure 3. Comparison of the spreading phase of the cylinder impacting a rigid surface. Experiments by L. - H. Luu and Y. Forterre (2009) are shown on the top, computed results are shown on the bottom.

Bibliography

[1] R. Abgrall (1996) How to prevent pressure oscillations in multicomponent flow calculations : A quasi-conservative approach. J. Comput. Phys. **125**, 150-160.

[2] B. Alvarez-Samaniego and D. Lannes (2008) Large time existence for 3D water waves and asymptotics, Invent. math. v. 171, 485-541.

[3] M. Baer and J. Nunziato (1986) A two-phase mixture theory for the deflagration-to-detonation transition (DDT) in reactive granular materials. Int. J. Multiphase Flows, **12**, 861-889.

[4] V. L. Berdichevsky (2009) *Variational Principles of Continuum Mechanics*, Springer-Verlag Berlin Heidelberg.

[5] A. D. Drew and S. L. Passman (1996) *Theory of multicomponent fluids*, Springer-Verlag New York Inc.

[6] N. Favrie, S. L. Gavrilyuk and R. Saurel (2009) Diffuse solid-fluid interface model in cases of extreme deformations, Journal of Computational Physics, **228**, 6037-6077.

[7] G. A. El, R. H. J. Grimshaw and N. F. Smyth (2006), Unsteady undular bores in fully nonlinear shallow-water theory, *Physics of Fluids* 18:027104 (17 pages).

[8] S. L. Gavrilyuk and H. Gouin (1999), A new form of governing equations of fluid arising from Hamilton's principle, *Int. J. Eng. Sci.* 37:1495–1520.

[9] S. L. Gavrilyuk and V. M. Teshukov (2001), Generalized vorticity for bubbly liquid and dispersive shallow water equations, *Continuum Mech. Thermodyn.* 13:365–382.

[10] S.L. Gavrilyuk and R. Saurel (2002), Mathematical and Numerical Modeling of Two-Phase Compressible Flows with Microinertia, Journal of Computational Physics, **175**, 326-360.

[11] S. L. Gavrilyuk and V. M. Teshukov (2004), Linear stability of parallel inviscid flows of shallow water and bubbly fluid, *Stud. Appl. Math.* 113:1–29.

[12] S.L. Gavrilyuk, N. Favrie and R. Saurel (2008) Modeling wave dynamics of compressible elastic materials. Journal of Computational Physics, **227**, 2941-2969.

[13] S. K. Godunov (1978) Elements of Continuum Mechanics, Nauka, Moscow (in Russian).

[14] S. K. Godunov and E. I. Romenskii (2003) Elements of Continuum Mechanics and Conservation Laws, Kluwer Academic Plenum Publishers, NY.

[15] A. E. Green, N. Laws and P. M. Naghdi (1974), On the Theory of Water Waves, *Proceedings of the Royal Society of London A* 338: 43–55.

[16] A. E. Green and P. M. Naghdi (1976), A derivation of equations for wave propagation in water of variable depth, *J. Fluid Mech.* 78:237–246.

[17] S. V. Iordansky (1960) On the equations of motion of the liquid containing gas bubbles. *Zhurnal Prikladnoj Mekhaniki i Tekhnitheskoj Fiziki* N3, 102–111 (in Russian).

[18] M. Ishii and T. Hibiki (2006), Thermo-fluid dynamics of two-phase flow, Springer.

[19] A. K. Kapila, R. Menikoff, J. B. Bdzil, S. F. Son, D. S. Stewart (2001) Two-phase modeling of deflagration-to-detonation transition in granular materials : reduced equations, Phys. Fluids, **13** (10), 3002-3024.

[20] S. Karni (1994) Multi-component flow calculations by a consistent primitive algorithm, Journal of Computational Physics, **112**, 31-43.

[21] B. S. Kogarko (1961) On the model of cavitating liquid. *Dokl. AN SSSR* **137**, 1331–1333 (in Russian).

[22] Y. A. Li (2001) Linear stability of solitary waves of the Green-Naghdi equations, *Comm. Pure Appl. Math.* 54:501–536.

[23] L.-H. Luu and Y. Forterre (2009) Drop impact of yield-stress fluids, J. Fluid Mechanics, **632**, 301-327.

[24] N. Makarenko (1986) A second long-wave approximation in the Cauchy-Poisson problem, Dynamics of Continuous Media, v. 77, pp. 56-72 (in Russian).

[25] J.-C. Micaelli (1982), Propagation d'ondes dans les écoulements diphasiques à bulles à deux constituants. Etude théorique et expérimentale. Thèse, Université de Grenoble, France.

[26] J. Miles and R. Salmon (1985), Weakly dispersive nonlinear gravity waves, *J. Fluid Mech.* 157:519–531.

[27] J. N. Plohr and B. J. Plohr (2005) Linearized analysis of Richtmyer-Meshkov flow for elastic materials, J. Fluid Mech., **537**, 55-89.

[28] R. Saurel and R. Abgrall (2001), A multiphase Godunov method for compressible multifluid and multiphase flows, J. Comput. Physics, **150**, 425-467.

[29] R. Saurel, S. L. Gavrilyuk and F. Renaud (2003) A multiphase model with internal degrees of freedom: application to shock-bubble interaction, Journal of Fluid Mechanics, **495**, 283-321.

[30] R. Saurel, O. Le Metayer, J. Massoni and S. Gavrilyuk (2007), Shock jump relations for multiphase mixtures with stiff mechanical relaxation, Shock Waves, **16**, 209-232.

[31] R. Salmon (1988), Hamiltonian Fluid Mechanics, *Ann. Rev. Fluid Mech* 20:225–256.

[32] R. Salmon (1998), *Lectures on Geophysical Fluid Dynamics*, Oxford University Press, New York, Oxford.

[33] C. H. Su and C. S. Gardner (1969), Korteweg-de Vries Equation and Generalizations. III. Derivation of the Korteweg-de Vries Equation and Burgers Equation, *J. Math. Phys.* 10:536–539.

[34] V. M. Teshukov and S. L. Gavrilyuk (2006), Three-Dimensional Nonlinear Dispersive waves on Shear Flows, *Stud. Appl. Math.* 116:241–255.

[35] L. van Wijngaarden (1968) On the equations of motion for mixtures of liquid and gas bubbles. *J. Fluid Mech.* **33**, 465–474.

[36] G. B. Whitham (1974), *Linear and Nonlinear Waves*, John Wiley & Sons, New York.

Introduction to stochastic variational problems

Victor L. Berdichevsky
Mechanical Engineering, Wayne State University, Detroit MI 48202 USA

Abstract The lectures provide an introduction to the Chapters on stochastic variational problems from the author's book Variational Principles of Continuum Mechanics, Springer, 2009.

0.1 Reminder from complex analysis

The major goal of these lectures is to explain how to compute probabilities in some stochastic variational problems. The lecturers can be used as an introduction to the Chapters of my book Variational Principles of Continuum Mechanics, Springer-Verlag, 2009, which are concerned with stochastic variational problems. Before proceeding to stochastic variational problems we have to learn how to compute probability in much simpler cases, like, for example, how to find probability distribution of a sum of independent random variables. Unfortunately, these issues are discussed in engineering probability courses at the time when the students do not have enough mathematical background, and a simple and beautiful nature of the classical results of probability theory, like the central limit theorem or Gauss distribution, remain unrevealed. Therefore, I will spend the first part of the course to cover these issues.

To do the calculations we will use complex analysis. I would assume with a great deal of certainty, that not everyone in this room was exposed to properly taught complex analysis. In high school you, perhaps, were taught that the complex unity i is a square root from -1:

$$i = \sqrt{-1}.$$

What does that mean? Of course, you can solve the equation $x^2 + 1 = 0$ in terms of i and write $x = \pm i$. So what? Well, you can then write the solution of any polynomial equation in the form $\alpha + i\beta$, where α and β are real numbers, besides, the number of roots is equal to the power of the polynom. This is nice indeed. This was the way in which the complex numbers were introduced in XVI century by Italian mathematicians

Cardano and Bombelli. This way penetrated in modern text books without much change. Unfortunately, one key word is missing in such a treatment, the word without which the real understanding of complex numbers is hardly possible. I begin with an explanation of what the complex numbers are (note that complex numbers are not numbers!) and why they are needed for the problems under consideration. Besides, I will review the basics of complex analysis. Then we spend two lectures for classical results of probability theory, and then go on to stochastic variational problems.

Complex numbers. We know two basic operations with vectors: We can multiply vector, $\vec{a}$, by a number, λ; if a^i $(i = 1, ..., n)$ are the components of a vector $\vec{a}$ in some basis $\hat{e}_i$[1],

$$\vec{a} = a^i \hat{e}_i,$$

then vector $\lambda\vec{a}$ has the components λa^i :

$$\lambda\vec{a} = \lambda a^i \hat{e}_i.$$

We also can sum vectors; if a^i and b^i are the components of vectors $\vec{a}$ and $\vec{b}$, then vector $\vec{a} + \vec{b}$ has the components $a^i + b^i$:

$$\vec{a} + \vec{b} = \left(a^i + b^i\right) \hat{e}_i.$$

The latter definition corresponds to the parallelogram rule (Fig. 1).

Now we wish more. We wish to operate with vectors as we do with numbers. We wish to introduce multiplication of vectors, $\vec{a}$ and $\vec{b}$, in such a way that the product of vectors is a vector, and, as for numbers,

$$\vec{a} \cdot \vec{b} = \vec{b} \cdot \vec{a}, \qquad \left(\vec{a} + \vec{b}\right) \cdot \vec{c} = \vec{a} \cdot \vec{c} + \vec{b} \cdot \vec{c}, \qquad \left(\vec{a} \cdot \vec{b}\right) \cdot \vec{c} = \vec{a} \cdot \left(\vec{b} \cdot \vec{c}\right).$$

Besides, we would like to be able to divide vectors, i.e., for each vectors $\vec{a}$ and $\vec{b}$, we should be able to compute their ratio, a vector $\vec{c}$,

$$\vec{c} = \vec{a} \,/ \vec{b}.$$

For given $\vec{a}$ and $\vec{b}$, vector $\vec{c}$ must be determined uniquely from the equation

$$\vec{b} \cdot \vec{c} = \vec{a}.$$

[1] In all formulas summation over repeated indices is assumed, e.g.

$$a^i \hat{e}_i \equiv \sum_{i=1}^{n} a^i \hat{e}_i.$$

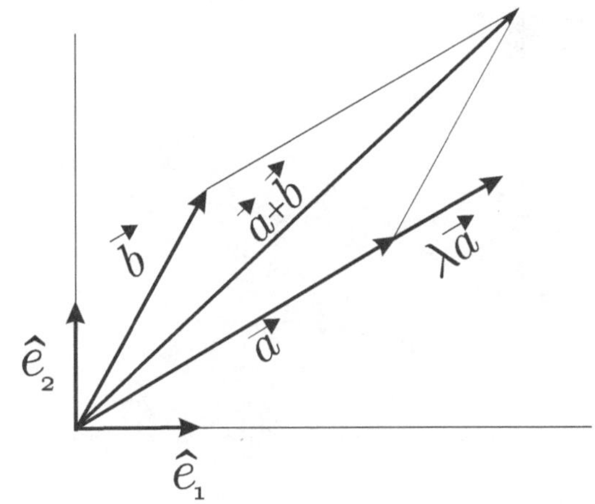

Figure 1. The definition of a sum of two vectors and multiplication of a vector by number.

Clearly, the products of vectors which are introduced in vector analysis, the vector product, $\vec{a} \times \vec{b}$, and the dot product, $\vec{a} \cdot \vec{b}$, do not fit: $\vec{a} \times \vec{b}$ is anti-symmetric $\left(\vec{a} \times \vec{b} = -\vec{b} \times \vec{a}\right)$, the dot product is a scalar, not a vector. The multiplication operation we are looking for must be a new one. It turns out that such product can be introduced only in three spaces: two-dimensional, four-dimensional and eight-dimensional. The vectors of the corresponding two-dimensional space are called complex numbers, four-dimensional space quaternions and eight-dimensional space octonions. Quaternions were first conceived by W.R. Hamilton, octonions by A. Caley (Caley numbers). Only vectors of two-dimensional space, complex numbers, possess all the features of usual numbers; quaternions and octonions do not.

Now we have to define what is the product of two vectors in two-dimensional space. To this end, it is enough to define the products of the basic vectors. Then the product of any two vectors can be computed from the relation

$$\left(a^1\hat{e}_1 + a^2\hat{e}_2\right) \cdot \left(b^1\hat{e}_1 + b^2\hat{e}_2\right) = a^1b^1\hat{e}_1 \cdot \hat{e}_1 + \left(a^1b^2 + a^2b^1\right)\hat{e}_1 \cdot \hat{e}_2 + a^2b^2\hat{e}_2 \cdot \hat{e}_2.$$

Let us denote the two basic vectors by the symbols $\hat{e}_1 = \hat{1}$, $\hat{e}_2 = \hat{\imath}$. Then any vector z has the form $z = x \cdot \hat{1} + y\hat{\imath}$. The products of basic vectors are defined by the following rule:

$$\hat{1} \cdot \hat{1} = \hat{1}, \qquad \hat{1} \cdot \hat{\imath} = \hat{\imath} \cdot \hat{1} = \hat{\imath}, \qquad \hat{\imath} \cdot \hat{\imath} = -\hat{1}.$$

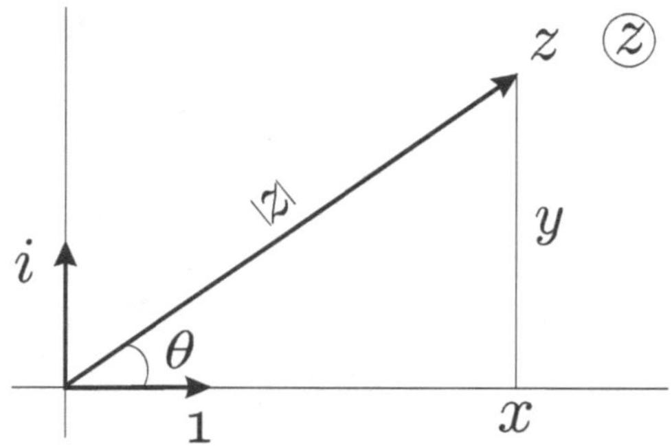

Figure 2. Notations in complex plane.

$$z_1 = x_1 \cdot \hat{1} + y_1 \hat{\imath}, \qquad z_2 = x_2 \cdot \hat{1} + y_2 \hat{\imath},$$

their product is

$$z_1 \cdot z_2 = (x_1 x_2 - y_1 y_2)\, \hat{1} + (x_1 y_2 + x_2 y_1)\, \hat{\imath}.$$

One can prove that such a product has all the features of the product of usual numbers.

The tradition is not to write the hat at the basic vector $\hat{\imath}$: one writes for this basic vector $i \equiv \hat{\imath}$. Moreover, without confusion the basic vector $\hat{1}$ in the expression $x\hat{1} + iy$ can be dropped. So the complex number takes the form

$$z = x + iy.$$

Such form defines the components of vector z, x and y, uniquely. They are called real and imaginary parts of z and denoted by Rez and Imz, respectively. The complex numbers with zero imaginary part are called real. The two-dimensional space with such defined vector product is called complex plane. We arrive at the usual definition of complex numbers, which you can find e.g. in Wikipedia: complex numbers are the numbers of the form $x + iy$, where x, y are real numbers while $i^2 = -1$. What is missing in this definition is the key word: vector. Complex numbers are not numbers, they are vectors.

E x e r c i s e s. Complex conjugate of $z = x + iy$ is, by definition, the number $\bar{z} = x - iy$. Show that $z\bar{z}$ is a real number equal to the squared length of vector z; the length of z, $|z|$ is defined as $|z| = \sqrt{x^2 + y^2}$ (see Fig.2). Find the number $1/z$.

Functions of complex variables. An advantage we gain working in complex plane is that we can operate with vectors in the same way we do with numbers. In particular, all functions we used in calculus, make sense being applied to vectors. For example, for numbers, function e^x can be defined as a sum

$$e^x = 1 + x + \frac{1}{2}x^2 + \frac{1}{6}x^3 + ... = \sum_{n=0}^{\infty} \frac{1}{n!}x^n.$$

In the same way it can be defined for vectors because the sum of vectors and the powers $z^n = \underbrace{z...z}_{n \text{ times}}$ are well defined:

$$e^z = 1 + z + \frac{1}{2}z^2 + ... = \sum_{n=0}^{\infty} \frac{1}{n!}z^n. \tag{1}$$

Of course, one has to complement this by a notion of convergence, but this is done in a natural way: sequence $z_n \to 0$ as $n \to \infty$, if $|z_n| \to 0$ as $n \to \infty$. By multiplying two series,

$$\sum_0^{\infty} \frac{1}{n!}z_1^n \text{ and } \sum_0^{\infty} \frac{1}{n!}z_2^n,$$

one can check that

$$e^{z_1+z_2} = e^{z_1}e^{z_2}. \tag{2}$$

An immediate consequence of (1) and (2) is Euler formula

$$e^{iy} = \cos y + i \sin y. \tag{3}$$

Indeed, plugging in (1) $z = iy$, and using that

$$i^n = \{.(-1)^k i \text{ if } n = 2k + 1(-1)^k \text{ if } n = 2k,$$

we get

$$e^{iy} = \sum_{\text{even } n} \frac{1}{n!}z^n + \sum_{\text{odd } n} \frac{1}{n!}z^n = \sum_{k=0}^{\infty} \frac{(-1)^k}{(2k)!}y^{2k} + i\sum_{k=0}^{\infty} \frac{(-1)^k}{(2k+1)!}y^k. \tag{4}$$

Euler formula follows from the Taylor expansion of sin and cos :

$$\cos y = \sum_{k=0}^{\infty} \frac{(-1)^k}{(2k)!}y^{2k}, \quad \sin y = \sum_{k=0}^{\infty} \frac{(-1)^k}{(2k+1)!}y^k.$$

Denote by θ the angle between the vector z and positive real axis. We use the convention, that angle θ increases in counter clockwise direction and changes within the limits $-\pi, \pi$. According to Euler formula one can write z in the polar form:

$$z = |z| \, e^{i\theta}. \tag{5}$$

Another important function of complex variable to be encountered further is $\ln z$. It is defined as inversion of exponential function, i.e.

$$e^{\ln z} = z. \tag{6}$$

Since, according to (5),

$$z = e^{\ln|z|+i\theta},$$

we can rewrite (6) as

$$e^{\ln z} = e^{\ln|z|+i\theta}$$

from which

$$\ln z = \ln |z| + i\theta. \tag{7}$$

Construction of functions by means of a series makes a class of functions defined by a converging Taylor series especially important; they are called analytic functions. More precisely, $f(z)$ is an analytic function in a region D if it can be presented in D by a converging series. We drop further all mathematical details, they can be found in numerous sources, and focus only on the basic ideas.

Derivative of a function of complex variable is defined in terms of its Taylor series[2]

$$f(z) = \sum_{k=0}^{\infty} a_k z^k,$$

as

$$f'(z) = \sum_{k=0}^{\infty} k a_k z^{k-1}.$$

This definition is equivalent to the usual one

$$f'(z) = \lim_{\Delta z \to 0} \frac{f(z + \Delta z) - f(z)}{\Delta z}$$

[2]Here we consider a function which is analytical in a vicinity of the point $z = 0$. If a function is analytical in a vicinity of some point z_0, Taylor series is written with respect to the shift $z - z_0$:

$$f(z) = \sum_{k=0}^{\infty} a_k (z - z_0)^k .$$

provided the convergence of all series, which are involved.

Integral over a contour γ in z-plane,

$$\int_{\gamma} f(z)dz,$$

is defined in the same way as for a contour in two-dimensional real plane.

Integral of an analytic function over a closed contour in a simply-connected region[3] is zero. Indeed, let a function be analytic in a vicinity of the point $z = 0$. The integral over closed contour of z^n is zero:

$$\oint_{\gamma} z^n dz = \oint d\frac{z^{n+1}}{n} = 0.$$

Thus, integrating the Taylor series of an analytic function we get zero for any analytic function. Consideration of functions analytical in vicinity of non-zero points is similar.

An example of non-analytic function in vicinity of $z = 0$ is the function $f(z) = 1/z$; it is analytic in any region with excluded point $z = 0$ and non-analytic in any vicinity of point $z = 0$.

E x e r c i s e s. 1. Write down polar forms for $1/z$, $\bar{z}$, $1/\bar{z}$, -1, z^n.

2. Find $(1+i)^5$, $(1+i)^{100}$.

3. Derive from the definitions of trigonometric functions of complex variables,

$$\sin z = \frac{1}{2i}\left(e^{iz} - e^{-iz}\right), \qquad \cos z = \frac{1}{2}\left(e^{iz} + e^{-iz}\right),$$

that

$$\sin^2 z + \cos^2 z = 1.$$

4. Hyperbolic sin and cos are defined by the formulas:

$$\sinh z = \frac{1}{2}\left(e^{z} - e^{-z}\right), \qquad \cosh z = \frac{1}{2}\left(e^{z} + e^{-z}\right).$$

Show that

$$\cosh^2 z - \sinh^2 z = 1.$$

5. Show that

$$\sin 2z = 2\sin z \cos z.$$

[3] A region is simply-connected if any closed contour can be shrunk by a continuous transformation to a point.

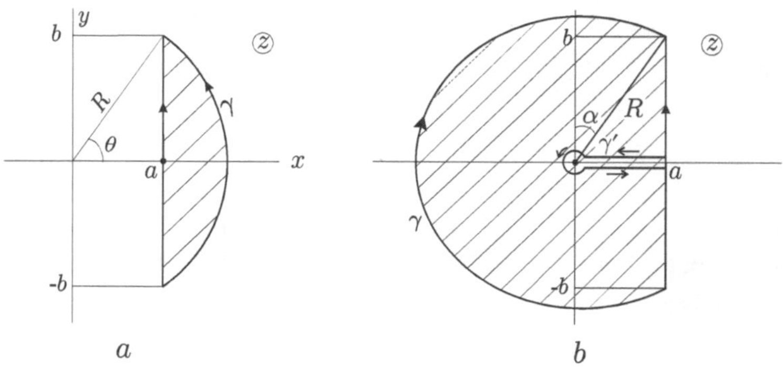

Figure 3. To justification of formula (10).

6. Show that for functions $f(z)$ with real coefficients of Taylor expansion

$$\overline{f(z)} = f(\bar{z}).$$

7. Write Taylor series for the function $(e^z - 1)/z$.

Step function. The only reason why we need complex numbers in what follows is the presentation of the step function by means of an integral in complex plane. By the step function, $\theta(t)$, we mean the following function

$$\theta(t) = \{.1 \quad t > 00 \quad t < 0. \tag{8}$$

Usually, the value of this function at $t = 0$ is not essential. We will set for definiteness

$$\theta(0) = \frac{1}{2}. \tag{9}$$

We need the following integral presentation of $\theta(t)$:

$$\theta(t) = \frac{1}{2\pi i} \int\limits_{a-i\infty}^{a+i\infty} \frac{e^{tz}}{z} dz, \qquad a \geqslant 0, \tag{10}$$

where the integration contour is a straight line in the right half-plane z.

Since formula (10) will play a central role in further calculations, we outline here its derivation.

First of all we need to specify the meaning of the integral in (10). The integral has infinite limits, and, in general, one can write

$$\int\limits_{a-i\infty}^{a+i\infty} \frac{e^{tz}}{z} dz = \lim_{b_1, b_2 \to \infty} \int\limits_{a-ib_1}^{a+ib_2} \frac{e^{tz}}{z} dz. \tag{11}$$

Such a formula would be meaningful if the integral converges absolutely, i.e. there exist an integral

$$\int_{a-i\infty}^{a+i\infty} \left|\frac{e^{tz}}{z}\right| |dz| .$$

The role of absolute convergence follows from the inequality: for any contour γ,

$$\left|\int_{\gamma} f(z)dz\right| \leqslant \int_{\gamma} |f(z)|\,|dz| .$$

This inequality is a consequence of the inequality

$$|z_1 + ... + z_n| \leqslant |z_1| + ... + |z_n| ,$$

which obviously follows from the triangle inequality,

$$|z_1 + z_2| \leqslant |z_1| + |z_2| .$$

If an integral with infinite limits converges absolutely, it does not matter in which way the infinite limits are approached. The integral in (11) does not converge absolutely. Indeed

$$\int_{a-i\infty}^{a+i\infty} \left|\frac{e^{tz}}{z}\right| |dz| = \int_{-\infty}^{+\infty} \frac{\left|e^{t(a+iy)}\right|}{|(a+iy)|} dy = \int_{-\infty}^{+\infty} \frac{e^{ta}}{\sqrt{a^2+y^2}} dy.$$

This integral diverges logarithmically. Therefore, we have to specify what is meant by integral in (11). We set

$$\int_{a-i\infty}^{a+i\infty} \frac{e^{tz}}{z} dz = \lim_{b\to\infty} \int_{a-ib}^{a+ib} \frac{e^{tz}}{z} dz. \tag{12}$$

Let us compute this limit. First of all, for $t = 0$

$$\theta(0) = \frac{1}{2\pi i} \lim_{b\to\infty} \int_{a-ib}^{a+ib} \frac{1}{a+iy} idy = \frac{1}{2\pi i} \lim_{b\to\infty} \left(\ln(a+ib) - \ln(a-ib)\right).$$

From (7)

$$\ln(a+ib) = \ln\sqrt{a^2+b^2} + i\arctan\frac{b}{a}$$

$$\ln(a-ib)=\ln\sqrt{a^2+b^2}-i\arctan\frac{b}{a}.$$

Since

$$\lim_{b\to\infty}\arctan\frac{b}{a}=\frac{\pi}{2},$$

we have

$$\theta(0)=\frac{1}{2\pi}\left[\frac{\pi}{2}-\left(-\frac{\pi}{2}\right)\right]=\frac{1}{2}.$$

This is in accord with (9).

Let now $t<0$. We aim to show that the limit in (12) is zero. Function e^{tz}/z is analytic in the shadowed region in Fig. 3a. Therefore, integral of e^{tz}/z over the boundary of the shadowed region is zero. Hence, the integral of e^{tz}/z over the straight segment $[a-ib,\ a+ib]$ is equal to the integral over the contour γ. Let γ be an arc of a circle with radius $R=\sqrt{a^2+b^2}$ and the center at $z=0$. Then the integral is bounded from above:

$$\left|\int\limits_{\gamma}\frac{e^{tz}}{z}dz\right|\leqslant\int\frac{|e^{tz}|}{|z|}|dz|=\int\limits_{-\theta}^{\theta}e^{tR\cos\theta}d\theta.$$

Since $t<0$, this integral does not exceed the integral,

$$J(R)=\int\limits_{-\pi/2}^{\pi/2}e^{-|t|R\cos\theta}d\theta=2\int\limits_{0}^{\pi/2}e^{-|t|R\cos\theta}d\theta.$$

We are going to show that function $J(R)$ tends to zero as $R\to+\infty$. This is the first point in these lectures where we encounter the necessity to find the asymptotics of an integral; we will have several such problems later on. If the function in the exponent, $\cos\theta$, were strictly positive everywhere on the integration interval, i.e. $\cos\theta\geqslant\min\cos\theta=c>0$, then the integrand does not exceed $e^{-|t|R\min(\cos\theta)}$, and the integral tends to zero as $R\to+\infty$. In our case, however, $\min\cos\theta=0$, (see Fig. 4a), it is achieved at $\theta=\pi/2$.

We have to study the contribution to the integral of the vicinity of the point $\theta=\pi/2$. Let us split the integral in two parts

$$\frac{1}{2}J(R)=\int\limits_{0}^{\pi/2-\alpha}e^{-|t|R\cos\theta}d\theta+\int\limits_{\frac{\pi}{2}-\alpha}^{\pi/2}e^{-|t|R\cos\theta}d\theta.\tag{13}$$

The first integral in (13) does not exceed

$$e^{-|t|R\cos\left(\frac{\pi}{2}-\alpha\right)}\left(\frac{\pi}{2}-\alpha\right)$$

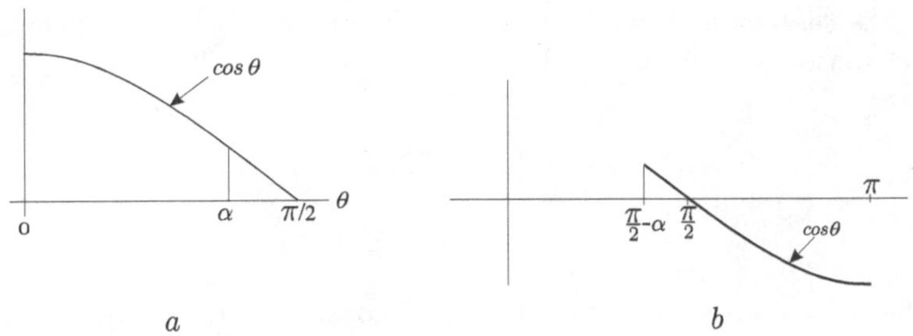

Figure 4. Graph of $\cos\theta$.

and tends to zero as $R \to +\infty$. For sufficiently small α, in the second integral we can replace $\cos\theta$ by the first terms of Taylor expansion

$$\cos\theta = \cos\theta\,\Big|_{\theta=\pi/2} + \frac{d\cos\theta}{d\theta}\Big|_{\theta=\pi/2}\left(\theta - \frac{\pi}{2}\right) = \left(\frac{\pi}{2} - \theta\right).$$

Then the second integral takes the form

$$\int\limits_{\frac{\pi}{2}-\alpha}^{\pi/2} e^{-|t|R\left(\frac{\pi}{2}-\theta\right)} d\theta = \int\limits_{0}^{\alpha} e^{-|t|R\xi} d\xi.$$

We can increase this integral replacing the upper limit by $+\infty$. Then we get the integral which is computed analytically

$$\int\limits_{0}^{\infty} e^{-|t|R\xi} d\xi = \frac{1}{|t|\,R}.$$

Hence, $J(R) \to 0$ as $R \to +\infty$, and $\theta(t) = 0$ for $t < 0$ indeed.

Consider now the case $t > 0$. In this case we introduce a region shadowed in Fig. 3b. Function e^{tz}/z is analytic in a shadowed region, and integral over the boundary of the shadowed region is zero. Therefore, the integral over the segment $[a - ib;\ a + ib]$ is equal to the sum of integrals over contours γ and γ', γ being a circle of radius R. The integral over γ can be estimated from above as

$$\left|\int\limits_{\gamma} \frac{e^{tz}}{z} dz\right| = \int\limits_{\gamma} \frac{|e^{tz}|}{|z|}\,|dz| = \int\limits_{\frac{\pi}{2}-\alpha}^{\frac{3\pi}{2}+\alpha} e^{tR\cos\theta} d\theta = 2\int\limits_{\frac{\pi}{2}-\alpha}^{\pi} e^{tR\cos\theta} d\theta. \qquad (14)$$

Here α is the angle shown in Fig. 3b. The integral in the right hand side of (14) can be written as a sum

$$\int_{\frac{\pi}{2}-\alpha}^{\pi} e^{tR\cos\theta} d\theta = \int_{\frac{\pi}{2}}^{\pi} e^{tR\cos\theta} d\theta + \int_{\frac{\pi}{2}-\alpha}^{\frac{\pi}{2}} e^{tR\cos\theta} d\theta.$$

Function $\cos\theta$ on the segment $[\pi/2-\alpha, \pi]$ is shown in Fig. 4b. The first member of the sum coincides with $J(R)$ considered above and thus tends to zero as $R \to \infty$. The behavior of the second integral is not immediately clear. Let us estimate this integral using that $R\cos\theta \leqslant a$:

$$\int_{\frac{\pi}{2}-\alpha}^{\pi/2} e^{tR\cos\theta} d\theta \leqslant e^{ta} \int_{\frac{\pi}{2}-\alpha}^{\pi/2} d\theta = \alpha e^{ta}.$$

As follows from Fig. 3b, $\alpha \to 0$ as $R \to \infty$. Hence, the second integral tends to zero as well. So, as $b \to \infty$, the integral of e^{tz}/z over the segment $[a-ib;\ a+ib]$ tends to the integral over γ', which, in turn, is equal to the integral over a circle of small radius r,

$$\frac{1}{2\pi i}\oint \frac{e^{tz}}{z} dz = \frac{1}{2\pi i}\int \frac{e^{tre^{i\theta}}}{re^{i\theta}} rie^{i\theta} d\theta$$
$$= \frac{1}{2\pi}\oint e^{tre^{i\theta}} d\theta|_{\cdot r\to 0} \to \frac{1}{2\pi}\oint d\theta = 1,$$

as claimed.

Sometimes, it is convenient to have in formula (10) the integral over a line in the left half-plane. The corresponding relation is obtained from (10), if we notice that the integral of e^{tz}/z over the boundary of the shadowed region in Fig. 5 is zero.

Therefore, the integral of e^{tz}/z over the segment $[a-ib;\ a+ib]$ is equal to the sum of integrals over $[-a-ib;\ -a+ib]$, γ^+, γ^- and γ'. The integral over γ^+ tends to zero as $b \to \infty$:

$$\left| \int_{\gamma^+} \frac{e^{tz}}{z} dz \right| \leqslant \int_{-a}^{a} \frac{|e^{tz}|}{|z|} dx = \int_{-a}^{a} \frac{e^{tx}}{\sqrt{x^2+b^2}} dx \leqslant e^{ta}\frac{1}{b}\cdot 2a.$$

Similarly, integral over γ^- tends to zero. Integral over γ' is equal to 1. Hence,

$$\theta(t) = 1 + \frac{1}{2\pi i}\int_{-a-i\infty}^{-a+i\infty} \frac{e^{tz}}{z} dz, \qquad a \geqslant 0. \qquad (15)$$

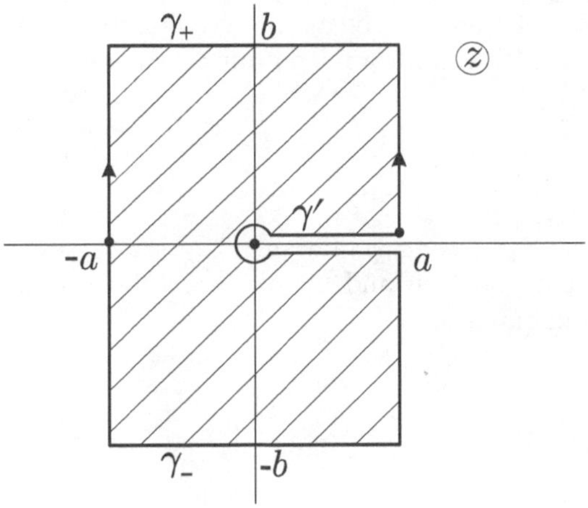

Figure 5. To the move of integration contour to the left half-plane.

It is known, that derivative of the step function is $\delta-$function:

$$\frac{d\theta(t)}{dt} = \delta(t). \tag{16}$$

Differentiating (10) over t we formally obtain a presentation of $\delta-$function

$$\delta(t) = \frac{1}{2\pi i} \int_{a-i\infty}^{a+i\infty} e^{tz} dz. \tag{17}$$

This relation, strictly speaking, does not make sense, because the integral in (17) does not converge. However, $\delta-$function is used only in the form of integrals with smooth functions. In such cases, one can write

$$\int \delta(t) \varphi(t) dt = \frac{1}{2\pi i} \int_{a-ib}^{a+ib} \int \varphi(t) e^{tz} dtdz, \tag{18}$$

and, if the function of z,

$$\int \varphi(t) e^{tz} dt,$$

decays fast enough as $|z| \to \infty$, (18) holds true.

E x e r c i s e s. 1. Let $f(z)$ be an analytic function in a region D. Show that for a point $z \in D$ and for any contour C, surrounding z,

$$f(z) = \frac{1}{2\pi i} \int\limits_C \frac{f(\zeta)d\zeta}{\zeta - z}.$$

2. Let $f(z) = u(x,y) + iv(x,y)$ be an analytic function. Show that real and imaginary part of f, u and v, are linked by a system of equations (Cauchy-Riemann equations)

$$\frac{\partial u}{\partial x} = \frac{\partial v}{\partial y}$$

$$\frac{\partial u}{\partial y} = -\frac{\partial v}{\partial x}.$$

3. Derive from Cauchy-Riemann equations that $u(x,y)$ and $v(x,y)$ are harmonic functions, i.e.

$$\Delta u\,(x,y) = 0, \qquad \Delta v\,(x,y) = 0,$$

where Δ is Laplace's operator

$$\Delta = \frac{\partial^2}{\partial x^2} + \frac{\partial^2}{\partial y^2}.$$

0.2 Some facts about integrals

Gauss formula. Consider in some finite-dimensional space, R_n, a quadratic form,

$$(Au, u) = A_{ij}u^i u^j, \tag{19}$$

The form is assumed to be positive,

$$(Au, u) > 0 \ \text{ if } u \neq 0.$$

Then the Gauss formula holds true:

$$\int\limits_{R_n} e^{-\frac{1}{2}(Au,u)} du = \frac{1}{\sqrt{\det A}}. \tag{20}$$

Here

$$\det A \equiv \det \|A_{ij}\|\,, \quad du = \frac{du^1}{\sqrt{2\pi}} \cdots \frac{du^n}{\sqrt{2\pi}}.$$

The Gauss formula can be proved by changing the variables, $u \to \mathring{u}$,

$$u^i = \lambda^i_j \mathring{u}^j, \qquad \det \left\| \lambda^i_j \right\| = 1,$$

$\mathring{u}^j$ being the coordinates in which the tensor A_{ij} is diagonal,

$$(Au, u) = A_{ij} \lambda^i_{i'} \lambda^j_{j'} \mathring{u}'^i \mathring{u}'^j = A_1 \left(\mathring{u}^1\right)^2 + \ldots + A_n \left(\mathring{u}^n\right)^2 . \tag{21}$$

In the new variables,

$$\begin{aligned} \int\limits_{R_n} e^{-\frac{1}{2}(Au,u)} du &= \int\limits_{R_n} e^{-\frac{1}{2}\left(A_1\left(\mathring{u}^1\right)^2 + \ldots + A_n\left(\mathring{u}^n\right)^2\right)} d\mathring{u} \\ &= \frac{1}{\sqrt{A_1 \ldots A_n}} = \frac{1}{\sqrt{\det A}} \end{aligned}$$

Here we used that[4]

$$\int\limits_{-\infty}^{+\infty} e^{-\frac{1}{2}x^2} dx = \sqrt{2\pi}. \tag{22}$$

The Gauss formula admits the following generalization: for any linear function of u, $(l, u) = l_i u^i$,

$$\sqrt{\det A} \int\limits_{R_n} e^{-\frac{1}{2}(Au,u) + (l,u)} du = e^{\frac{1}{2}\left(A^{-1}l, l\right)} \tag{23}$$

where A^{-1} is the inverse matrix to the matrix A. Formula (23) follows from (20) and the identity,

$$\frac{1}{2}(Au, u) - (l, u) = \frac{1}{2}\left(A\left(u - A^{-1}l\right), \left(u - A^{-1}l\right)\right) - \frac{1}{2}\left(A^{-1}l, l\right). \tag{24}$$

Plugging (24) in (23), changing the variables of integration, $u \to u + A^{-1}l$, and using (20) we obtain the right hand side of (23).

[4]The integral (22) can be found by a witty trick suggested by Poisson:

$$\left(\int\limits_{-\infty}^{+\infty} e^{-\frac{1}{2}x^2} dx\right)^2 = \int\limits_{-\infty}^{+\infty} e^{-\frac{1}{2}x^2} dx \int\limits_{-\infty}^{+\infty} e^{-\frac{1}{2}y^2} dy = \int\limits_{-\infty}^{+\infty}\int\limits_{-\infty}^{+\infty} e^{-\frac{1}{2}x^2} e^{-\frac{1}{2}y^2} dx dy$$

$$= \int\limits_{-\infty}^{+\infty}\int\limits_{-\infty}^{+\infty} e^{-\frac{1}{2}(x^2+y^2)} dx dy = \int\limits_{0}^{+\infty}\int\limits_{0}^{2\pi} e^{-\frac{1}{2}r^2} r dr d\theta = 2\pi \int\limits_{0}^{+\infty} e^{-\frac{1}{2}r^2} d\frac{1}{2}r^2 = 2\pi.$$

E x e r c i s e. Let (Au, u) be a positive quadratic form of variables $u = (u^1, ..., u^n)$, and u be subject to s linear constraints,

$$(l_\alpha, u) = c_\alpha, \qquad \alpha = 1, ..., s; \qquad s < n.$$

Consider an integral,

$$J = \int\limits_{R_n} e^{-\frac{1}{2}(Au,u)} \prod_{\alpha=1}^{s} \delta\left(c_\alpha - (l_\alpha, u)\right) du^1 ... du^n.$$

Show that

$$J = \sqrt{\frac{(2\pi)^n}{\det A}} \sqrt{\frac{(2\pi)^s}{\det \mathcal{A}}} e^{-\frac{1}{2}\left(\mathcal{A}^{-1}c, c\right)} \tag{25}$$

where $\mathcal{A}$ is the matrix with components

$$\mathcal{A}_{\alpha\beta} = \left(A^{-1} l_\alpha, l_\beta\right).$$

H i n t. Use the presentation of δ–function (17) and formula (20).

Laplace's asymptotics. Consider an integral which depends on a parameter, λ, in the following way:

$$I(\lambda) = \int_V f(x) e^{\lambda S(x)} dV,$$

where V is a bounded region of n-dimensional space, $f(x)$ and $S(x)$ are some smooth functions. We wish to find the asymptotics of this integral as $\lambda \to \infty$. Laplace suggested that the leading terms of the asymptotics of $I(\lambda)$ are the same as that of the integral over the vicinities of the points where the function, $S(x)$, has the maximum value. Then the asymptotics can be easily found. Indeed, let $S(x)$ achieve its maximum value only at one point, $\hat{x}$, this point is an internal point of V, and the matrix of the second derivatives, $\left\|\partial^2 S(\hat{x})/\partial x^i \partial x^j\right\|$ is non-degenerated, i.e. its determinant, Δ, is non-zero. We can write,

$$I(\lambda) = f(\hat{x}) e^{\lambda S(\hat{x})} \int_V \frac{f(x)}{f(\hat{x})} e^{-\lambda[S(\hat{x}) - S(x)]} dV.$$

In a small vicinity of the point, $\hat{x}$, we can replace $S(\hat{x}) - S(x)$ by the non-degenerated quadratic form,

$$S(\hat{x}) - S(x) \approx -\frac{1}{2} S_{ij}(x^i - \hat{x}^i)(x^j - \hat{x}^j), \tag{26}$$

where $S_{ij} = \partial^2 S(\hat{x})/\partial x^i \partial x^j$. Note that the quadratic form (26) is positive because $\hat{x}$ is the point of maximum of $S(x)$. In a small vicinity of $\hat{x}$ we can replace $f(x)/f(\hat{x})$ by unity thus obtaining

$$I(\lambda) \approx f(\hat{x})e^{\lambda S(\hat{x})} \int_{\text{small vicinity of } \hat{x}} e^{-\lambda[-\frac{1}{2}S_{ij}(x^i-\hat{x}^i)(x^j-\hat{x}^j)]} dV. \tag{27}$$

Since $\lambda \to \infty$, the function, $\exp\left[-\lambda[-\frac{1}{2}S_{ij}(x^i - \hat{x}^i)(x^j - \hat{x}^j)]\right]$, decays very fast away from $\hat{x}$. We do not pause to justify that the expansion of the integration region from a small vicinity of $\hat{x}$ to the entire space, R,causes only exponentially small corrections in (27). Thus, we can write:

$$I(\lambda) \approx f(\hat{x})e^{\lambda S(\hat{x})} \int_R e^{-\lambda[-\frac{1}{2}S_{ij}(x^i-\hat{x}^i)(x^j-\hat{x}^j)]} dV.$$

The integral here, according to the Gauss formula (20), is equal to $\sqrt{(2\pi)^n/\lambda^n\,|\Delta|}$. Finally, the leading term of the asymptotics is

$$I(\lambda) \approx \sqrt{\frac{(2\pi)^n}{\lambda^n\,|\Delta|}}\, f(\hat{x})e^{\lambda S(\hat{x})}. \tag{28}$$

As a more elaborated derivation shows, the error of the formula (28) is on the order of $(e^{\lambda S(\hat{x})}/\lambda^{n/2})/\lambda$. If $S(x)$ achieves its maximum at several internal points, one should sum the contributions (28) of all points. One can check that in the cases of the point of maximum lying on the boundary and/or degeneration of the quadratic form $-\frac{1}{2}S_{ij}(x^i - \hat{x}^i)(x^j - \hat{x}^j)$ the asymptotics remains qualitatively the same,

$$I(\lambda) \approx \text{prefactor}(\lambda)e^{\lambda S(\hat{x})}, \tag{29}$$

with the prefactor being a decaying power function of λ.

The prefactor is a constant independent on λ, if $S(x)$ has maximum value on a set with non-zero volume. By Laplace's asymptotics we mean further the asymptotics of the form (29) where the prefactor changes slower than the exponential function of λ:

$$\frac{1}{\lambda} \ln \text{prefactor}(\lambda) \to 0 \qquad \text{as } \lambda \to \infty.$$

Changing in the previous consideration $S(x)$ by $-S(x)$ we obtain the asymptotics,

$$\int_V f(x)e^{-\lambda S(x)} dV \approx \text{prefactor}(\lambda)e^{-\lambda S(\check{x})}, \tag{30}$$

where $\check{x}$ is the point of minimum of $S(x)$.

In applications to the variational problems, we need also to know the asymptotics of integrals of the form (30) for complex values of λ. In this case we denote the parameter by z,

$$I(z) = \int_V f(x) e^{zS(x)} dV,$$

and consider the asymptotics of $I(z)$ as $|z| \to \infty$. Note first of all that $I(z)$ is an analytical function of z at any finite point, z, if the integral, as we accept, converges absolutely, i.e.

$$\int_V |f(x)| \, e^{Re z S(x)} dV < \infty.$$

The point $z = \infty$ can be, however, the singular point of $I(z)$. Usually, the singularity is essential, i.e. the asymptotics of $I(z)$ along different paths, $z \to \infty$, are different. It turns out that Laplace's asymptotics,

$$I(z) \approx \sqrt{\frac{(2\pi)^n}{z^n |\Delta|}} f(\hat{x}) e^{zS(\hat{x})} \left(1 + O\left(\frac{1}{z}\right)\right), \tag{31}$$

holds true for all paths, $z \to \infty$, such that $|\text{Arg}\, z| \leq \pi/2 - \varepsilon$, for some small ε. For other paths, this asymptotics does not hold. This is seen from studying the asymptotics when $z \to \infty$ along the imaginary axis, $z = iy$, $|y| \to \infty$. It turns out that in this case the leading contribution to the asymptotics is provided by not only the point of maximum of $S(x)$, but by all stationary points of $S(x)$, in particularly, by all points of local maxima and minima. This asymptotics is called the stationary phase asymptotics; we do not dwell on it here since it will not be used further.

0.3 Reminder from probability theory

The random variables which we will be dealing with are the points, x, of some finite-dimensional region, V. Probability of the event that x belongs to a set A, $A \subset V$, is, by definition,

$$p(A) = \int_A f(x) dx.$$

Non-negative function $f(x)$ is called the probability density. Since $p(V) = 1$,

$$\int_V f(x) dx = 1.$$

Average value of function $\varphi(x)$ is defined as

$$M\varphi = \int_V \varphi(x) f(x) dx.$$

Here M stands for "mathematical expectation". This operation is often denoted by the symbol E (for expectation), but we reserve E to be the symbol of energy.

For any function φ, $\varphi(x)$ is a random variable. Probability density of $\varphi(x)$, f_φ, is defined in terms of probability of the event $\{\xi \leqslant \varphi(x) \leqslant \xi + \Delta\xi\}$ for small $\Delta\xi$:

$$f_\varphi(\xi)\Delta\xi = \text{Prob}\left\{\xi \leqslant \varphi(x) \leqslant \xi + \Delta\xi\right\}.$$

It is convenient to introduce the distribution function,

$$F(\xi) = \text{Prob}\left\{\varphi(x) \leqslant \xi\right\}.$$

If the distribution function is smooth, then

$$f_\varphi\left(\xi\right) = \frac{dF(\xi)}{d\xi}.$$

It is convenient to write the distribution function in terms of the step function

$$F(\xi) = M\theta\left(\xi - \varphi\left(x\right)\right) = \int \theta\left(\xi - \varphi\left(x\right)\right) f(x) dx. \tag{32}$$

This formula explains how the step function, to which we have paid already much attention, enters our consideration.

Consider a random variable, which is a couple (x, y). Its probability density is a function of x and y, $f(x, y)$. Let y take values only in some region B. What would be a probability density of x? It is natural to define probability density of x under condition that $y \in B$ as

$$f(x) = \int_B f\left(x, y\right) dy \Big/ \int_V \int_B f\left(x, y\right) dx dy.$$

Obviously,

$$\int f(x) dx = 1.$$

In general, the conditional probability density of x depends on the choice of B. By definition the random variables x and y are statistically independent, if the conditional probability does not depend on B. This is possible only if $f(x,y)$ is the product of two functions

$$f(x,y) = f(x)g(y), \quad \int_V f(x)dx = 1, \quad \int_V g(y)dy = 1. \tag{33}$$

The notion of statistical independence is the central one in probability theory. All most important facts of probability theory are concerned with the sets of independent events. The major applications of probability theory are based to the possibility to identify the independent (or slightly dependent) events[5].

We prepared everything to solve the central for applications problem of probability theory. Let $x_1, ..., x_n$ be independent identically distributed random variables. Find probability distribution of the sum

$$\varphi(x_1) + ... + \varphi(x_N)$$

where φ is a given function. Of course, we expect to get an analytical answer only in the limit of large N.

Let us find the probability distribution of the arithmetic average

$$E = \frac{1}{N}(\varphi(x_1) + ... + \varphi(x_N)).$$

We have

$$F_N(E) = M\theta\left(E - \frac{1}{N}(\varphi(x_1) + ... + \varphi(x_N))\right).$$

We use for the sum the symbol E because in similar problems to be considered later, it has the meaning of energy. Using the presentation of the step function (10) we have

$$F_N(E) = M\frac{1}{2\pi i}\int_{a-\infty}^{a+i\infty} \frac{dz}{z} e^{Ez - z\frac{1}{N}\varphi(x_1) - ... - z\frac{1}{N}\varphi(x_N)}. \tag{34}$$

[5] By the way, the recent financial crisis was caused in part by a wrong identification of independent events. The hedge fund traders believed that combining mortgage loans into large packages for sale to banks and pension funds reduces the risk of default. This is true if the defaults of individual loans were independent. However, as we have experienced, this is not always the case: there are rare catastrophic events when probabilities of defaults become strongly correlated. This is what happened in the recent economic crisis: simultaneous default of many mortgage loans along with other negative events drove the economy down, which, in turn, resulted in more mortgage defaults and bankruptcy of financial institutions.

It is convenient to change variable z by zN. Then

$$F_N(E) = M\frac{1}{2\pi i}\int\limits_{a-i\infty}^{a+i\infty}\frac{dz}{z}e^{NEz - z\varphi(x_1) - \ldots - z\varphi(x_N)}.$$

In such change, the constant a in the integral limit must be replaced by Na, but, since this constant is arbitrary, we keep the some notations for the integral limit. The operation of mathematical expectation is, in essence, integration. The order of integrals can be changed almost always; we do not pay attention to degenerated cases. Since the variables $x_1, ..., x_N$ are statistically independent and identically distributed,

$$Me^{-z\varphi(x_1) - \ldots - z\varphi(x_N)} = \left(Me^{-z\varphi(x)}\right)^N,$$

and we get

$$F_N(E) = \frac{1}{2\pi i}\int\limits_{a-i\infty}^{a+i\infty}\frac{dz}{z}e^{NEz}\left(Me^{-z\varphi(x)}\right)^N$$

$$= \frac{1}{2\pi i}\int\limits_{a-i\infty}^{a+i\infty}\frac{dz}{z}e^{NEz + N\ln Me^{-z\varphi(x)}} = \frac{1}{2\pi i}\int\limits_{a-i\infty}^{a+i\infty}\frac{dz}{z}e^{NS(E,z)} \tag{35}$$

where we introduced a function of E and z

$$S(E,z) = Ez + \ln Q(z), \quad Q(z) \equiv \int e^{-z\varphi(x)}f(x)\,dx. \tag{36}$$

In physical applications, functions S has the meaning of entropy, and we will call S the entropy of the problem.

The integral (35) contains a large parameter, N. Therefore, probability distribution $F_N(E)$ for large N can be found by studying the asymptotics of the integral (35) as $N \to \infty$. This idea is in the core of all further examples considered.

Usually, the integral in (35) can be differentiated over E. After differentiation we obtain formula for probability density of the normalized sum,

$$f_N(E) = \frac{N}{2\pi i}\int\limits_{a-i\infty}^{a+i\infty} e^{NS(E,z)}dz. \tag{37}$$

Note that the integrand, which is equal to

$$e^{NEz}\left(\int e^{-z\varphi(x)}f(x)\,dx\right)^N,$$

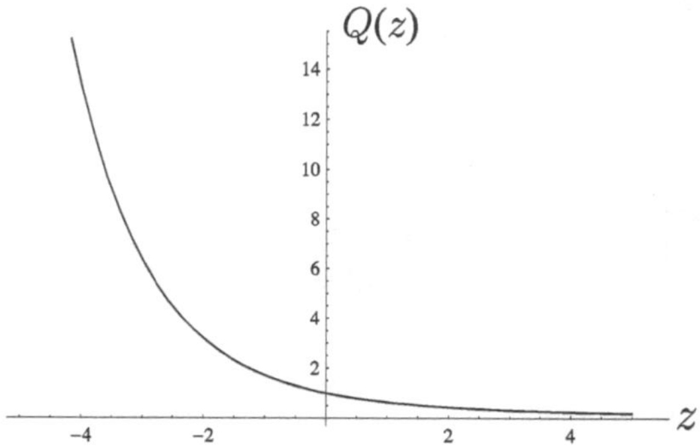

Figure 6. Plot of $Q(z)$.

does not have singularities for finite z, and, if it decays as $|Imz| \to \infty$, the line of integration can be moved to the left half-plane. Therefore, a in (37) can take both positive and negative values.

We consider examples in the next lecture.

0.4 The central limit theorem and the law of large numbers

We begin with the following example. Let x be a random number on the segment [0,1], which is homogeneously distributed, i.e. $f(x) \equiv 1$. Then $Q(z)$ is computed analytically:

$$Q(z) = \int_0^1 e^{-zx} dx = \frac{1}{z}\left(1 - e^{-z}\right).$$

Graph of $Q(z)$ is shown in Fig. 6.

Function $S(E, z)$ is shown in Fig. 7.

Function $\ln Q(z)$ is a convex function of z (we will prove it in a more general case later). Therefore, $S = Ez + \ln Q$ is also convex and has a minimum. To find the minimizer we have to solve the equation

$$\frac{\partial S(E, z)}{\partial z} = E + \frac{Q'(z)}{Q(z)} = 0.$$

It can be written as

$$-\frac{Q'(z)}{Q(z)} = \frac{1}{z} - \frac{e^{-z}}{1 - e^{-z}} = E. \tag{38}$$

Function $-Q'(z)/Q(z)$ takes the values between 0 and 1 (again, we show this later in a general case). Therefore, for $0 \leqslant E \leqslant 1$, equation (38) has a

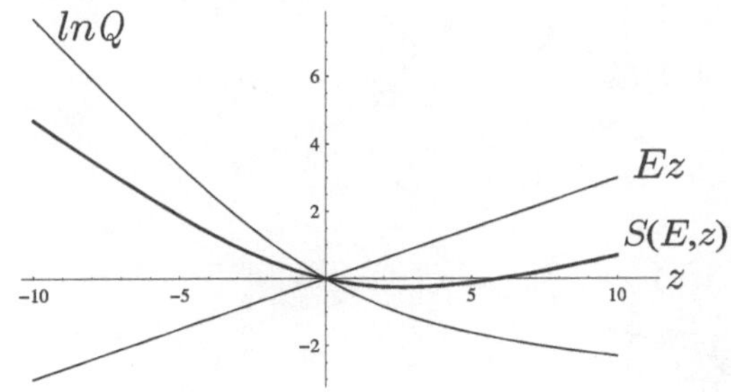

Figure 7. Plot of $\ln Q(z)$, Ez and $S(E,z)$ for $E = 0.3$.

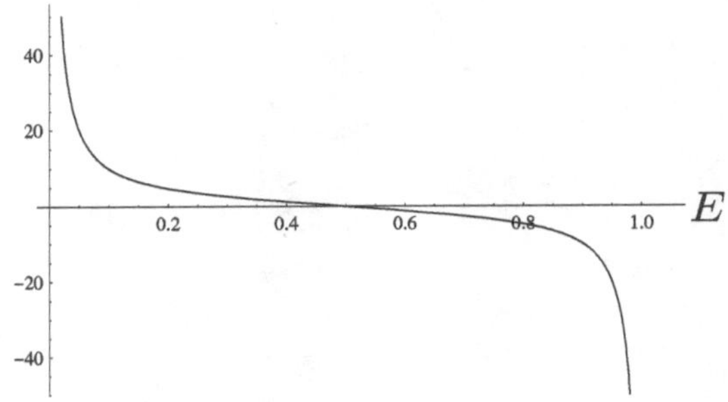

Figure 8. Dependence of inverse temperature β on energy E.

unique solution, which we denote $\beta(E)$. This notation is motivated again by physical problems: in similar physical problems β plays the role of inverse temperature. Function $\beta(E)$ is shown in Fig. 8.

If we tend z to zero in (38), we find

$$\frac{1}{z} - \frac{1 - z + \frac{1}{2}z^2}{1 - \left(1 - z + \frac{1}{2}z^2\right)} \rightarrow \frac{1}{2}.$$

Therefore, the value of E at which $\beta = 0$, is $1/2$. If $E < 0$, then $S(E,z)$ has the minimum at $z = +\infty$, if $E > 1$, $S(E,z)$ has the minimum at $z = -\infty$ (Fig. 9).

Denote by $S(E)$ the value of $S(E,z)$ at the point of minimum over z,

$$S(E) = S(E, \beta(E)).$$

The graph of function $S(E)$ is shown in Fig. 10.

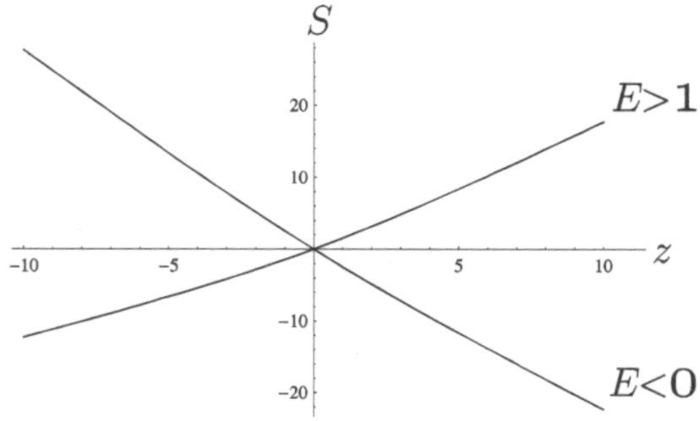

Figure 9. A qualitative graph of $S(E,z)$ as a function of z for $E>1$ and $E<0$.

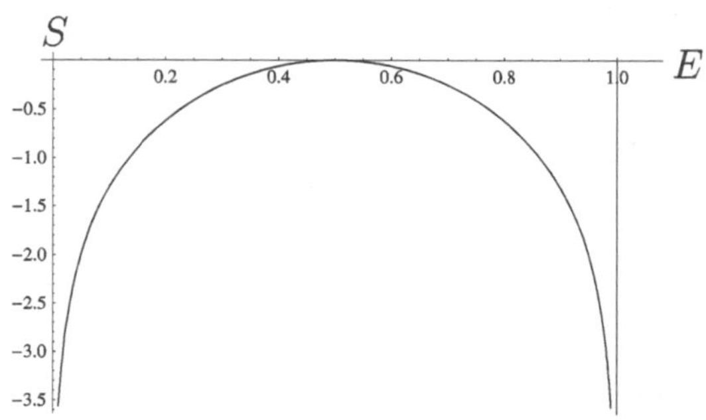

Figure 10. Dependence of $S(E)$ on E.

E x e r c i s e. Show that

$$\beta(E) = \frac{dS(E)}{dE}.$$

In the sum, $x_1 + ... + x_N$, each member takes the values between 0 and 1. Therefore, the admissible values of the sum lie between 0 and N, while the values of E are between 0 and 1. Probability that $E<0$ or $E>1$ is zero. This fact can be derived directly from (35), (36), but we do not pause on this derivation and focus on the values of E from 0 to 1. We set a in (37) equal to $\beta(E)$. Formula (37) takes the form

$$f_N(E) = \frac{N}{2\pi} \int_{-\infty}^{+\infty} e^{NS(E,\beta+iy)} dy \tag{39}$$

$$S(E,z) = Ez + \ln\left[\frac{1-e^{-z}}{z}\right].$$

Since $f_N(E)$ is real, the imaginary part of the integral (39) is zero.

Consider function $S(E,\beta+iy)$ for small y. Since $\partial S(E,z)/\partial z = 0$ at $y=0$, we have

$$S(E,z) = S(E) + \frac{1}{2}S''(E)(iy)^2 = S(E) - \frac{1}{2}S''(E)y^2, \tag{40}$$

$$S''(E) \equiv \left.\frac{\partial^2 S(E,z)}{\partial z^2}\right|_{z=\beta(E)} = \left.\frac{\partial^2 \ln Q(z)}{\partial z^2}\right|_{z=\beta(E)}.$$

Function $S(E,z)$ is strictly convex on real axis, thus $S''(E) > 0$, and $S(E,\beta+iy)$ has a local maximum on $y-$axis at $y=0$. For finite y, $ReS(E,\beta+iy)$ decays. Indeed,

$$\left|\ln\left[\frac{1-e^{-\beta-iy}}{\beta+iy}\right]\right| = \ln\frac{\left|1-e^{-\beta-iy}\right|}{\sqrt{\beta^2+y^2}}$$

$$= \ln\left(e^{-\beta}\left|e^{\beta}-e^{-iy}\right|\right) - \ln\sqrt{\beta^2+y^2}.$$

The first term here is bounded, while the second one goes to $-\infty$. So, the major contribution to this integral as $N\to\infty$ is provided by a vicinity of the point $y=0$. Replacing $S(E,z)$ by (40) we have

$$f_N(E) = e^{NS(E)}\frac{N}{2\pi}\int_{-\infty}^{+\infty} e^{-\frac{N}{2}S''(E)y^2}dy = \sqrt{\frac{N}{2\pi S''(E)}}e^{NS(E)}. \tag{41}$$

Here we made the change of variable $y \to y\big/\sqrt{NS''}$ and used (22).

Formula (41) is an asymptotic formula as $N\to\infty$. The normalization condition

$$\int_{-\infty}^{+\infty} f_N(E)\,dE = \int_0^1 f_N(E)\,dE = 1 \tag{42}$$

is satisfied asymptotically. Indeed, in the integral

$$\int_{-1}^{1}\frac{1}{\sqrt{S''(E)}}e^{NS(E)}dE$$

the major contribution is provided by the vicinity of the point of maximum of $S(E)$, which is $\hat{E} = \frac{1}{2}$. At this point $S = 0$; thus

$$\int_{-1}^{1} \frac{1}{\sqrt{S''(E)}} e^{NS(E)} dE \simeq \int_{-\infty}^{+\infty} \frac{1}{\sqrt{S''(E)}} e^{\frac{1}{2} N S_{EE}(\hat{E})(E-\hat{E})^2} dE. \quad (43)$$

Here we denoted by $S_{EE}(E)$ the second derivative of $S(E)$:

$$S_{EE}(E) = \frac{d^2 S(E)}{dE^2}.$$

The derivative $S_{EE}(E)$ is negative, because $S(E)$ is a concave function. For the integral (43) we have in the leading approximation,

$$\frac{\sqrt{2\pi/N}}{\sqrt{-S_{EE}(\hat{E})}} \frac{1}{\sqrt{S''\left(\hat{E}\right)}}. \quad (44)$$

From (38) and (40)

$$S''\left(\hat{E}\right) = \left.\frac{d^2 S(E,z)}{dz^2}\right|_{z=\beta(\hat{E})} = \left.\frac{d^2 \ln Q}{dz^2}\right|_{z=\beta(\hat{E})} = -\left.\frac{dE}{d\beta}\right|_{z=\beta(\hat{E})}.$$

Hence

$$S''\left(\hat{E}\right) \cdot S_{EE}(\hat{E}) = S''\left(\hat{E}\right) \left.\frac{d\beta}{dE}\right|_{\hat{E}} = -1. \quad (45)$$

Combining (44), (45) and (41) we obtain (42).

Our asymptotic result converges to the exact one very fast. For $N = 10$, the exact and asymptotic results are shown in Fig. 11.

For $N = 100$ the exact and asymptotic results are practically indistinguishable.

As N increases, probability density converges to δ–function, concentrated at the point $E = \frac{1}{2}$. This value, $E = \frac{1}{2}$, is the mathematical expectation of each of the members of the sum. We obtain *the law of large numbers:*

$$\frac{1}{N}(x_1 + ... + x_N) \to Mx = \frac{1}{2} \quad \text{as} \quad N \to \infty. \quad (46)$$

Equation (46) can be interpreted in the following way. Let we do some experiments and the outcome of the experiment is a number, x, $0 \leqslant x \leqslant 1$. All values of x on the segment $[0,1]$ are equiprobable. The outcomes of the experiments are independent. Then the arithmetic average of all outcomes for large N is approximately equal to the mathematical expectation of x.

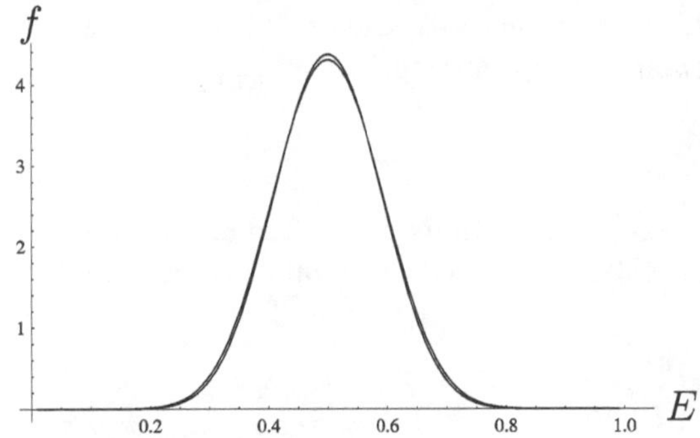

Figure 11. Exact probability density of E compared with the asymptotic formula when $N = 10$; the exact and asymptotic results are hardly distinguishable.

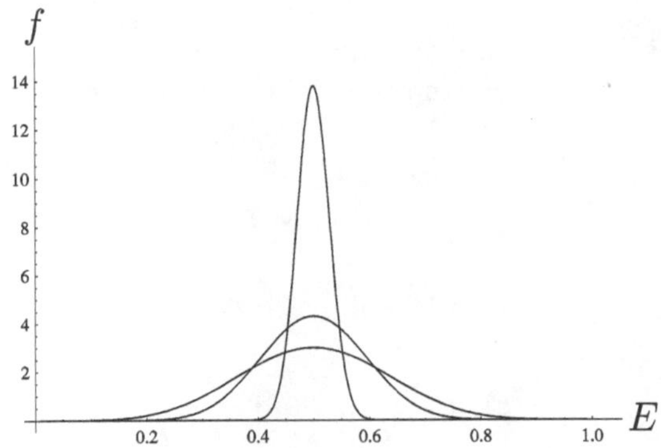

Figure 12. Exact probability densities of E for $N = 5$, 10 and 100.

It is interesting to characterize the deviations of the arithmetic average from the mathematical expectation,

$$\frac{1}{N}\left(x_1+...+x_N\right)-Mx=E'.$$

To estimate the order of deviations, let us find the mathematical expectation of $E^{'2}$. We can do that using the probability density of E (41). From (41) we have

$$ME^{'2}=M\left(E-\frac{1}{2}\right)^2=\int_0^1\left(E-\frac{1}{2}\right)^2\sqrt{\frac{N}{2\pi S''(E)}}e^{NS(E)}dE$$

$$=\sqrt{\frac{N}{2\pi S''\left(\frac{1}{2}\right)}}\int_{-\infty}^{+\infty}E^{'2}e^{NS_{EE}\left(\frac{1}{2}\right)E'^2}dE'=\frac{1}{N}\frac{S''\left(\frac{1}{2}\right)}{\sqrt{2\pi}}\int_{-\infty}^{+\infty}x^2e^{-\frac{1}{2}x^2}dx.$$

Here we used that $S(E)$ has maximum at $E=\frac{1}{2}$ and equal to zero at this point, besides, equation (45) was also employed.

We see that $ME^{'2}$ is of the order $1/N$. Hence, E' is of the order $1/\sqrt{N}$. This suggests that the scaled deviations,

$$\xi=\sqrt{N}E',$$

are of the order of unity and can have a non-singular probability distribution. Let us find it. Denote the probability density function of ξ by $g(\xi)$. Since,

$$E=\frac{1}{2}+\frac{\xi}{\sqrt{N}}$$

and

$$f_N\left(E\right)dE=g(\xi)d\xi,$$

we have

$$g(\xi)=\frac{1}{\sqrt{N}}f_N\left(\frac{1}{2}+\frac{\xi}{\sqrt{N}}\right).$$

Plugging here (41) we find for large N

$$g(\xi)=\frac{1}{\sqrt{2\pi S''\left(\frac{1}{2}\right)}}e^{\frac{1}{2}S_{EE}\left(\frac{1}{2}\right)\xi^2}.$$

Denoting $S''\left(\frac{1}{2}\right)$ by σ^2 and using (45) we obtain

$$g(\xi)=\frac{1}{\sqrt{2\pi}\sigma}e^{-\frac{1}{2\sigma^2}\xi^2}. \tag{47}$$

This is Gaussian distribution. The constant σ has the meaning of variance of ξ,

$$M\xi^2 = \sigma^2.$$

Formula (47) expresses the so-called *central limit theorem.* The law of large numbers and the central limit theorem are simple consequences of (41). Formula (41) provides much more information: it determines the probability of large deviations of the arithmetical average from the average value. This probability is exponentially small because $S(E)$ is negative.

Exercises. 1. Let u_a be non-negative numbers, $a = 1, ..., N$. All points of the space $\{u_1, ..., u_N\}$ are equiprobable. Note that we cannot introduce probability in a usual sense because the volume of the admissible values is infinite. Let u_a be constrained by the condition

$$\frac{1}{N}\sum_{a=1}^{N} u_a = 1.$$

This condition makes the volume of the admissible values finite. Show that in the limit $N \to \infty$ the values of any two numbers (say, u_1 and u_2) become statistically independent, and each number has the probability distribution

$$f(u) = e^{-u}.$$

2. Let $a_1, ..., a_n, ...$ be an infinite sequence of numbers and $x_1, ..., x_n, ...$ a sequence of independent identically distributed variables with mathematical expectation Mx, variance Mx^2 and probability density function $f(x)$. Consider a random number

$$\xi = a_1 x_1 + ... + a_n x_n + ...$$

show that

$$M\xi = Mx \sum_{k=1}^{\infty} a_k$$

$$M\xi^2 = Mx^2 \sum_{k=1}^{\infty} a_k^2$$

while the probability density function of ξ, $f_\xi(y)$ is given by the integral

$$f_\xi(y) = \frac{1}{2\pi i} \int_{a-i\infty}^{a+i\infty} e^{zy - g(y)} dz,$$

$$g(y) = \sum_{k=1}^{\infty} \ln \int e^{-ya_k x} f(x)\, dx.$$

It is assumed that the sums and integrals in these formulas exist. A necessary condition for convergence of series is the decay a_k as $k \to \infty$.

0.5 Poisson distribution

Now we return to a more general case we have started with, when x was a point of some region distributed with some probability density $f(x)$. As we will see all results we obtained for numbers are extended to this case. We have obtained already the formulas for probability density (37), (36). Consider the properties of $S(E, z)$ on real axis. This function is a convex function of z. Indeed,

$$\frac{\partial S(E,z)}{\partial z} = E - \frac{\int \varphi e^{-z\varphi} f dx}{\int e^{-z\varphi} f dx} \tag{48}$$

$$\frac{\partial^2 S(E,z)}{\partial z^2} = \frac{\int \varphi^2 e^{-z\varphi} f dx \int e^{-z\varphi} f dx - \left(\int \varphi e^{-z\varphi} f dx\right)^2}{\left(\int e^{-z\varphi} f dx\right)^2}. \tag{49}$$

Using Cauchy inequality,

$$\left(\int f \cdot g dx\right)^2 \leqslant \int f^2 dx \int g^2 dx,$$

we have

$$\left(\int \varphi e^{-z\varphi} f dx\right)^2 = \left(\int \varphi e^{-\frac{1}{2}z\varphi} f^{\frac{1}{2}} \cdot e^{-\frac{1}{2}z\varphi} f^{\frac{1}{2}} dx\right)^2$$
$$\leqslant \int \varphi^2 e^{-z\varphi} f dx \int e^{-z\varphi} f dx.$$

Therefore,

$$\frac{\partial^2 S(E,z)}{\partial z^2} \geqslant 0,$$

and $S(E, z)$ is a convex function of z. Hence, it may have only one local minimum at a finite z. It may have also minimum at $z = +\infty$ or $z = -\infty$. Consider the case when minimum is achieved at a finite point, $\check{z}$. According to (48), z is the solution of the equation

$$\frac{\int \varphi e^{-z\varphi} f dx}{\int e^{-z\varphi} f dx} = E. \tag{50}$$

We assume that function $\varphi(x)$ is piecewise continuous and bounded in V and has the minimum and maximum values, φ_- and φ_+. Since

$$\varphi_- \int e^{-z\varphi} f dx \leqslant \int \varphi e^{-z\varphi} f dx \leqslant \varphi_+ \int e^{-z\varphi} f dx,$$

the left hand side of (50) is within the limits

$$\varphi_- \leqslant \frac{\int \varphi e^{-z\varphi} f dx}{\int e^{-z\varphi} f dx} \leqslant \varphi_+.$$

Therefore, the solution of equation (50) exists only for the values of E belonging to the segment

$$\varphi_- \leqslant E \leqslant \varphi_+.$$

According to Laplace asymptotics, the left hand side of (50) tends to φ_- as $z \to +\infty$ and φ_+ as $z \to -\infty$. So, the picture is completely similar to that of the case of random numbers.

Consider one special case which has a lot of applications. Let us choose $\varphi(x)$ to be a characteristic function of some subregion B of volume V, i.e.

$$\varphi(x) = \{.1 \quad \text{if} \quad x \in B0 \text{ otherwise.}$$

Then the sum,

$$\mathcal{N} = \varphi(x_1) + ... + \varphi(x_N),$$

has the meaning of the number of points which are in the region B. This number is random and takes the values $0, 1, ...N$. We wish to find the probabilities that $\mathcal{N}$ has values $0, 1, ...$ To this end, we have to rewrite formula (35) in terms of probability distribution of the non-scaled sum, $\mathcal{N} = NE$. Denotes its values by k. Repeating the derivation from (34) to (35) we have

$$F_N(k) = M \frac{1}{2\pi i} \int_{a-\infty}^{a+\infty} \frac{dz}{z} e^{kz - z\varphi(x_1) - ... - z\varphi(x_N)}$$

$$= \frac{1}{2\pi i} \int_{a-\infty}^{a+\infty} \frac{dz}{z} e^{kz} \left(\int e^{-z\varphi(x)} f(x) dx \right)^N. \tag{51}$$

Let all points be homogeneously distributed over V, i.e.

$$f(x) = \frac{1}{|V|} = const.$$

Then

$$\int e^{-z\varphi(x)} f(x)\, dx = \frac{1}{|V|}\left(|V| - |B| + |B|\, e^{-z}\right). \tag{52}$$

Note that region B is not necessarily simply connected and may consist of many pieces, but formula (52) contains only the volume of region B. We consider the "thermodynamic limit", when $|V| \to \infty$, $N \to \infty$, while the number of points per unit volume,

$$n = \frac{N}{|V|},$$

remains finite. We have

$$\left(\int e^{-z\varphi(x)} f(x)\, dx\right)^N = \left(1 - \frac{|B|\, n}{N}\left(1 - e^{-z}\right)\right)^N.$$

In the limit $N \to \infty$

$$\left(\int e^{-z\varphi(x)} f(x)\, dx\right)^N = e^{-|B|n\left(1-e^{-z}\right)}.$$

Hence, (51) takes the form

$$F_\infty(k) = \frac{1}{2\pi i}\int_{a-i\infty}^{a+i\infty} \frac{dz}{z} e^{kz-|B|n\left(1-e^{-z}\right)}. \tag{53}$$

We replace N by ∞ in notation of distribution function because the right hand side of (53) is the limit as $N \to \infty$. One can show that (53) can be differentiated over k. We get for probability density

$$\frac{dF_\infty(k)}{dk} = f_\infty(k) = e^{-|B|n}\frac{1}{2\pi i}\int_{a-i\infty}^{a+i\infty} e^{kz+|B|ne^{-z}}\, dz.$$

This integral can be computed analytically. Indeed, let us present $\exp\left(|B|\, ne^{-z}\right)$ as the series

$$e^{|B|ne^{-z}} = 1 + |B|\, ne^{-z} + \frac{1}{2}\left(|B|\, n\right)^2 e^{-2z} + \ldots \frac{1}{s!}\left(|B|\, n\right)^s e^{-sz} + \ldots$$

According to (17),

$$\frac{1}{2\pi i}\int e^{kz}\frac{1}{s!}\left(|B|\, n\right)^s e^{-sz} dz = \delta(k-s)\frac{1}{s!}\left(|B|\, n\right)^s.$$

So,

$$f_\infty(k) = e^{-|B|n}\left(\delta(k) + |B|\, n\delta(k-1) + \frac{1}{2}(|B|\, n)^2\,\delta(k-2) + ...\right),$$

i.e. $f_\infty(k)$ is a sum of $\delta-$functions, concentrated at the points $k = 0, 1, 2....$ This means that the region B contains k points with probability

$$p_k = \frac{1}{k!}(|B|\, n)^k\, e^{-|B|n}.$$

The sum of all probabilities is equal to 1,

$$\sum_0^\infty p_k = 1,$$

as follows from Taylor expansion of exponential function

$$e^{|B|n} = \sum_{k=0}^\infty \frac{1}{k!}(|B|\, n)^k.$$

We arrived at the so-called *Poisson distribution.*

0.6 Stochastic variational problems

Many problems of physics and mechanics can be formulated as variational problems, i.e. as problems of minimization of some functional, $I(u)$, on a set of elements, u. We will consider the simplest case, when $I(u)$ is a quadratic functional, i.e. the functional of the form

$$I(u) = \frac{1}{2}(Au, u) - (l, u). \tag{54}$$

By u one can mean a point of a multidimensional space, $u = \left(u^1, ..., u^n\right)$, (l, u) a linear function

$$(l, u) = l_i u^i, \tag{55}$$

and (Au, u) a quadratic function

$$(Au, u) = A_{ij} u^i u^j. \tag{56}$$

In continuum mechanics problems, one considers the limit $n \to \infty$, but a finite-dimensional truncation of continuum mechanics problems (for example, by the finite-element method) returns us to the finite-dimensional case

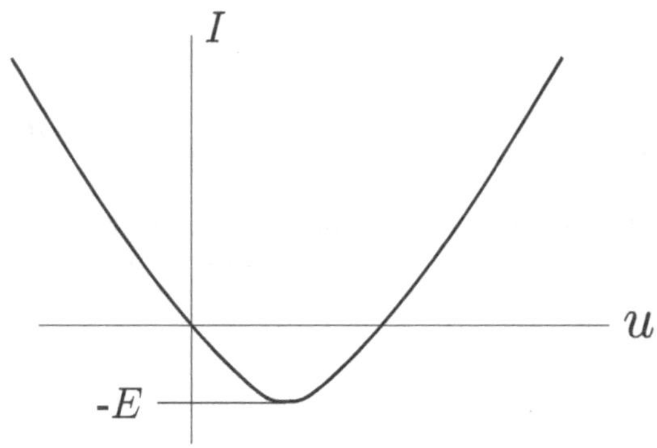

Figure 13. Minimum value of $I(u)$ is negative; it has the meaning of negative energy.

(55), (56). The minimum value of $I(u)$ is always negative; in 1D case this is seen from Fig. 13.

In physical problems the minimum value has the meaning of negative energy of the system.

If the properties of the system are random, so are the matrix (operator) $A = (A_{ij})$ and the vector (linear functional) $l = (l_i)$. We consider the simplest "probabilistic" question: What is the probability distribution of minimum values of $I(u)$ (i.e. probability distribution of energy)? It is enough to discuss the finite-dimensional case; the results for continuum mechanics are obtained in the limit $n \to \infty$.

If the matrix A does not depend on the event, ω, the problem is called weakly stochastic, otherwise the variational problem is called strongly stochastic. Many physical theories provide examples of such type of problems. The analytical results can be obtained mostly for weakly stochastic problems.

As is usual in the probabilistic approach, the probabilistic modeling is especially effective, if one can identify in the phenomenon to be modeled the statistically independent (or slightly dependent) events. Analytical investigation can be advanced considerably, if there are many statistically independent events involved. We focus here on a special case when (l, u) is a sum of small independent linear functionals. More precisely, there is a large number, N, of independent identically distributed random variables, $r_1, ..., r_N$, and a given random linear functional, $(l_0(r), u)$. Then the linear functional of the variational problem, (l, u), is defined as an "empirical average" of N values of $(l_0(r), u)$:

$$(l, u) = \frac{1}{N} \sum_{a=1}^{N} (l_0(r_a), u)\,. \tag{57}$$

Consider a quadratic function of a finite number of variables (54). The minimum value of this function is

$$\check{I} = -\frac{1}{2}\left(A^{-1}l, l\right) \tag{58}$$

where A^{-1} is the inverse matrix for the matrix A. We wish to find probability distribution of energy,

$$f(E) = M\delta\left(E + \min_u I(u)\right). \tag{59}$$

Following the same path as for a sum of independent random variables we plug in (59) the presentation of δ–function (17),

$$f(E) = M\frac{1}{2\pi i}\int_{-i\infty}^{i\infty} e^{Ez + z\min\limits_u I(u)}dz = \frac{1}{2\pi i}\int_{-i\infty}^{i\infty} e^{Ez} M e^{z\min\limits_u I(u)}dz. \tag{60}$$

It would be desirable to change somehow the operations of mathematical expectation and minimization: then we would arrive to some deterministic problem. This can be achieved by presenting $\exp\left[-z\min I(u)\right]$ by an integral of some function of u over u. Since mathematical expectation is, in fact, also integration, the order of integrals can be changed, and we obtain an integral of the mathematical expectation of the function of u, which can be found explicitly in some cases. Now let us discuss precisely what this trick means.

According to (58) formula (23) can be also written as

$$e^{-\min\limits_u I(u)} = \sqrt{\det A}\int e^{-I(u)}du \tag{61}$$

We see that this relation reduces the computation of the minimum value to integration indeed. Since any quadratic functional in variational problems of continuum mechanics admits a finite-dimensional truncation, one can write formula (61) for a finite-dimensional truncation, and then consider the limit when the dimension of the truncation tends to infinity. In the limit, in the right hand side of (61) we obtain what is called the functional integral. We include $\sqrt{\det A}$ in the definition of the "volume element" in the functional space,

$$\mathcal{D}_A u = \sqrt{\det A}du \tag{62}$$

and write (61) as

$$e^{\min_u I(u)} = \int e^{-I(u)} \mathcal{D}_A u. \tag{63}$$

The notation, $\mathcal{D}_A u$, emphasizes that the volume element depends on the operator A.

E x e r c i s e. Consider a variational problem with a set of constraints: minimize a quadratic functional,

$$I(u) = \frac{1}{2}(Au, u)$$

on all u obeying to linear constraints

$$(l_\alpha, u) = c_\alpha, \qquad \alpha = 1, ..., s.$$

Show that

$$e^{-\min I(u)} = \int e^{-I(u)} \prod_\alpha \delta\left(c_\alpha - (l_\alpha, u)\right) \mathcal{D}u$$

where

$$\mathcal{D}u = \sqrt{\frac{(2\pi)^n}{\det A}} \sqrt{\frac{(2\pi)^s}{\det \mathcal{A}}} du^1 ... du^n$$

and $\mathcal{A}$ is a matrix with the components

$$\mathcal{A}_{\alpha\beta} = \left(A^{-1} l_\alpha, l_\beta\right).$$

H i n t. Use (25).

In Section 5.12 of the above-cited book Variational Principles of Continuum Mechanics one can find various generalizations of (63) involving non-positive quadratic functionals and complex-valued functionals. We illustrate the idea using one of such generalizations,

$$e^{z \min_u I(u)} = \int_{-i\infty}^{i\infty} e^{z[\frac{1}{2}(Au,u)-(l,u)]} \mathcal{D}_{zA} u, \qquad \text{for } Re z > 0. \tag{64}$$

In (64) the parameter, z, is also included in the volume element: for $m-$dimensional truncation, $\mathcal{D}_{zA} u = \sqrt{z^m \det A} du$.

If we plug in (64) the linear functional (57), we get

$$M e^{z \min_u I(u)} = \int_{-i\infty}^{i\infty} M e^{z[\frac{1}{2}(Au,u)-(l,u)]} \mathcal{D}_{zA} u$$

$$= \int_{-i\infty}^{i\infty} e^{z\frac{1}{2}(Au,u)} M e^{-z\frac{1}{N}\sum_{a=1}^{N}(l_0(r_a),u)} \mathcal{D}_{zA}u$$

$$= \int_{-i\infty}^{i\infty} e^{z\frac{1}{2}(Au,u)} \left(Me^{-z\frac{1}{N}(l_0(r),u)}\right)^N \mathcal{D}_{zA}u.$$

If we change variable $z \to Nz$, then probability density $f(E)$ (60) takes the form

$$f(E) = \frac{N}{2\pi i} \int_{-i\infty}^{i\infty} e^{NS(E,z,u)} \mathcal{D}_{NzA}udz, \tag{65}$$

where $S(E, z, u)$ is a function that is independent on N,

$$S(E, z, u) = Ez + \frac{z}{2}(Au, u) + \ln Me^{-z(l_0,u)}. \tag{66}$$

The functional integral (65) depends on a large parameter N. In many cases the asymptotics of this integral can be studied. It is determined by the stationary points of the entropy functional (66). Examples and further details can be found in the above-cited book.

New concepts in damping generation and control: theoretical formulation and industrial applications

A. Carcaterra*

Department of Mechanics and Aeronautics
University La Sapienza
Rome, Italy

1 A New Point of View about Dissipation: Introduction

These notes are finalized to a particular study of the **damping mechanism** in Hamiltonian systems, characterized indeed by the **absence of any energy dissipation** effect. It is important to make a clear distinction between the two previous concepts, since they seem to be somehow contradictory. A Hamiltonian system is characterized by an invariant total energy (the Hamiltonian H) that is equivalent to state any energy dissipation process is absent. This circumstance, especially from an engineering point of view, leads to the wrong expectation that the motion of any part of such a dissipation-free system, subjected to some initial conditions, maintains a sort of constant amplitude response. This is, although unexpectedly, a wrong prediction and the "mechanical intuition" leads, in this case, to a false belief. It is indeed true the converse: **even in the absence of any energy dissipation mechanisms, mechanical systems can exhibit damping, i.e. a decay amplitude motion.**

This statement makes clear how the two concepts rely on completely different properties of the Hamiltonian systems.

This fact is the root of very fundamental physical properties of mechanical systems, and touches charming and thorny questions about them. It is a matter of fact that looking at the atomic scale of mechanical systems, they can be described through the use of Hamiltonian equations: an atomic lattice vibrates and moves energy along its structure without any dissipation mechanisms. But it is a trivial consideration about every day physical world, that any energy released at macro-scale to an atomic lattice tends to disappear at that scale, to be moved into the micro-scale vibrations of the atoms. Damping is observed. This process is the base for converting mechanical energy (large scale motion) into heat (small scale thermal vibration), and without any energy loss. The natural tendency to produce this process is at the base of the **macroscopic irreversibility** and at the root of the second principle of thermodynamics itself.

As it appears, the questions behind the distinction between damping and dissipation involves fundamental aspects of physical systems, that have been the subject of diatribes, debates and fascinating investigations starting from the end of the nineteenth century up to the present days, in the field of theoretical physics as well as in mathematical physics. Famous names are involved as those of Boltzmann, Loschmidt, Curvelwell, the Ehrenfests, Poincarè, Zermelo, Prigogine and many others, turning around the Boltzmann's *H-theorem.*

The investigation contained in the present notes does not attack directly the previous questions (wisely), but a problem somehow close to the previous, and the related possibilities for engineering applications(!). As a consequence, the obtained results seem to be theoretically intriguing (aesthetic in science is a luxury relatively down market) and practically usable (engineers are sensitive). Optimists would say a good compromise.

But, aside these considerations, let us illustrate the main point we have here.

More precisely, we consider a partition of the Hamiltonian system S into two subsystems: we select, among its N degrees of freedom, one of them x, indicated as **master**. The remaining part of the system, consisting of $N-1$ degrees of freedom $\tilde{\mathbf{x}}$, is called the **hidden** or **unmonitored part** of the system. With this view of the problem, we are interested in the following analysis.

Let the Lagrangian function L of S be:

$$L\left(\mathbf{x},\, \dot{\mathbf{x}}\right) = L_x\left(x,\, \dot{x}\right) + L_{mix}\left(x, \dot{x}, \tilde{\mathbf{x}}, \dot{\tilde{\mathbf{x}}}\right) + L_{hid}\left(\tilde{\mathbf{x}}, \dot{\tilde{\mathbf{x}}}\right)$$

The motion of the whole system S is governed by the minimum principle:

$$\delta \int_{t_1}^{t_2} L\left(\mathbf{x}, \dot{\mathbf{x}}\right)\, dt = 0$$

to which correspond the Euler-Lagrange equations:

$$\frac{d}{dt}\frac{\partial L\left(\mathbf{x}, \dot{\mathbf{x}}\right)}{\partial \dot{\mathbf{x}}} - \frac{\partial L\left(\mathbf{x}, \dot{\mathbf{x}}\right)}{\partial \mathbf{x}} = 0 \qquad (1)$$

providing the solutions $x*$, $\tilde{\mathbf{x}}^*$.

The motion $x*$ would be here derived directly from the solution of a reduced problem, by means of a new equation of motion written in terms of x only in the form:

$$\frac{d}{dt}\frac{\partial L_x(x,\dot{x})}{\partial \dot{x}} - \frac{\partial L_x(x,\dot{x})}{\partial x} + Q(x,\dot{x},\ddot{x}) = 0 \tag{2}$$

where the form for Q depends on the structure of the Lagrangian L.

Comparing the complete set of equations for S and that for x, the following replacement is operated:

$$V\left(x,\dot{x},\tilde{\mathbf{x}},\dot{\tilde{\mathbf{x}}}\right) = \frac{d}{dt}\frac{\partial L_{mix}\left(x,\dot{x},\tilde{\mathbf{x}},\dot{\tilde{\mathbf{x}}}\right)}{\partial \dot{x}} - \frac{\partial L_{mix}\left(x,\dot{x},\tilde{\mathbf{x}},\dot{\tilde{\mathbf{x}}}\right)}{\partial x}$$

$$\overset{replaced\ by}{\Longrightarrow} \quad Q(x,\,\dot{x},\,\ddot{x})$$

Thus, we desire to determine, if possible, the form of Q such that equations (1) and (2) provide the same solution x^* for x. It is shown in the sections ahead that in general it is not strictly possible but, under some restrictive hypotheses, the difference between Q and V can be small, and negligible, at least in a prescribed time window. The nature of the term Q under such hypotheses, can be also nicely interpreted as a dissipative effect for the system x; it means:

$$L_x(x,\dot{x}) = T_x(x,\,\dot{x}) - U_x(x) \;\Rightarrow\; \dot{H}_x = \frac{d}{dt}\left[T_x(x,\,\dot{x}) + U_x(x)\right] < 0 \tag{3}$$

Therefore, such analysis explains well the chance of observing damping in the motion of x even if dissipation in the system S is absent i.e.:

$$\dot{H} = \frac{d}{dt}\left[T(\mathbf{x},\dot{\mathbf{x}}) + U(\mathbf{x})\right] = 0 \tag{4}$$

being clear how equations (3) and (4)are not contradictory.

The question of the form of Q is approached in Section 2, namely subsections 1, 2 and 3, where it is illustrated how Q can include a dissipation term. Subsection 4 studies the time window within which the substitution of V by Q is permitted because of the small error.

Subsection 5 approach an inverse problem through a variational technique: design the system S such that Q generates the fastest energy transfer from x to the unmonitored system. Subsection 6 studies examples of application of the theory presented in 5.

Finally, last but not least, Section 3 describes an industrial application of the theory outlined, showing how the theoretical speculation can dress the clothes of a design procedure for the production of a real innovative category of damping devices.

Therefore, the main point we have here is that very effective **damping properties can be indeed obtained by coupling a main structure to a secondary structure that serves as energy storage**. This opportunity opens a new way to damping control: damping is produced not acting on a local dissipation process, a weak controllable phenomenon, but it can be shaped by the designer modifying the purely elastic response of a secondary part of the structure. This result can be achieved by **purely mechanical means or by electromechanical devices**. The theory developed here applies to both, and in both cases the ability in controlling the damping properties of the primary structure does not rely on any local dissipation mechanism.

More precisely, it has been shown how the motion of a principal structure, called master, can be damped as an effect amounting to an energy transfer process by which the master energy is moved to a set of resonators attached to it (1; 2; 3; 4; 5; 6; 7; 8). The phenomenon, sometime called apparent damping, is intriguing considering that this spontaneous energy transfer can have irreversible characteristics, i.e. the energy can be permanently transferred from the master to the set of resonators. This result is always observed when the number of degrees of freedom within the set is infinite, or actually very large (9; 10; 11; 20; 21; 22; 23; 26; 27; 28) and not for small number of oscillators, where recursive phenomena are observed (8; 12). In some cases, the same irreversible energy transfer can be indeed predicted considering statistical ensemble average over a population of similar structures (13). Moreover, a quasi-irreversible energy transfer can be produced even with a finite number of resonators within the attachment, when selecting special distribution of the natural frequencies of the attached resonators, or alternatively, introducing non-linear or parametric effects within the set (14; 15; 16; 17; 27; 28). In some cases the attachment can have an electrical nature (24; 25). In (38; 39; 40; 41; 42; 43; 44; 45; 46; 47) it is shown how an electrical passive network, a sort of electrical-double of the master mechanical structure, can absorb very effectively the energy pumped in it. An energy transfer from the mechanical to the electrical part of the system initially takes place; later, energy is actually dissipated by resistive effects into the electrical energy storage. Also in this case, the observed damping is not a direct consequence of a local dissipation, but rather it is due to a fast energy displacement form the master part of the system to a sacrificial attachment.

Further interesting readings on the subject can be found in (18; 19; 29; 30; 31; 32; 33; 34; 35; 36; 37).

2 Pseudo-Dissipative Systems: an Outline of the Theory

This section outlines an original theory of pseudo-dissipative structures, *N-dimensional* Hamiltonian systems for which a certain degree of freedom x, called **master**, is separated from the remaining $N-1$ *dofs*, the **hidden variables**, and studied apart from them. The effect of the interaction between x and the hidden *dofs* is taken into account by terms dependent only on x, following the procedure outlined in these notes. Namely the attention is focused on three main aspects: (i) collapse this interaction effect into simple terms, (ii) provide for them a physical interpretation, (iii) discuss in depth the limit of these simplifications.

The analysis starts with static systems. For them the effect of the hidden variables is not a pseudo-dissipation but rather an additional elastic restoring force depending on the displacement of x and on the system of forces acting on the hidden variables. However, the static system allows to understand simply the way the hidden variables play their effect in the equation of motion of the master variable, especially looking at the Hamiltonian formulation of the reduced problem. This presents a formal analogy with the dynamic case, the one of main interest for the theory presented in this course.

The Hamiltonian presentation of the results permits to understand how the terms depending on the hidden variables and on the mixed terms, are indeed replaced by simpler equivalent terms in the new Lagrangian function of the reduced system. On the other hand, the variational approach plays a key role in determining the conditions for a permanent energy storage within the master variable x.

2.1 Static systems: the effect of the hidden variables

A static system S is described through the set of variables

$$x \equiv x_0, x_1, \ldots, x_N$$
$$x = \{x_0, x_1, \ldots, x_N\}^T$$

Assume S is controlled by the set of equations:

$$\frac{\partial L(\mathbf{x})}{\partial x_k} = 0, \qquad k = 0, 1, 2, \ldots, N$$

the Euler-Lagrange equations related to the variational principle:

$$\delta L(\mathbf{x}) = 0$$

where the characteristic lagrangian function L for S is dependent on $\mathbf{x}$ in a quadratic fashion

$$L(\mathbf{x}) = \frac{1}{2}\mathbf{x}^T\mathbf{K}\,\mathbf{x} - \mathbf{f}^T\mathbf{x}, \qquad \mathbf{K} = \begin{bmatrix} k_{00} & \dots & k_{0N} \\ \dots & \dots & \dots \\ k_{N0} & \dots & k_{NN} \end{bmatrix} = \begin{bmatrix} k_{00} & \tilde{\mathbf{k}}_0^T \\ \tilde{\mathbf{k}}_0 & \widetilde{\mathbf{K}} \end{bmatrix}$$

$$\widetilde{\mathbf{K}} = \begin{bmatrix} k_{11} & \dots & k_{1N} \\ \dots & \dots & \dots \\ k_{N1} & \dots & k_{NN} \end{bmatrix} = \begin{bmatrix} \tilde{\mathbf{k}}_1^T \\ \dots \\ \tilde{\mathbf{k}}_N^T \end{bmatrix}, \quad \tilde{\mathbf{k}}_0 = \begin{Bmatrix} k_{01} \\ \dots \\ k_{0N} \end{Bmatrix} = \begin{Bmatrix} k_{10} \\ \dots \\ k_{N0} \end{Bmatrix}$$

Accordingly to our assumption the equations for S are:

$$\frac{\partial L(\mathbf{x})}{\partial x_k} = 0, \quad \rightarrow \quad \mathbf{K}\,\mathbf{x} = \mathbf{f}$$

These can be written separating the degree of freedom x, the master, from the others as:

$$\begin{cases} k_{00}\,x + \tilde{\mathbf{k}}_0^T\tilde{\mathbf{x}} = f_0 \\ k_{10}\,x + \tilde{\mathbf{k}}_1^T\tilde{\mathbf{x}} = f_1 \\ \vdots \\ k_{N0}\,x + \tilde{\mathbf{k}}_N^T\tilde{\mathbf{x}} = f_N \end{cases} \qquad \text{or} \qquad \begin{cases} k_{00}\,x + \tilde{\mathbf{k}}_0^T\tilde{\mathbf{x}} = f_0 \\ \tilde{\mathbf{k}}_0\,x + \widetilde{\mathbf{K}}\,\tilde{\mathbf{x}} = \tilde{\mathbf{f}} \end{cases}$$

where:

$$\tilde{\mathbf{x}} = \begin{Bmatrix} x_1 \\ \dots \\ x_N \end{Bmatrix}, \qquad \tilde{\mathbf{f}} = \begin{Bmatrix} f_1 \\ \dots \\ f_N \end{Bmatrix}$$

We can eliminate from the equation for x the set of other variables (hidden) derived as the solution of the matrix equation, obtaining a single equation for x, where the hidden variables disappear:

$$\begin{cases} k_{00}\,x - \tilde{\mathbf{k}}_0^T\widetilde{\mathbf{K}}^{-1}\tilde{\mathbf{k}}_0\,x = f_0 - \tilde{\mathbf{k}}_0^T\widetilde{\mathbf{K}}^{-1}\tilde{\mathbf{f}} \\ \tilde{\mathbf{x}} = \widetilde{\mathbf{K}}^{-1}\left(\tilde{\mathbf{f}} - \tilde{\mathbf{k}}_0\,x\right) \end{cases}$$

The determined equation for x deserves some comments.

Consider explicitly the lagrangian of the original problem, decoupling the dependency of it upon x and the other variables:

$$L(\mathbf{x}) = \frac{1}{2}\mathbf{x}^T\mathbf{K}\mathbf{x} - \mathbf{f}^T\mathbf{x} = \frac{1}{2}k_{00}x^2 + \frac{1}{2}x\tilde{\mathbf{k}}_0^T\tilde{\mathbf{x}} + \frac{1}{2}x\tilde{\mathbf{k}}_0^T\tilde{\mathbf{x}} + \frac{1}{2}\tilde{\mathbf{x}}^T\widetilde{\mathbf{K}}\tilde{\mathbf{x}} - f_0x - \tilde{\mathbf{f}}^T\tilde{\mathbf{x}}$$

We can distinguish three kind of terms: terms directly dependent on x, terms dependent on the hidden variables, and mixed terms:

$$
\begin{aligned}
L(\mathbf{x}) &= L_X(x) + L_{mix}(x, \tilde{\mathbf{x}}) + L_{hid}(\tilde{\mathbf{x}}) \\
L_X(x) &= \frac{1}{2} k_{00}\, x^2 - f_0\, x \\
L_{mix}(x, \tilde{\mathbf{x}}) &= x\, \tilde{\mathbf{k}}_0^T \tilde{\mathbf{x}} \\
L_{hid}(\tilde{\mathbf{x}}) &= \frac{1}{2} \tilde{\mathbf{x}}^T \widetilde{\mathbf{K}}\, \tilde{\mathbf{x}} - \tilde{\mathbf{f}}^T \tilde{\mathbf{x}}
\end{aligned}
$$

It comes out the equation for x can be determined by modifying the lagrangian function L of the complete system retaining only the direct terms and substituting the mixed terms and the terms depending on the hidden variables by an **interaction potential** $D(x)$ and an **external force potential** Nx:

$$
L'_X(x) = L_X(x) + D(x) + N\, x \qquad \text{where:} \qquad
\begin{aligned}
D(x) &= \frac{1}{2} \tilde{\mathbf{k}}_0^T \widetilde{\mathbf{K}}^{-1} \tilde{\mathbf{k}}_0 x^2 \\
N &= \tilde{\mathbf{k}}_0^T \widetilde{\mathbf{K}}^{-1} \tilde{\mathbf{f}}
\end{aligned}
$$

2.2 Dynamic systems

The Hamiltonian system S is described through the set of variables

$$
\begin{aligned}
x(t) &\equiv x_0(t), x_1(t), \ldots, x_N(t) \\
\mathbf{x}(t) &= \{x_0(t), x_1(t), \ldots, x_N(t)\}^T
\end{aligned}
$$

and governed by the set of equations:

$$
\frac{d}{dt} \frac{\partial L(\mathbf{x}, \dot{\mathbf{x}})}{\partial \dot{x}_k} - \frac{\partial L(\mathbf{x}, \dot{\mathbf{x}})}{\partial x_k} = 0, \qquad k = 0, 1, 2, \ldots, N
$$

the Euler-Lagrange equations related to the well-known variational Hamilton principle:

$$
\delta \int_{t_1}^{t_2} L(\mathbf{x}, \dot{\mathbf{x}})\, dt = 0
$$

where L is the quadratic form:

$$L(\mathbf{x}, \dot{\mathbf{x}}) = \frac{1}{2}\dot{\mathbf{x}}^T \mathbf{M}\dot{\mathbf{x}} + \frac{1}{2}\mathbf{x}^T \mathbf{K}\mathbf{x} - \mathbf{f}^T \mathbf{x}$$

$$\mathbf{K} = \begin{bmatrix} k_{00} & \dots & k_{0N} \\ \dots & \dots & \dots \\ k_{N0} & \dots & k_{NN} \end{bmatrix} = \begin{bmatrix} k_{00} & \tilde{\mathbf{k}}_0^T \\ \tilde{\mathbf{k}}_0 & \widetilde{\mathbf{K}} \end{bmatrix}$$

$$\widetilde{\mathbf{K}} = \begin{bmatrix} k_{11} & \dots & k_{1N} \\ \dots & \dots & \dots \\ k_{N1} & \dots & k_{NN} \end{bmatrix} = \begin{bmatrix} \tilde{\mathbf{k}}_1^T \\ \dots \\ \tilde{\mathbf{k}}_N^T \end{bmatrix}, \quad \tilde{\mathbf{k}}_0 = \begin{Bmatrix} k_{01} \\ \dots \\ k_{0N} \end{Bmatrix} = \begin{Bmatrix} k_{10} \\ \dots \\ k_{N0} \end{Bmatrix},$$

$$\mathbf{M} = \begin{bmatrix} m_0 & 0 & \dots & 0 \\ 0 & m_1 & \dots & 0 \\ 0 & 0 & \dots & m_N \end{bmatrix}$$

This partitioning of matrices is useful ahead.

Accordingly to our assumption, the equations for S are:

$$\frac{d}{dt}\frac{\partial L(\mathbf{x}, \dot{\mathbf{x}})}{\partial \dot{x}_k} - \frac{\partial L(\mathbf{x}, \dot{\mathbf{x}})}{\partial x_k} = 0, \quad \rightarrow \quad \mathbf{M}\ddot{\mathbf{x}} + \mathbf{K}\mathbf{x} = \mathbf{f}$$

We can separate x from the other degrees of freedom as:

$$\begin{cases} m_0 \ddot{x} + k_{00}\, x + \tilde{\mathbf{k}}_0^T \tilde{\mathbf{x}} = f_0 \\ m_1 \ddot{x}_1 + k_{10}\, x + \tilde{\mathbf{k}}_1^T \tilde{\mathbf{x}} = f_1 \\ \dots \\ m_N \ddot{x}_N + k_{N0}\, x + \tilde{\mathbf{k}}_N^T \tilde{\mathbf{x}} = f_N \end{cases} \quad \text{or} \quad \begin{cases} m_0 \ddot{x} + k_{00}\, x + \tilde{\mathbf{k}}_0^T \tilde{\mathbf{x}} = f_0 \\ \widetilde{\mathbf{M}}\ddot{\tilde{\mathbf{x}}} + \tilde{\mathbf{k}}_0\, x + \widetilde{\mathbf{K}}\,\tilde{\mathbf{x}} = \tilde{\mathbf{f}} \end{cases}$$

where:

$$\tilde{\mathbf{x}} = \begin{Bmatrix} x_1 \\ \dots \\ x_N \end{Bmatrix}, \quad \widetilde{\mathbf{M}} = \begin{bmatrix} m_1 & 0 & \dots & 0 \\ 0 & m_2 & \dots & 0 \\ 0 & 0 & \dots & m_N \end{bmatrix}, \quad \tilde{\mathbf{f}} = \begin{Bmatrix} f_1 \\ \dots \\ f_N \end{Bmatrix}$$

The lagrangian function can be written separating the contribution of x from the others as:

$$L(\mathbf{x}, \dot{\mathbf{x}}) = \frac{1}{2}\dot{\mathbf{x}}^T \mathbf{M}\dot{\mathbf{x}} + \frac{1}{2}\mathbf{x}^T \mathbf{K}\mathbf{x} - \mathbf{f}^T \mathbf{x} =$$

$$= \frac{1}{2}m_0\,\dot{x}^2 + \frac{1}{2}k_{00}x^2 + \frac{1}{2}x\tilde{\mathbf{k}}_0^T \tilde{\mathbf{x}} + \frac{1}{2}x\tilde{\mathbf{k}}_0^T \tilde{\mathbf{x}} + \frac{1}{2}\dot{\tilde{\mathbf{x}}}^T \widetilde{\mathbf{M}}\dot{\tilde{\mathbf{x}}} +$$

$$+ \frac{1}{2}\tilde{\mathbf{x}}^T \widetilde{\mathbf{K}}\tilde{\mathbf{x}} - f_0 x - \tilde{\mathbf{f}}^T \tilde{\mathbf{x}}$$

The Laplace domain form of the equations of motion reads (capital symbols for transformed quantities):

$$\begin{cases} m_0 s^2 X + k_{00} X + \tilde{\mathbf{k}}_0^T \widetilde{\mathbf{X}} = F_0 \\ m_1 s^2 X_1 + k_{10} X + \tilde{\mathbf{k}}_1^T \widetilde{\mathbf{X}} = F_1 \\ \dots \\ m_N s^2 X_N + k_{N0} X + \tilde{\mathbf{k}}_N^T \widetilde{\mathbf{X}} = F_N \end{cases} \quad \text{or} \quad \begin{cases} m_0 s^2 X + k_{00} X + \tilde{\mathbf{k}}_0^T \widetilde{\mathbf{X}} = F_0 \\ s^2 \widetilde{\mathbf{M}} \widetilde{\mathbf{X}} + \tilde{\mathbf{k}}_0 X + \widetilde{\mathbf{K}} \widetilde{\mathbf{X}} = \widetilde{\mathbf{F}} \end{cases}$$

As for the static case, eliminating the hidden variables from the equation for x yields:

$$\begin{cases} m_0 s^2 X + k_{00} X - \tilde{\mathbf{k}}_0^T \left(s^2 \widetilde{\mathbf{M}} + \widetilde{\mathbf{K}} \right)^{-1} \tilde{\mathbf{k}}_0 X = F_0 - \tilde{\mathbf{k}}_0^T \left(s^2 \widetilde{\mathbf{M}} + \widetilde{\mathbf{K}} \right)^{-1} \widetilde{\mathbf{F}} \\ \widetilde{\mathbf{X}} = \left(s^2 \widetilde{\mathbf{M}} + \widetilde{\mathbf{K}} \right)^{-1} \left(\tilde{\mathbf{f}} - \tilde{\mathbf{k}}_0 X \right) \end{cases}$$

Transforming the first equation back to the time domain, one obtains the three equivalent forms:

$$m_0 \ddot{x}(t) + k_{00} x(t) - \frac{1}{2\pi} \int_0^\infty \tilde{\mathbf{k}}_0^T \mathbf{G}(s) \tilde{\mathbf{k}}_0 X(s) e^{st} dt = f_0 - \frac{1}{2\pi} \int_0^\infty \tilde{\mathbf{k}}_0^T \mathbf{G}(s) \widetilde{\mathbf{F}}(s) e^{st} dt$$

$$m_0 \ddot{x}(t) + k_{00} x(t) + \int_{-\infty}^{\infty} \tilde{\mathbf{k}}_0^T \mathbf{g}(t-\tau) \tilde{\mathbf{k}}_0 x(\tau) d\tau = f_0 + \int_{-\infty}^{\infty} \tilde{\mathbf{k}}_0^T \mathbf{g}(t-\tau) \tilde{\mathbf{f}}(\tau) d\tau$$

$$m_0 \ddot{x}(t) + k_{00} x(t) + \left(\tilde{\mathbf{k}}_0^T \mathbf{g}(t) \tilde{\mathbf{k}}_0 \right) * x(t) = f_0 + \left(\tilde{\mathbf{k}}_0^T \mathbf{g}(t) \right) * \tilde{\mathbf{f}}(t)$$

where

$$\mathbf{g}(t) = -\frac{1}{2\pi} \int_0^\infty \left(s^2 \widetilde{\mathbf{M}} + \widetilde{\mathbf{K}} \right)^{-1} e^{st} dt; \quad \mathbf{G}(s) = \left(s^2 \widetilde{\mathbf{M}} + \widetilde{\mathbf{K}} \right)^{-1}$$

These equations produce a clear qualitative picture of the nature of the interaction effect between x and the hidden variables.

More precisely, note that:

1. The reduced equation of the motion for x becomes an integral-differential equation, meaning the interaction with the remaining part of the system, the hidden part, amount to a **memory effect** (integral term); this is the dynamic counterpart of the terms related to the **interaction potential *D(x)*** appearing in the static case;

2. Additionally, the forces applied to the hidden part of the system appears as a known forcing term, that amounts to a **noise effect**; this is the dynamic counterpart of the static **external force potential** $\boldsymbol{Nx}$;
3. The kernel $\mathbf{G}$ apparently is a rational function in terms of s; namely it can be expressed as a ratio of two polynomials P and Q (the degree of P being larger than that of Q, because of the causality principle):
$$[\mathbf{G}(s)]_{ij} = \left[\left(s^2\widetilde{\mathbf{M}} + \widetilde{\mathbf{K}}\right)^{-1}\right]_{ij} = \left[\frac{P_{ij}(s)}{Q(s)}\right]$$
4. The kernel $\mathbf{G}$, since the system is Hamiltonian, is a real function, and does not contain any imaginary part, as an effect of the absence of any real dissipation in the motion of the system.
5. In the previous analysis, using the Laplace transform, it has been tacitly assumed that all initial conditions are set to zero (generalization is relatively easy).

As the simplest example, a nice exercise it to apply the previous analysis to a two degrees of freedom system; one can easily verify that the previously discussed equations becomes:

$$m_0\ddot{x}(t) + (k_1 + k_2)x(t) + k_2^2 g(t) * x(t) = f_0 = k_2 g(t) * f_1(t)$$

$$\mathbf{M} = \begin{bmatrix} m_0 & 0 \\ 0 & m_1 \end{bmatrix}, \quad \mathbf{K} = \begin{bmatrix} k_1 + k_2 & -k_2 \\ -k_2 & k_2 \end{bmatrix}$$

$$\mathbf{G}(s) = \frac{1}{m_1 s^2 + k_1}, \quad \mathbf{g}(t) = H(t)\sin\sqrt{\frac{k_2}{m_2}}\,t$$

2.3 Pseudo-dissipative effects: motion about ω_0

The previous analysis shows how following the motion of only one selected degree of freedom of S, its motion becomes controlled by an integral-differential equation, the integral part amounting to the interaction of x with the unmonitored part of the system.

However, a more clear and physically interesting picture of the problem arises from a more detailed analysis of this integral terms. If some approximations are made, it discloses sharp characteristics about the nature of the forces they represent.

To enlightening these properties, our following analysis offers two distinct mathematical approaches, that leads to look at different perspectives for the interaction terms. Both the approaches are based on an approximation about the integral term involving $\mathbf{G}$. **The idea is to consider**

systems having a dominant frequency in their response, say ω_0. This happens when the natural frequencies of S are all located in a rather narrow frequency bandwidth. In physical systems this case is actually met in many cases of interest and it is also a good approximation to approach the energy sharing process in the general case. For engineering system this can indeed intended as a design configuration purposely obtained to determine a desired energy sharing effect.

However, both of the proposed techniques permit to extract information about the energy exchange between x and the hidden variables. More precisely the first approach passes through a **Taylor expansion of the kernel G** in terms of s, while the second uses a less intuitive strategy, but more powerful, we named **integral-Padè expansion of the kernel**.

In the next sections the two mentioned approaches are developed in detail.

Pseudo-dissipative effects: Taylor expansion of the kernel G To carry on our point of view about unmonitored dynamic systems, it helps introduce some additional considerations and then some additional hypotheses on our system.

The kernel **G** can be written using a Taylor expansion in the complex plane in terms of s, valid within a circle around the complex s_0:

$$\mathbf{G}(s) = \mathbf{G}(s_0) + \mathbf{G}'(s_0)(s - s_0) + \mathbf{R}(s)$$

$$\mathbf{R}(s) = \frac{(s - s_0)^N}{N!}\left[\frac{d^{(N)}\mathbf{G}}{ds^N}\right]_{s=\zeta} = \frac{1}{2\pi i}\oint_{\gamma(s_0)} \frac{\mathbf{G}(s)}{(s - s_0)^{N+1}} ds$$

Note that this expansion of **G** produces, neglecting the remainder, a linear term $\mathbf{G}'(s_0)s$ plus a constant term $\mathbf{G}(s_0) - \mathbf{G}'(s_0)s_0$. In general $\mathbf{G}(s_0)$, $\mathbf{G}'(s_0)$ are complex quantities and it is useful to analyse them in depth.

We assume:

$$s_0 = \sigma_0 + j\omega_0, \qquad s = j\omega$$

This means: the center of the series expansion is in general on the complex plane, while the series is evaluated only along the imaginary axis. Since the expansion is made in the frequency domain (ω) about ω_0, we call the investigated response "**motion about ω_0**". The physical sense of this assumption and several cases of applications will be the subject of following subsections.

Neglecting the remainder one gets the equation:

$$\mathbf{G} \approx \mathrm{Re}(\mathbf{G}_0 - \sigma_0\mathbf{G}_0') - (\omega - \omega_0)\mathrm{Im}(\mathbf{G}_0') + j\left[\mathrm{Im}(\mathbf{G}_0 - \sigma_0\mathbf{G}_0') + (\omega - \omega_0)\mathrm{Re}(\mathbf{G}_0')\right]$$

This expression leads, in general, to four different force contributions in the equation of motion, that can be categorized as:
Dissipative actions:
(i) viscous damping:

$$j\omega \mathrm{Re}\left(\mathbf{G}_0'\right)$$

(ii) hysteretic damping:

$$j\left[\mathrm{Im}\left(\mathbf{G}_0-\sigma_0\mathbf{G}_0'\right)-\omega_0\mathrm{Re}\left(\mathbf{G}_0'\right)\right]$$

Conservative actions:
(iii) elastic:

$$\mathrm{Re}\left(\mathbf{G}_0-\sigma_0\mathbf{G}_0'\right)+\omega_0\mathrm{Im}\left(\mathbf{G}_0'\right)$$

(iv) gyroscopic effect:

$$-\omega\mathrm{Im}\left(\mathbf{G}_0'\right)$$

As it is clear two of them are dissipative, the other two conservative.

The last term is here called gyroscopic because this is a velocity dependent conservative contribution, in that the net energy added to the system along one period is zero, and it adds and subtracts the same amount of energy in half of the period, that is the characteristic of gyroscopic effects. For example, in the Euler equations for the rigid body motion, the components of the angular velocity are conservative terms coupling the differential equations in terms of angular velocity components derivatives.

The hysteretic damping is indeed a velocity independent effect frequently met (at least in the frequency domain) in structural dynamics, to represent inherent dissipation of the material.

To put the equation for x into a more concise form, let:

$$c_{eq}=-\tilde{\mathbf{k}}_0^T\mathrm{Re}(\mathbf{G}_0')\tilde{\mathbf{k}}_0 \qquad \eta_{eq}=-\frac{\tilde{\mathbf{k}}_0^T\left[\mathrm{Im}(\mathbf{G}_0-\sigma_0\mathbf{G}_0')-\omega_0\mathrm{Re}(\mathbf{G}_0')\right]}{k_{00}}$$

$$k_{eq}=-\tilde{\mathbf{k}}_0^T\mathrm{Re}(\mathbf{G}_0'-\sigma_0\mathbf{G}_0'+\omega_0\mathrm{Im}(\mathbf{G}_0'))\tilde{\mathbf{k}}_0 \qquad g_{eq}=-\tilde{\mathbf{k}}_0^T\mathrm{Im}(\mathbf{G}_0')\tilde{\mathbf{k}}_0$$

$$F_+(\cdot)=\int_0^\infty(\cdot)e^{j\omega t}dt \qquad \text{one sided Fourier tr.}$$

Considering again the Laplace domain equation for x:

$$m_0s^2X+k_{00}X-\tilde{\mathbf{k}}_0^T\mathbf{G}\tilde{\mathbf{k}}_0X=F_0-\tilde{\mathbf{k}}_0^T\mathbf{G}\widetilde{\mathbf{F}}$$

Assume for the sake of simplicity the forces on the hidden variables zero (zero noise); with the previous determined expression for $\mathbf{G}$, and with $s=j\omega$ it becomes:

$$\left[-m_0\omega^2+(k_{00}+k_{eq})+j\omega c_{eq}+jk_{00}\eta_{eq}+\omega g_{eq}\right]X(j\omega)=F_0$$

It is useful to treat the gyroscopic term in a slightly different manner, as:

$$\left[-m_0\omega^2 + (k_{00} + k_{eq}) + j\omega c_{eq} + jk_{00}\eta_{eq} + j(j\omega g_{eq})\right] X(j\omega) = F_0$$

One can notice that, under the assumptions made, the interaction between x and the others degrees of freedom generates: (i) **pseudo-damping viscous and hysteretic effects**, that amount to an energy release (or energy absorption, depending on the algebraic sign) from the coordinate x to the hidden variables, (ii) an **additional restoring elastic force**, (iii) a **gyroscopic term.**

Let now go back to the time domain representation of this equation. Introduce the relationships:

$$x(t) = \frac{1}{2\pi}\int_{-\infty}^{+\infty} X(j\omega)\, e^{j\omega t} d\omega = F^{-1}(X)$$

$$\tilde{x}(t) = \frac{1}{2\pi}\int_{-\infty}^{+\infty} -j\,\mathrm{sign}(\omega)\; X(j\,\omega)\, e^{j\omega t} d\omega$$

$$x_+(t) = \frac{1}{2\pi}\int_{0}^{+\infty} X(j\omega)\, e^{j\,\omega t} d\omega = F_+^{-1}(X) = \frac{1}{2}\left[x(t) + j\tilde{x}(t)\right]$$

where the last two equations stands for the so called **analytic signal** and **Hilbert transform** of x, respectively. If the one sided inverse Fourier transform is applied to the equation of motion, one has:

$$m_0\ddot{x}_+ + (k_{00} + k_{eq})\, x_+ + c_{eq}\dot{x}_+ + jk_{00}\eta_{eq}x_+ - jg_{eq}\dot{x}_+ = f_{0+}$$

Taking for this complex valued equation only the real part, one finally obtains:

$$\boxed{m_0\ddot{x} + (k_{00} + k_{eq})\; x + c_{eq}\dot{x} + k_{00}\eta_{eq}\tilde{x} + g_{eq}\dot{\tilde{x}} = f_0} \tag{5}$$

Equation (5) is one of the central result of the present theory.

This shows that, under some hypotheses related to the remainder term, a simple and strict form for the interaction between the variable x and the hidden part of the system is born. A nice distinction can be operated between reactive forces, elastic and gyroscopic, and active forces, viscous and hysteretic. Note that among the terms depending on x, the one involving the derivative is related to dissipation, while the opposite is true for those depending on the Hilbert transform.

This result is derived under the hypothesis of keeping only the first order term in the Taylor expansion for **G**. This can be a reasonable approximation when the system response presents a rather narrowband spectrum concentrated around a certain reference frequency ω_0.

Under a substantially equivalent hypothesis, a different approach can be developed that, although less intuitive, leads to a better development of our theory.

Pseudo-dissipative effects: integral-Padè expansion of the kernel G It is interesting to start a different approach to the analysis of **G**. The one presented in the previous section, based on the Taylor expansion, can nicely represent the interaction terms in a suggestive physical manner, but the analysis of the remainder $\boldsymbol{R}(s)$ in that form, does not allow a straightforward prediction of the effects produced when neglecting it, especially in the time domain. In other words we can expect that x is not controlled strictly by equation (1), and it can depart from it as an effect of the presence of $\boldsymbol{R}(s)$.

The present approach, based on a completely different idea, produces similar but simpler results in terms of the physical picture of the interaction term, but, additionally, it provides a new point of view about the pseudo-dissipative effect based on the analysis of the remainder term.

More precisely, we can put a list of key questions about the remainder:

Q1. What is the effect of the remainder on the motion of x?

Q2. Can the time domain counterpart of the remainder disclose some elements answering Q1?

Q3. In which manner the form of the remainder depends on the system physical properties?

Q4. The presence of the pseudo-dissipative terms, i.e. the energy released from x to the hidden variables, is a permanent property of the interaction or the existence of a not negligible remainder confines this effect in some restricted time window?

Q5. There exist physical systems, purposely selected or designed, for which the effect of the remainder can be controlled, for example making it as small as possible?

Q6. There is any chance to use answer to Q5 to design special engineering devices?

The compiled list opens probably the core question of our problem. In fact the results we found about pseudo-dissipation have their fundament only when the effect of the remainder term is, in some sense known as unimportant. Moreover, there are two additional points that can produce a breakthrough in our theory: (i) the possible existence of particular systems exhibiting **permanent pseudo-dissipative effects,** as those described for equation (1), pushes our analysis toward a new border: that of the irreversible process in Hamiltonian systems, a subject of great charm as it has been outlined in the initial survey; at the same time, and not surprisingly, this circumstance drives us towards engineering design of systems able to pilot their energy to enhance their dynamic performances.

For these reasons we go more in depth in analyzing the nature of the remainder term. Starting from the Fourier domain equation for x:

$$-m_0\omega^2 X + k_{00}X - \tilde{\mathbf{k}}_0^T \mathbf{G}(\omega)\, \tilde{\mathbf{k}}_0 X = F_0$$

and using the Padè expansion for $\mathbf{G}$ (a real valued function of ω):

$$G_{ij} = \sum_{n=1}^{2N} \frac{\vartheta_n^{(i,j)}}{j\omega - j\omega_n}, \qquad \mathbf{G} = \sum_{n=1}^{2N} \frac{\mathbf{\Theta}_n}{j\omega - j\omega_n}$$

$$\tilde{\mathbf{k}}_0^T \mathbf{G} \tilde{\mathbf{k}}_0 = \sum_{n=1}^{2N} \frac{\tilde{\mathbf{k}}_0^T \mathbf{\Theta}_n \tilde{\mathbf{k}}_0}{j\omega - j\omega_n} = \sum_{n=1}^{N} \frac{\alpha_n}{j\omega - j\omega_n} + \frac{\alpha_n^*}{j\omega + j\omega_n}$$

$$\alpha_n = \lim_{\omega \to \omega_n} j(\omega - \omega_n)\, \tilde{\mathbf{k}}_0^T \mathbf{G}(\omega)\, \tilde{\mathbf{k}}_0 = j\beta_n$$

where the poles of the transfer function occur in imaginary conjugate pairs because of the absence of any real dissipative effect in the (Hamiltonian) system.

The equation of motion becomes:

$$\left(-m_0\omega^2 + k_{00}\right) X(\omega) - X(\omega) \sum_{n=1}^{N} \frac{\alpha_n}{j\omega - j\omega_n} + \frac{\alpha_n^*}{j\omega + j\omega_n} = F_0$$

Taking the inverse Fourier transform and introducing the Heavisdie function $H(t)$:

$$m_0\ddot{x}(t) + k_{00}x(t) \; - x(t) * \left[H(t) \sum_{n=1}^{N} \alpha_n e^{j\omega_n t} + \alpha_n^* e^{-j\omega_n t}\right] = F_0$$

$$m_0\ddot{x}(t) + k_{00}x(t) - 2x(t) * \left[H(t) \sum_{n=1}^{N} \mathrm{Re}\left\{\alpha_n e^{j\omega_n t} + \alpha_n^* e^{-j\omega_n t}\right\}\right] = F_0$$

$$m_0\ddot{x}(t) + k_{00}x(t) + 2x(t) * \left[H(t) \sum_{n=1}^{N} \beta_n \sin \omega_n t\right] = F_0$$

Let:

$$S(t) = \frac{1}{N} \sum_{n=1}^{N} \beta_n \sin \omega_n t$$

With this positions, the equation for x becomes:

$$m_0\ddot{x}(t) + k_{00}x(t) + 2Nx * [HS] = F_0$$

$$m_0\ddot{x}(t) + k_{00}x(t) + 2N\int_0^{\infty} x(\tau)\, H(t-\tau)\, S(t-\tau)\, d\tau = F_0$$

The nature of the interaction between x and the hidden variables expressed through $x * [HS]$, is difficult to handle because of the discrete summation appearing in S. An interesting chance comes indeed from the substitution of S by a suitable integral, that makes the physical interpretation of the interaction term easier. In fact, **the core of our analysis assumes the system's eigenfrequencies are "dense enough" over the frequency axis**; this means that all the natural frequencies belong to the same narrow frequency bandwidth. Under this condition it is reasonable to replace the summation over n of the Padè expansion by an integral $I(t)$:

$$I(t) = \int_0^1 \beta(\xi)\sin\Omega(\xi)\, t\, d\xi \qquad S(t) = \sum_{n=1}^{N} \beta_n \sin\omega_n t\Delta\xi$$

$$\Delta\xi = \frac{1}{N} \qquad I(t) \approx S(t)$$

where ξ is a dummy variable that belongs to the interval [0,1]. Introducing the remainder *r(t)*:

$$I(t) = S(t) + r(t)$$

the equation of motion takes the form:

$$m_0\ddot{x}(t) + k_{00}x(t) + 2Nx * [HI] + x * [Hr] = F_0$$

$$m_0\ddot{x}(t) + k_{00}x(t) + 2N\int_0^{\infty} x(\tau)\, H(t-\tau)\, I(t-\tau)\, d\tau +$$

$$+ 2N\int_0^{\infty} x(\tau)\, H(t-\tau)\, r(t-\tau)\, d\tau = F_0$$

meaning the equation is rewritten as:

$$\boxed{m_0\ddot{x}(t) + k_{00}x(t) + 2Nx * [HI] + \varepsilon(t) = F_0} \qquad (6)$$

Equation (2) is the new starting station for our analysis. A physical interpretation of the term $[HI]$ should be provided, together with an estimate for ε that is the force generated by the remainder term.

From the mathematical and physical point of view, the use of $[HI]$ instead of $[HS]$ brings a great benefit to our analysis. In fact its Fourier transform highlights clearly the effect of this interaction term; moreover an estimate of some properties of ε can follow.

Thus, move again to the frequency domain:

$$F\{HI\} = F\{I\} * F\{H\}$$

The first factor in the convolution is:

$$F\{I\} = \int_0^1 \beta(\xi) \int_{-\infty}^{+\infty} e^{-j\omega t} \sin\Omega(\xi)t \, dt \, d\xi =$$

$$= -j\frac{\pi}{2} \int_0^1 \beta(\xi) \left[\delta\left(\omega + \Omega(\xi)\right) + \delta\left(\omega - \Omega(\xi)\right)\right] d\xi$$

Let $d\Omega = \Omega'(\xi)\, d\xi$:

$$F\{I\} = -j\frac{\pi}{2} \int_0^1 \frac{\beta(\xi)}{\Omega'} \left[\delta\left(\omega + \Omega\right) + \delta\left(\omega - \Omega\right)\right] d\Omega$$

Writing the distribution of natural frequency $\Omega(\xi)$ as the solution of the differential equation $\Omega' = f(\Omega)$, for an arbitrary function f, the previous integral produces the expression:

$$F\{I\} = -j\omega \left[\frac{\pi}{2} \frac{\beta(\omega)}{\omega f(\omega)}\right] \qquad \text{for } \omega \in [\Omega(0); \Omega(1)]$$

$$F\{I\} = 0 \qquad \text{elsewhere}$$

The second factor in the convolution is:

$$F\{H\} = \frac{1}{2}\delta(\omega) + \frac{1}{j\omega}$$

Therefore:

$$\boxed{F\{HI\} = -j\omega \left[\frac{\pi}{4} \frac{\beta(\omega)}{\omega\, f(\omega)}\right] + \int_{-\infty}^{+\infty} \frac{\pi}{2} \frac{\beta(\zeta)}{f(\zeta)\,(\zeta - \omega)} \, d\zeta}$$

This expression provides the frequency domain counterpart of the term $x * [HI]$. This result shows a double contribution in the interaction term between x and the hidden variables: one is imaginary, one is real.

The imaginary dissipative term is a frequency dependent damping controlled by an equivalent viscous coefficient $c_{eq}(\omega) = \frac{\pi}{4}\frac{\beta(\omega)}{\omega f(\omega)}$. The function f **depends on the natural frequency distribution within the hidden part of the system; more precisely the inverse of f ($1/f = n/N$) is proportional to the density n of the natural frequencies over the frequency axis**, so that the more the natural frequencies are dense, the more the damping effect is large.

The coefficient β is indeed:

$$\beta(\Omega) = \lim_{\omega \to \Omega} (\omega - \Omega)\, \tilde{\mathbf{k}}_0^T \mathbf{G}(\omega)\, \tilde{\mathbf{k}}_0$$

and corresponds to the reactive part of the interaction term.

For the second contribution, introduced the function $\Gamma = \pi\beta/2f$ one has:

$$\int_{-\infty}^{+\infty} \frac{\pi}{2}\frac{\beta(\zeta)}{f(\zeta)(\zeta - \omega)} d\zeta = \Gamma(\omega) * \frac{1}{\omega} = \pi\tilde{\Gamma}(\omega)$$

where the tilde stands, as in the previous section, for Hilbert transform.

Without any further simplification, the equation of motion exhibits an interaction force that is directly the inverse transform of the previous determined term $[HI]$. However, accordingly with the analysis developed in the previous section, and on the same line that led to replace S by I under the hypothesis of eigenvalues "dense enough", one is pushed to **estimate the determined frequency dependent terms at a given frequency ω_0**, that is the characteristic frequency around which the natural frequencies of the hidden system are located. Therefore, the imaginary part can be approximated by a viscous force represented by an equivalent viscous damper as:

$$c_{eq}(\omega) = \left[\frac{\pi}{4}\frac{\beta(\omega)}{\omega f(\omega)}\right] \approx \left[\frac{\pi}{4}\frac{\beta(\omega_0)}{\omega_0 f(\omega_0)}\right] = c_{eq}$$

The real part can be indeed roughly estimated by identifying *1/f* as a delta function centered at ω_0 so that:

$$\int_{-\infty}^{+\infty} \frac{\pi}{2}\frac{\beta(\zeta)}{f(\zeta)\,(\zeta - \omega)}\, d\zeta \approx \frac{\pi}{2}\frac{\beta(\omega_0)}{\omega_0} = k_{eq}$$

With these expressions, the time domain equation for x becomes:

$$\boxed{m_0\ddot{x} + (k_{00} + k_{eq})\, x + c_{eq}\dot{x} + \varepsilon(t) = f_0}$$

For the reasons arising clearly from the previous analysis, we name again the motion of x determined by the previous equation, and for $\varepsilon = 0$, **motion of x about ω_0**. A comparison with the result obtained in the previous section shows strong analogies, even in the form of the equivalent damping and equivalent stiffness, with the equation obtained using the Taylor expansion of $\mathbf{G}$.

As a final point let us summarize the obtained result under a **Hamiltonian point of view.**

If a Hamiltonian system S has stiffness and mass matrices:

$$\mathbf{K} = \begin{bmatrix} k_{00} & \tilde{\mathbf{k}}_0^T \\ \tilde{\mathbf{k}}_0 & \widetilde{\mathbf{K}} \end{bmatrix}, \qquad \mathbf{M} = \begin{bmatrix} m_0 & 0 \\ 0 & \widetilde{\mathbf{M}} \end{bmatrix}$$

and lagrangian function:

$$\begin{aligned} L(\mathbf{x}, \dot{\mathbf{x}}) &= L_x(x, \dot{x}) + L_{mix}(x, \tilde{\mathbf{x}}) + L_{hid}\left(\tilde{\mathbf{x}}, \dot{\tilde{\mathbf{x}}}\right) \\ L_x(x, \dot{x}) &= \frac{1}{2} m_0 \dot{x}^2 + \frac{1}{2} k_{00} x^2 - f_0 x \\ L_{mix}(x, \tilde{\mathbf{x}}) &= \frac{1}{2} x \tilde{\mathbf{k}}_0^T \tilde{\mathbf{x}} + \frac{1}{2} X \tilde{\mathbf{k}}_0^T \tilde{\mathbf{x}} \\ L_{hid}\left(\tilde{\mathbf{x}}, \dot{\tilde{\mathbf{x}}}\right) &= \frac{1}{2} \dot{\tilde{\mathbf{x}}}^T \widetilde{\mathbf{M}} \dot{\tilde{\mathbf{x}}} + \frac{1}{2} \tilde{\mathbf{x}}^T \widetilde{\mathbf{K}} \tilde{\mathbf{x}} - \tilde{\mathbf{f}}^T \tilde{\mathbf{x}} \end{aligned}$$

then, the motion of x about ω_0 (in the sense previously specified), when the other variables are unmonitored, is controlled by the new lagrangian

$$L'_x(x, \dot{x}) = L_x(x, \dot{x}) + \frac{\pi}{4} \frac{\beta(\omega_0)}{\omega_0} x^2$$

and by the Rayleigh dissipation function:

$$D(\dot{x}) = \frac{\pi}{8} \frac{\beta(\omega_0)}{\omega_0 f(\omega_0)} \dot{x}^2$$

where:

$$\beta(\omega_0) = \lim_{\omega \to \omega_0} (\omega - \omega_0) \tilde{\mathbf{k}}_0^T \mathbf{G}(\omega) \tilde{\mathbf{k}}_0$$

$$f(\omega_0) = \left. \frac{\Delta\omega}{\Delta N} \right|_{\omega_0} = \frac{N}{n(\omega_0)}$$

$\Delta\omega$ in a bandwidth around ω_0 and ΔN the number of eigenvalues of $\mathbf{G}$ that falls within $\Delta\omega$, or if $n(\omega)$ is the so called modal density, $\boldsymbol{n(\omega_0)}$ is the maximum of the modal density of $\mathbf{G}$.

This is the fundamental result of the paper.

However, as mentioned at the beginning of this section, the present approach gives the additional chance of estimating interesting properties of the remainder effect ε. The next section attacks just this problem.

2.4 Remainder term: return time and energy transfer rate

We provide in this section a nice property for ε, that is:

> **for $t<t^*$ the contribution of ε is negligible; t^* , named return time, depends on the form of G.**

a. Lemma 1

As a first lemma, a simple asymptotic property for I is enlightened:

$$\lim_{t\to\infty} I(t) = 0$$

This property follows from an asymptotic expansion (integration by parts) for I. Precisely (Watson's Lemma):

If $\Omega(\xi) \neq 0,\ \forall\xi \in [0,1]$:

$$I(t) = \int_0^1 \beta(\xi) \sin\Omega(\xi)\, t d\xi = \frac{1}{t}\left[\frac{\beta(0)}{\Omega'(0)}\cos\Omega(0)\, t - \frac{\beta(1)}{\Omega'(1)}\cos\Omega(1)\, t\right] + o\left(\frac{1}{t}\right)$$

If $\exists\xi_0 \in [0,1] : \Omega'(\xi_0) = 0$:

$$I(t) = \int_0^1 \beta(\xi)\sin\Omega(\xi)\, t\, d\xi = \sqrt{\frac{2\pi}{t\,|\Omega''(\xi_0)|}}\,\beta(\xi_0)\cos\Omega(\xi_0)\, t + o\left(\frac{1}{\sqrt{t}}\right)$$

This property implies that for t large enough, $S(t)$ is well represented by the remainder $r(t)$:

$$S(t) \approx r(t)\,, \qquad \text{for large } t$$

that provides the **behaviour of S at late time**.

b. Lemma 2

Classical results in numerical integration theory, shows that:

$$I(t) = S(t) + r(t)$$

$$r(t) = \frac{\Delta\xi}{2}\frac{d}{d\xi}\Big[\beta(\xi)\sin\Omega(\xi)\, t\Big]_{\xi=\bar{\xi}(t)}$$

And explicitly:

$$r(t) = \frac{\Delta\xi}{2}\Big[\beta'(\xi)\sin\Omega(\xi)\,t + \beta(\xi)\Omega'(\xi)t\cos\Omega(\xi)t\Big]_{\xi=\bar{\xi}(t)}$$
$$= \frac{\Delta\xi}{2}A(\xi)\sin\Big[\Omega(\xi)\,t + \varphi(\xi)\Big]_{\xi=\bar{\xi}(t)}$$
$$A(\xi) = \sqrt{\beta'^2 + \beta^2\Omega'^2 t^2}, \qquad \varphi(\xi) = \arctan\left(\frac{\beta\Omega'}{\beta'}t\right)$$

Thus, the late time behaviour for r may be represented by:

$$\lim_{t\to\infty} r(t) = \frac{\Delta\xi}{2}\beta(\xi)\,\Omega'(\xi)\,t\sin\Big[\Omega(\xi)\,t + \varphi(\xi)\Big]_{\xi=\bar{\xi}(t)}$$

while it appears that:

$$\lim_{t\to 0} r(t) = 0$$

This permits to conclude that the **behaviour of S** at **early time** is well represented by I:

$$S(t) \approx \int_0^1 \beta(\xi)\sin\Omega(\xi)\,t\,d\xi, \qquad \text{at early time}$$

c. Estimate of the return time

The results provided by the two previous lemmas can be used to give an estimate for the return time t^*. The general picture for S, I and r is summarized: at early time, S and I are close because r is small, and the perturbation ε is small too; as time increases, at late time, I departs from S: in fact I vanishes and it is not for S, that is indeed close to r, and the perturbation ε becomes large.

We are interested in providing the order of magnitude of the time needed for S to depart from I, that provides also an estimate of the time scale over which the perturbation ε holds small. The strategy is simple: (i) estimate the peak value S_{peak} for both I and S reached at early time, (ii) then it is expected that I decreases (because of lemma 1), and that S initially holds close to I because r is small, (iii) at some late time t^*, it is indeed expected that S increases departing from I reaching an amplitude close to its initial peak.

Let estimate S_{peak}:

$$S(t) \approx \int_0^1 \beta(\xi) \sin\Omega(\xi)\, t\, d\xi \Longrightarrow$$

$$\frac{dS}{dt} \approx \int_0^1 \beta(\xi)\,\Omega(\xi) \cos\Omega(\xi)\, t\, d\xi \approx \int_0^1 \beta(\xi)\ \Omega(\xi) \left[1 - \frac{1}{2}\Omega^2(\xi)\, t^2\right] d\xi$$

$$\frac{dS}{dt}(t_{peak}) = 0 \Longrightarrow \int_0^1 \beta(\xi)\,\Omega(\xi)\, d\xi - t_{peak}^2 \frac{1}{2} \int_0^1 \beta(\xi)\,\Omega^3(\xi)\, d\xi = 0$$

$$t_{peak} \approx \sqrt{\frac{2\int_0^1 \beta(\xi)\,\Omega(\xi)\, d\xi}{\int_0^1 \beta(\xi)\,\Omega^3(\xi)\, d\xi}}, \qquad S_{peak} \approx \frac{\sqrt{2}\left[\int_0^1 \beta(\xi)\,\Omega(\xi)\, d\xi\right]^{3/2}}{\left[\int_0^1 \beta(\xi)\,\Omega^3(\xi)\, d\xi\right]^{1/2}}$$

Let finally estimate the time t^* for which S regain the value S_{peak}; this happens at late time, for which S can be approximated by r; evaluating all the frequency dependent functions at ω_0:

$$S \approx \frac{\Delta\xi}{2}\beta(\xi)\,\Omega'(\xi)\, t \sin\left[\Omega(\xi)\, t + \varphi(\xi)\right]_{\xi=\bar{\xi}(t)} \approx \frac{\Delta\xi}{2}\beta(\omega_0)\, f(\omega_0)\, t$$

$$t^* \approx \frac{2\sqrt{2}N}{f(\omega_0)} = 2\sqrt{2}n(\omega_0)$$

This is also a central result of the paper.

Note that the time t^* has not surprisingly this expression: in fact, considering the special case in which S is a periodic function (a Fourier series), t^* would be the period of S:

$$S = \sum_{k=1}^{N} \beta_k \sin\omega_k t = \sum_{k=1}^{N} \beta_k \sin\frac{2\pi k t}{t^*}$$

$$\omega_k = \frac{2\pi k}{t^*} \Longrightarrow \Delta\omega = \frac{2\pi}{t^*} \Longrightarrow n = \frac{1}{\Delta\omega} \Longrightarrow t^* = 2\pi n$$

Thus, the obtained expression for t^* provides essentially a **proportionality between the return time and the modal density n at ω_0.**

This makes clear how the interaction term in the found simplified form of an equivalent damping and stiffness is valid only when ε is small, that is up to time t^*.

A final interesting result correlates the time t^* to the equivalent damping and stiffness. In particular it is easy to find:

$$\boxed{t^* \approx \frac{4\sqrt{2}\,N}{k_{eq}}\,c_{eq}} \tag{7}$$

where an intriguing proportionality is found between the return time and the equivalent damping. This result states: **the faster the energy transfer from x to the hidden variables, the longer the energy storage within them.**

This result makes meaningful the search of special configurations of the hidden system that **optimize the energy suction and storage from x**: in fact if we are able to make the energy transfer as fast as possible, it comes together with the longest storage time for the moved energy.

The question is analysed in the next section.

2.5 A variational theorem for the minimum remainder term

The present section describes the conditions under which the series $S(t)$ approaches the integral $I(t)$ and shows that a criterion of minimum distance $D(t)$

$$D^2(t) = \int_C [S - I]^2\, W\, dC = \overline{(S-I)^2} = \overline{r^2}$$

can be satisfied with a suitable weighting function W in a prescribed domain C within a certain space Σ.

A method to find such a weighting function that depends on I, which depends on $\omega(\xi)$ is described by the use of a variational approach.

The results show there exists a class of functions $\omega(\xi)$ that minimizes the distance between $S(t)$ and $I(t)$. In such cases, the series $S(t)$ tends to match as close as possible the trend of the integral $I(t)$, producing closely the apparent damping effects previously discussed, and a nearly irreversible energy transfer processes in conservative linear systems. The next section reviews the definitions and properties necessary to form the basis for the ensuing theoretical development.

The following approach is based on an average defined in a multi-dimensional space C with the use of a non trivial weighting function W. The reason for this is that, despite the difficult mathematical form, as it appears at a first glance, it can indeed lead, surprisingly, to a closed form solution of our minimization problem.

In order to correctly formulate our minimization problem, and some related constraints, let introduce some preliminary considerations.

a. Definition of average
For a set of functions $s_i = \beta_i \sin \omega_i t$, $i = 1, 2, \ldots, N$ at any time t, $\mathbf{s} = [s_1, s_2, \ldots, s_N]^T$ defines a point (or a vector) in the space Σ of harmonics; $\mathbf{s}$ exists within the hypercube $C \equiv \{E \times E \times \cdots \times E\}$, with $E \equiv [-\beta_{\max}, \beta_{\max}]$, $\beta_{\max} = \max\{\beta_1, \beta_2, \ldots, \beta_N\}$.

Let $f(\mathbf{s})$ be an arbitrary function defined over $C \subset \Sigma$ with the vector $\mathbf{s} \in \Sigma$. In general, the average value $\bar{f}(t)$ of $f(\mathbf{s})$ over C can be expressed using a weighting function $P(\mathbf{s}, I)$ as:

$$\bar{f}(t) = \int_C f(\mathbf{s}) P(\mathbf{s}, I)\, dC, \qquad dC = \prod_{k=1}^{N} ds_k \tag{8}$$

where the weighting function is selected to depend on $\mathbf{s}$ and I as described below.

As a consequence, scalar product of the two functions $f(\mathbf{s})$ and $g(\mathbf{s})$ in C follows as

$$\overline{f \cdot g} = \int_C f(\mathbf{s}) g(\mathbf{s}) P(\mathbf{s}, I)\, dC \tag{9}$$

Similarly, the distance $D(t)$ between $f(\mathbf{s})$ and $g(\mathbf{s})$ follows form:

$$D^2(t) = \overline{(f-g)\cdot(f-g)} = \overline{(f-g)^2} = \int_C [f(\mathbf{s}) - g(\mathbf{s})]^2 P(\mathbf{s}, I)\, dC \tag{10}$$

b. Weighting Function
The weighting function $P(\mathbf{s}, I)$ in Eq. (10) is selected to have the form:

$$P(\mathbf{s}, I) = \prod_{k=1}^{N} p(s_k, I) \tag{11}$$

where $p(s_k, I)$ is an arbitrary function that satisfies the conditions:

$$\int_{-G_{\max}}^{G_{\max}} \sigma p(\sigma, I)\, d\sigma = I(t) \tag{12}$$

$$\int_{-G_{\max}}^{G_{\max}} p(\sigma, I)\, d\sigma = 1 \tag{13}$$

Equation (12) offers a comparison with the integral in Eq.(2) for $\sigma = \beta(\omega) \sin \omega t$ and through a change of integration variables in Eq. (12) (first from $d\sigma$ to $d\omega$, then to $d\xi$):

$$I(t) = \int_{-G_{\max}}^{G_{\max}} \sigma p(\sigma, I)\, d\sigma = \int_{\omega_{\min}}^{\omega_{\max}} \sigma(\omega) p[\sigma(\omega), I] \frac{d\sigma}{d\omega} d\omega$$

$$I(t) = \int_{0}^{\xi_{\max}} \sigma(\omega(\xi)] p\{\sigma[\omega(\xi)], I\} \frac{d\sigma}{d\omega} \frac{d\omega}{d\xi} d\xi \tag{14}$$

provided that in the interval $[\omega_{\min}, \omega_{\max}]$, $\sigma = \beta(\omega) \sin \omega t$ is single-valued and $\sigma \in [-\beta_{\max}, \beta_{\max}]$. A comparison of equations (2) and (14) implies that the function $p(\sigma, I)$ must satisfy the following compatibility condition :

$$p(\sigma, I) \frac{d\sigma}{d\omega} \frac{d\omega}{d\xi} = 1 \tag{15}$$

The condition expressed by Equation (15) also implies a dependence between $p\ (\sigma,\ I)$ and $\omega\ (\xi)$ for $\sigma = \beta(\omega) \sin \omega t$.

It follows that substituting equation (15) in equation (13) yields the upper bound of ξ as :

$$\int_{-G_{\max}}^{G_{\max}} p(\sigma, I) d\sigma = \int_{0}^{\xi_{\max}} p\left\{\sigma\left[\omega\left(\xi\right)\right], I\right\} \frac{d\sigma}{d\omega} \frac{d\omega}{d\xi}\, d\xi = \xi_{\max}$$

yielding $\xi_{\max} = 1$, which leads to the conclusion about the bounds of ξ as: $\xi \in [0, 1]$.

c. Average of S

The average of the function $S(t)$ in Eq. (1)can be expressed by substituting for $\bar{s}_i$ the averaging property —expressed for $\bar{f}(t)$— in Eq. (8):

$$\overline{S}(t) = \frac{1}{N} \sum_{i=1}^{N} \bar{s}_i = \frac{1}{N} \sum_{i=1}^{N} \int_C s_i P(\mathbf{s}, I)\, dC$$

Further substitution for $P(\mathbf{s}, I)$ from Eq. (11) and for dC from Eq. (8)yields:

$$\overline{S}(t) = \frac{1}{N} \sum_{i=1}^{N} \int_{-G_{\max s}}^{G_{\max}} s_i p(s_i, I)\, ds_i \int_{C^{(N-1)}} \prod_{k \neq i}^{N} p(s_k, I)\, ds_k$$

The condition (13) leads the mutiplication series in the second integral to produce identity and applying condition (12) to the first integral shows that:

$$\overline{S}(t) = I(t) \tag{16}$$

By invoking the definition of average value in (8) leads to the fundamental result:

$$\overline{S}\,(t) = \int_C SP(\mathbf{s}, I)\, dC = I(t) \tag{17}$$

provided that equation (15) is satisfied.

d. Constraints

The weigthing function, as it has been defined in subsection **b**, must satisfy the constraints (12) and (13). Equation (12) is automatically satisfied provided that equation (15) holds, as it will be used ahead. It is indeed suitable to introduce explicitly the constraint (13) in a form that makes easier to approach the following variational problem.

Substituting for $P\,(\mathbf{s}, I)$ from Eq. (11) and for dC from Eq. (8) and invoking the condition in Eq. (13) it can be show that:

$$\int_C P(\mathbf{s}, I)\, dC = 1 \tag{18}$$

The first derivative of Eq. (18) with respect to I can be expressed as:

$$\int_C \frac{\partial}{\partial I} P(\mathbf{s}, I)\, dC = 0 \quad \Longrightarrow \quad \int_C \left[\frac{1}{P(\mathbf{s}I)} \frac{\partial}{\partial I} P(\mathbf{s}, I)\right] P(\mathbf{s}, I)\, dC = 0$$

Thus:

$$\int_C \frac{\partial}{\partial I} \left[\log P(\mathbf{s}, I)\right] P(\mathbf{s}, I)\, dC = 0$$

which is equivalent to stating:

$$\overline{\frac{\partial}{\partial I} \log P} = 0 \tag{19}$$

that is a constraint derived from (12) or (18). Following the same approach, the first derivative of Eq. (17) with respect to I, produces a similar expression:

$$\int_C S \frac{\partial}{\partial I} P(\mathbf{s}, I)\, dC = 1 \quad \Longrightarrow \quad \int_C S \left[\frac{1}{P(\mathbf{s}, I)} \frac{\partial}{\partial I} P(\mathbf{s}, I)\right] P(\mathbf{s}, I)\, dC = 1$$

$$\overline{S \frac{\partial}{\partial I} \log P} = 1 \tag{20}$$

that is a consequence of (17). In order to find the distance between $S(t)$ and $I(t)$, an equivalent expression for Equation (20) is developed for $I(t)$ by multiplying Eq. (19) by the factor I, which is independent of the integration variable $\mathbf{s}$:

$$\overline{I\frac{\partial}{\partial I}\log P} = 0 \tag{21}$$

Finally the combination of (20) and (21) produces:

$$\boxed{\overline{(S-I)\frac{\partial}{\partial I}\log P} = 1}$$

e. The Euler-Lagrange equation minimizing $\mathbf{D^2} = (\mathbf{S}-\mathbf{I})^2$

The problem outlined at the beginning of this section 2.5 is solved indeed by a variational approach finding the minimum of the modified functional:

$$\delta\widetilde{D}^2 = 0\,, \qquad \tilde{D}^2 = \overline{(S-I)^2} + \lambda\left[\overline{(S-I)\left(\frac{\partial}{\partial I}\log P\right)}\right]$$

where λ is the Lagrange's multiplier to include the constraint derived in subsection **d**.

The variation of $\tilde{D}^2$ to be considered is with respect to I, that is, equivalently, with respect to $\omega\left(\xi\right)$ from which I depends, but it is easier not to consider explicitly. Therefore:

$$\boxed{-2\left(S-I\right)+\lambda\left[-\frac{\partial}{\partial I}\log P+\left(S-I\right)\frac{\partial}{\partial I}\left(\frac{\partial}{\partial I}\log P\right)\right]=0}$$

where the last is the Euler-Lagrange equation.

This differential equation, as it can be easily verified, admits a solution of the form:

$$\frac{\partial}{\partial I}\log P = -\frac{(S-I)}{\lambda} \tag{22}$$

Condition (22) represents a differential equation in terms of P and its solution leads to a family of exponential functions $P\left(\mathbf{s},I\right) = \prod_{k=1}^{N} p\left(s_k,I\right)$. The solution to Eq. (22), originally obtained by Pitman and Koopman in the context of the theory of estimators, is given as:

$$p(\sigma, I) = \exp\left\{A(I)B(\sigma) + C(\sigma) + D(I)\right\}$$

where $A(I), B(I), C(\sigma)$, and $D(\sigma)$ are arbitrary functions of their respective arguments.

Gauss function also belongs to this family of solutions and provides an excellent example that can be easily verified by substituting into equation (22):

$$p(\sigma, I) = \frac{1}{r\sqrt{2\pi}} \exp\left\{-\frac{1}{2}\frac{(\sigma - I)^2}{r^2}\right\} \tag{23}$$

The solution $p(\sigma, I)$, with $\sigma\,(\beta, \omega)$, has a shape that depends on the function $I(t)$ and on the parameter r.

Together with equation (23), the compatibility equation (15) becomes a nonlinear differential equation and its solution provides the frequency distribution $\omega(\xi)$ that minimizes the distance D:

$$\frac{1}{r\sqrt{2\pi}} \exp\left\{-\frac{1}{2}\frac{(\sigma - I)^2}{r^2}\right\} \frac{d}{d\omega}\left[\beta(\omega)\sin\omega t\right] \frac{d\omega}{d\xi} = 1 \tag{24}$$

Equation (24) can be solved for $\omega(\xi)$ numerically; however, an alternative approach using density of harmonic functions, analogous to modal density in a dynamical system, produces a closed-form expression. Since $\frac{d\xi}{d\omega}N$ represents the harmonic density, $\delta(\omega)$, that counts the number of harmonics, or modes, contained within the frequency band $d\omega$, Eq. (24) directly leads to an expression for $\delta(\omega)$. With $d\xi = \frac{1}{N}\,dn$, N being the total number of harmonics, s_i, for $\xi \in [0,\, 1]$, and dn the number of harmonics for $\xi \in [\xi, \xi + d\xi]$, it follows that $N\frac{d\omega}{d\xi} \propto \frac{d\omega}{dn} = \frac{1}{\delta(\omega)}$. Substituting in equation (24) produces:

$$\boxed{\delta_{\mathrm{opt}}(t) = \frac{1}{N}\frac{1}{r\sqrt{2\pi}} \exp\left\{-\frac{1}{2}\frac{(\sigma - I)^2}{r^2}\right\} \frac{d}{d\omega}\left[\beta(\omega)\sin\omega t\right]} \tag{25}$$

Equations (24) and (25) show that time appears as a parameter in the frequency distribution that minimizes the difference between S and I. Oscillators with time-dependent parameters or, equivalently, with time-varying natural frequencies, imply presence of parametrically controlled resonators or nonlinear resonators. The problem under consideration in this paper addresses linear time-invariant dynamical systems and, thus, equations (24) and (25) cannot be satisfied at all times t. Thus, the approach taken here uses the frequency distribution $\omega(\xi)$ that results from equation (24) or (25) for a particular time t_0 to solve the compatibility equation

$$p(\sigma, I)\frac{d\sigma}{d\omega}\frac{d\omega}{d\xi}\bigg|_{t=t_0} = 1 \tag{26}$$

The choice for t_0, selection of the frequency interval $[\omega_{\min}, \omega_{\max}]$ within which $\omega(\xi)$ falls, which also depends the choice of t_0, and the implication of their selection are discussed with examples in the next sections.

Normally, the form of Eq. (23) satisfies Eqs. (12) and (13) automatically for an integration domain $[-\infty, +\infty]$; however, since the actual domain is finite $E \equiv [-G_{\max}, G_{\max}]$, r and $I(t_0)$ must satisfy the additional constraints:

$$r \ll G_{\max}, \quad I(t_0) \in E \tag{27}$$

These constraints guarantee that the function represented by equation (23) has its primary distribution within the interval E and therefore (approximately) satisfying equations (9) and (10).

2.6 Examples of application: set of parallel resonators

The examples given in this section illustrate application of the theory described above. Each case demonstrates how to minimize the difference between a sum of harmonic functions and the corresponding integral summation. The first example consists of a simple summation of sine functions for which $\beta(\omega) \equiv 1$. In the second example, $\beta(\omega) \equiv \omega$ represents the reaction force of a set of undamped resonators on a common rigid base. Subsections c and d examine more complex examples.

a. Simple sine series $\beta(\omega) \equiv 1$

Summation of a series of $N = 100$ sine functions with frequencies ω_i results from Eq. (1)by substituting for $\beta(\omega) \equiv 1$:

$$S(t) = \frac{1}{N}\sum_{i=1}^{N} \sin \omega_i t$$

and the corresponding integral from Eq. (2)becomes:

$$I(t) = \int_0^1 \sin \omega(\xi)\, t\, d\xi$$

For this case, the nonlinear differential equation (24) becomes:

$$\frac{1}{r\sqrt{2\pi}} \exp\left\{-\frac{1}{2}\frac{(\sigma - I_0)^2}{r^2}\right\} t_0 \cos \omega t_0 \frac{d\omega}{d\xi} = 1 \tag{28}$$

where σ and I_0 from Eqs. (26) and (2)become

$$\sigma = \sin \omega(\xi)\, t_0, \qquad I_0 = \int_0^{\xi_{\max}} \sin \omega(\xi) t_0\, d\xi$$

In this case, $E \equiv [-1, 1]$. Restricting the selection to monotonic frequency distributions $\omega(\xi)$, so that $\frac{d\omega}{d\xi} > 0$ for $\omega \in [\omega_{\min}, \omega_{\max}]$, implies that according to Eq. (28), $\frac{d\sigma}{d\omega} = \cos\omega(\xi)\, t_0$ must always be positive for$\omega \in [\omega_{\min}, \omega_{\max}]$. It follows that assigning, for example, $t_0 = \frac{\pi}{4}$, yields $\frac{d\omega}{d\xi} > 0$ for $\omega \in [0, 1]$. Under these conditions, the values for I_0 may be arbitrary, except that they must satisfy the inequalities in (27) and p satisfies the conditions (11) and (15).

Figures 1-5 illustrate the results obtained for $N = 100$, $t_0 = \frac{\pi}{4}$, $r = 0.05$, $\omega \in [0, 1]$ and with the choice of $I_0 = 0.2$ and $r \ll 1$, both of which satisfy equation (27). Figure 1 represents the frequency distribution $\omega(\xi)$ determined by a numerical integration of equation (28) from which the set ω_i $(i = 1, \ldots, 100)$ is determined by sampling 100 points equally spaced alongξ. Figure 2 represents the harmonic density and Figure 3 the time history of $S(t)$. As shown in Figures 4 and 5 for different time scales, in the time history of the series obtained using the theory developed here the strong periodicity disappears when compared with the corresponding series consisting of a linear frequency distribution with period $2\pi N/\omega_{\max}$for two different time scales.

Figures 6-9 show a case analogous to the previous one except for $r = 0.01$, representing a higher harmonic density around its peak resulting in a somewhat better performance.

The third example, shown in Figures 10-12, uses $t_0 = \frac{\pi}{8}$ and $N = 100$, $r = 0.05$, $\omega \in [0, 1]$.

The results of the examples above show that the frequency distributions satisfying the minimum distance bound requirements produce time histories that bring the summation $S(t)$ very close to $I(t)$, without recurrence or periodicity in its time history unlike, say, the case of a linear frequency distribution. The envelope of the summation in Eq. (1)decays significantly with respect to its early oscillations and without regaining its initial amplitude, following closely the same trend that its integral counterpart $I(t)$ exhibits in Eq. (3).

b. Reaction force of a set of parallel oscillators on a rigid base, $\beta(\omega) \equiv \omega$

Consider a set of N parallel oscillators attached to a common rigid base. Oscillators have equal mass m and uncoupled natural frequencies $\omega_i = \sqrt{k_i/m}$, where k_i represents the stiffness of oscillator i. Impulse response of each oscillator is expressed as:

$$h_i(t) = \frac{1}{m\omega_i} \sin\omega_i t$$

The total reaction force exerted on the base by a set of N oscillators can be represented as:

$$S(t) = \sum_{i=1}^{N} k_i h_i(t) = \sum_{i=1}^{N} \omega_i \sin \omega_i t$$

$S(t)$ has the same form as the series in (1) with $G \equiv \omega$. In this case, for time $t = t_0$, Eq. (26) together with Eq. (23) for $p(\sigma)$, provides:

$$\frac{1}{r\sqrt{2\pi}} \exp\left\{-\frac{1}{2}\frac{(\sigma - I_0)^2}{r^2}\right\} [\sin \omega t_0 + t_0 \cos \omega t_0] \frac{d\omega}{d\xi} = 1$$
$$\sigma = \omega(\xi) \sin \omega(\xi) t_0, \qquad I_0 = \int_0^{\xi_{\max}} \omega(\xi) \sin \omega(\xi) t_0 \, d\xi \tag{29}$$

As before, restricting the analysis only to monotonic frequency distributions $\omega(\xi)$, such that $\frac{d\sigma}{d\omega} = \sin \omega t_0 + t_0 \cos \omega t_0 > 0$, and choosing, for example, $t_0 = \frac{\pi}{4}$, leads to the condition that in the frequency interval $\omega \in [0, 2]$, $\frac{d\sigma}{d\omega} > 0$ and $\sigma \in [0, 2]$. Again, r and I_0 can be assigned arbitrarily, but consistent with inequalities (27); in this case, $r = 0.1$ and $I_0 = 0.8$.

Figure 13 displays the frequency distribution obtained by solving Eq. (29) and Figure 14 shows the corresponding optimal modal density from equation (25). The time history of the reaction force on the rigid base, shown in Figure 15, exhibits a rapid decay and remains at a negligibly low amplitude.

c. Pseudo-damping in conservative continuous structures, $\beta(\omega) \equiv \frac{1}{\omega}$

A continuous linear undamped dynamic system, excited by a unit impulse at point x_0, satisfies the equation of motion:

$$L[w(x,t)] + m' \frac{\partial^2 w(x,t)}{\partial t^2} = 0$$

with initial conditions $w(x, 0) = 0$, $\dot{w}(x, 0) = \delta(x - x_0)$, where δ is the Dirac's distribution. $L[\ \ ]$, $w(x,t)$, m' represent the system operator, the displacement response and the mass density, respectively. The general response of such a linear system can be represented by its orthonormal modes $\Phi_i(x)$ and principal co-ordinates $q_i(t)$, as

$$w(x,t) = \sum_{i=1}^{N} \Phi_i(x), q_i(t)$$

$$q_i(t) = A_i \sin \omega_i t\,, \qquad A_i = \frac{m'}{\omega_i} \Phi_i(x_0)$$

Then its impulse response at x_0 can be represented by the series expression $S(t)$ in Eq. (1), with $G_i = \frac{m'}{\omega_i}\Phi_i^2(x_0)$.

In general, in the absence of damping, this finite series, a superposition of pure sine functions, exhibits an almost-periodic trend. For example, the case of a Fourier series of sine functions with linearly distributed frequencies $\omega_i = i\omega_0$, where ω_0 is the fundamental frequency, becomes periodic. As before, a decaying trend in $S(t)$ is expected only in the presence of energy dissipation. However, as shown in previous studies that in cases where condensation points exist within the frequency distribution or, equivalently, natural frequencies accumulate around a particular frequency, impulse response of that linear system exhibits a decaying characteristic even in the absence of dissipation sources, a phenomenon referred here as near-irreversibility or apparent-damping.

Application of the theory developed in this paper to the continuous system described above provides a theoretical basis to the numerically obtained results in earlier studies and demonstrates how a class of frequency distributions ω_i can produce apparent-damping.

As an example, consider a simply-supported beam as a prototypical linear system for which $\Phi_i(x_0) = \sqrt{\frac{2}{m'L}}\sin\frac{\pi i x_0}{L}$, $\frac{x_0}{L} = \frac{1}{2}$. Substituting for $G_i = \frac{1}{\omega_i}\frac{2}{L}\left(\sin\frac{\pi i}{2}\right)^2$ in Eq. (1), and retaining only the odd terms:

$$S(t) = \sum_{i=1}^{N/2} \frac{2}{L}\frac{1}{\omega_{2i-1}} \sin \omega_{2i-1} t$$

Selecting $t_0 = \frac{\pi}{4}$, the function $\sigma = \frac{1}{\omega}\sin\omega t_0$ has a monotonically increasing trend, for example, within the interval $\omega \in [8, 9]$. Choosing values $I_0 = 0.05$, $r = 0.005$, which satisfy the conditions (27), the frequency distribution and the corresponding modal density can be obtained from equations (24) and (25), as shown in Figures 16 and 17. The impulse response of the beam with such a frequency distribution is shown in Fig. 18 for $N = 200$ (but includes only the 100 odd modes). The impulse response shows a rapid decay reminiscent of the impulse response of a highly-damped system, being indeed the system is Hamiltonian.

d. Pseudo-irreversible energy transfer between a single dof resonator and a set of parallel oscillators, $\beta(\omega) \equiv \omega^3$

Figure 19 depicts the system under consideration in this section, which consists of set of resonators with natural frequencies ω_i $(i = 1, 2, \ldots, N)$, that are connected in parallel to a common principal structure. The system does not possess any means of energy dissipation. For a very large number of oscillators N, approaching infinity, but with a constant total mass, the attached oscillators can be considered as a continuous distribution with the equations of motion for the coupled system expressed as:

$$\begin{cases} M\ddot{x}_M(t) + K_M x_M(t) + \int_0^1 k(\xi)\left(x_M(t) - x(\xi,t)\right) d\xi = 0 \\ m\ddot{x}(\xi,t) - k(\xi)\left(x_M(t) - x(\xi,t)\right) = 0 \end{cases} \tag{30}$$

where M, K_M, x_M are the mass, stiffness and displacement of the master structure, respectively; m, k, x represent the same quantities of the distributed oscillators in the attached set.

Several studies have shown that such a distribution of oscillators, produce a damping effect on the principal mass (1; 2; 3; 4; 5; 6; 7; 8) as N approaches infinity.

An alternative derivation of this result, presented in the Appendix, shows that the impulse response of the principal oscillator progressively decays and vanishes asymptotically with time. Energy initially imparted to the principal structure migrates to the attached set of infinite number of oscillators that have frequencies that fall within a finite bandwidth, where it remains indefinitely. As discussed earlier, it is commonly accepted that, in general, such irreversible energy transfer does not hold for a finite N (8). However, as the following application of the theory developed in this paper shows, there exist particular frequency distributions which afford a nearly-irreversible energy transfer even for a finite set of oscillators.

Considering the second of equations (29), the displacement of the continuous set of resonators in terms of the master response can be expressed by the convolution integral:

$$x(\xi,t) = \omega_n(\xi) \int_0^t x_M(\tau) H\left(t-\tau\right) \sin\omega_n(\xi)\left(t-\tau\right) d\tau$$

where H is the Heaviside function. Introducing this expression into the first of Eq. (29), an integro-differential equation results in terms of x_M:

$$M\ddot{x}_M(t) + K_M x_M(t) + x_M(t) \int_0^1 k(\xi)\, d\xi +$$
$$- \int_0^t x_M(\tau) \int_0^1 m\omega_n^3(\xi) H(t-\tau) \sin\omega_n(\xi)(t-\tau)\, d\xi\, d\tau = 0$$

which can be also expressed as:

$$M\ddot{x}_M(t) + (K_M + \bar{k})x_M(t) - x_M(t) * [I(t)H(t)] = 0 \qquad (31)$$

where $I(t)H(t)$ is the kernel of the integral part of the previous equation and

$$\bar{k} = \int_0^1 k(\xi)\, d\xi, \qquad I(t) = \int_0^1 m\omega_n^3(\xi) \sin\omega_n(\xi)\, t\, d\xi \qquad (32)$$

In the case of a finite set of N resonators, the equation of motion takes a different form where integrals over ξ are substituted by summations. Thus, Eq. (31) remains applicable provided that $\bar{k} = \sum_{i=1}^{N} k_i$ and $I(t)$ is replaced by its discrete counterpart $S(t) = \frac{1}{N}\sum_{i=1}^{N} m\omega_i^3 \sin\omega_i t$:

$$M\ddot{x}_M(t) + (K_M + \bar{k})x_M(t) - x_M(t) * [S(t)H(t)] = 0 \qquad (33)$$

The apparent damping and near irreversibility as manifested by the decay characteristics of the impulse response result from the application of the present theory by considering $G(\omega) = m\omega^3$ with $\sigma = m\ \omega^3 \sin\omega t_0$ and searching for the optimum frequency distribution.

As an example, consider a master structure, with an uncoupled natural frequency $\omega_M = 1$, with $N = 100$ attached oscillators. Assuming $t_0 = \frac{\pi}{4}$ and searching for a monotonic frequency distribution $\omega(\xi)$, it follows that, within the frequency interval$\omega \in [0, 2]$, $\frac{d\sigma}{d\omega} > 0$ and $\sigma \in [0, 8]$. The values of r and I_0 ($r = 0.4$ and $I_0 = 0.6$) are selected to be consistent with inequalities (27), and to assure that the function represented by equation (23) has its peak around $\omega = \omega_M = 1$.

Figures 20 and 21 show the frequency distribution and the frequency density of the attached oscillators determined by solving equations (24) and (26). Figure 22 shows the master response following an impulse applied at $t = 0$, which illustrates how a significant part of its energy is transferred to the set of oscillators and remains there without returning back to the master, producing an irreversible energy transfer.

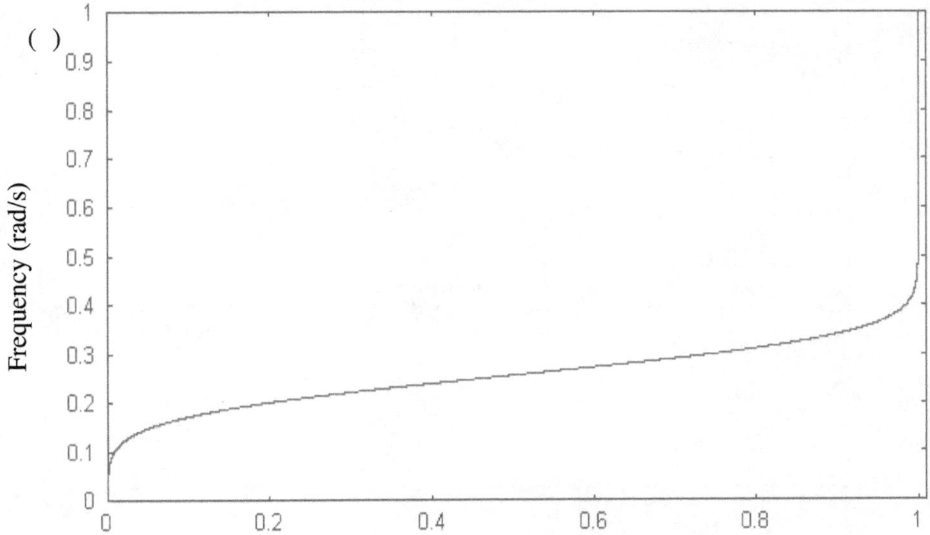

Figure 1

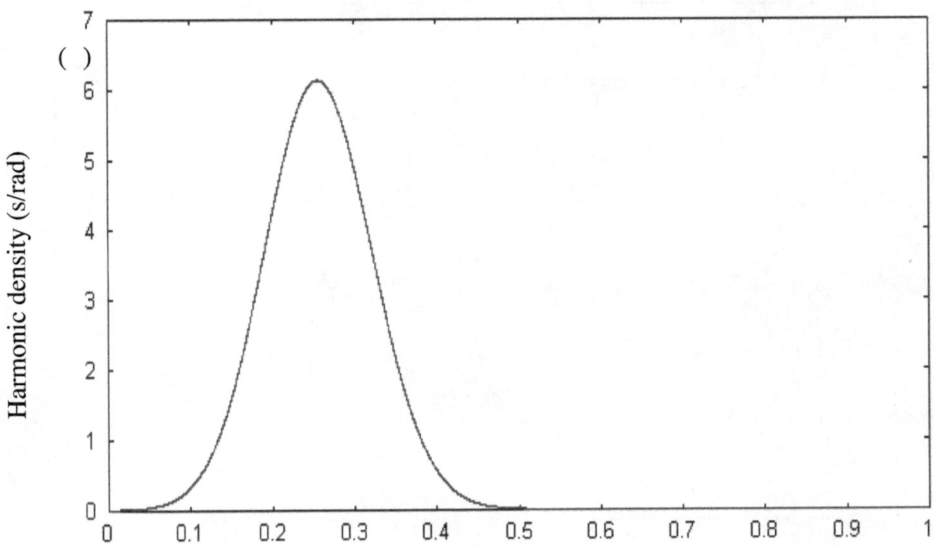

Figure 2

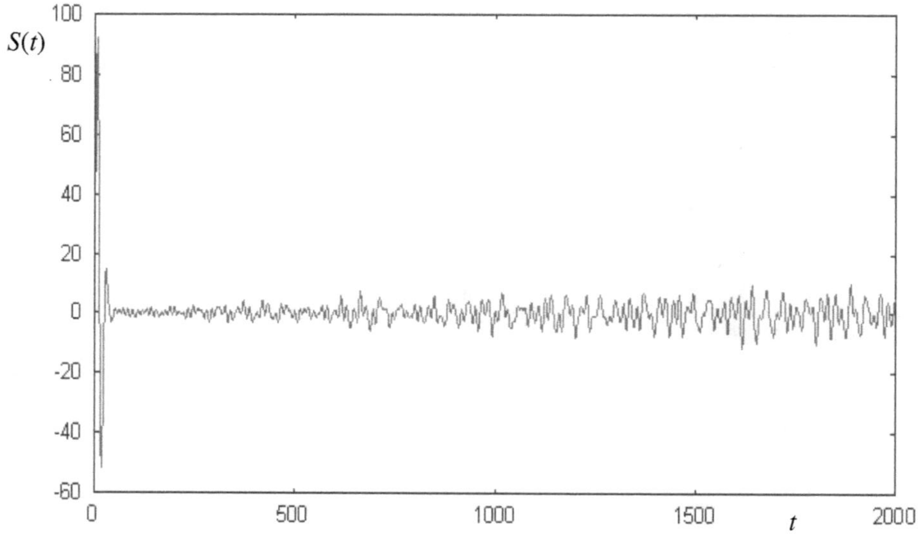

Figure 3

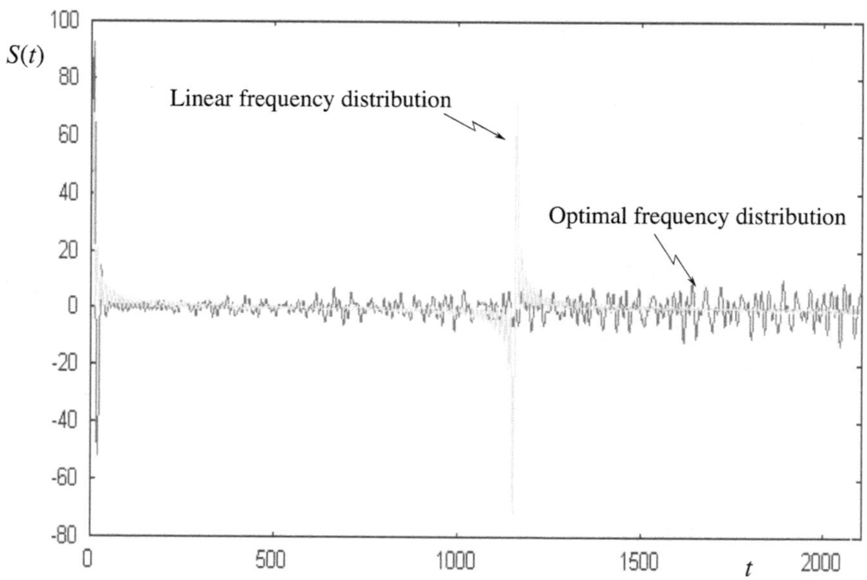

Figure 4

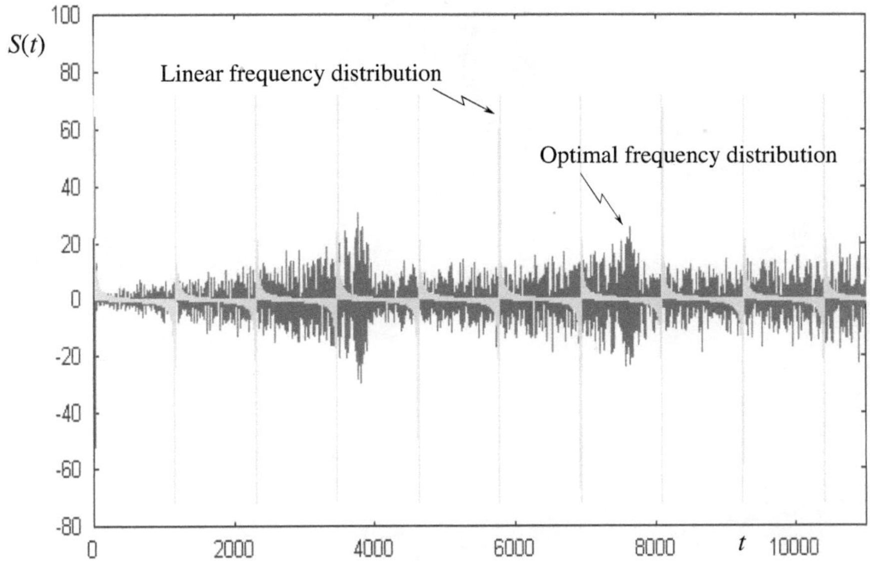

Figure 5

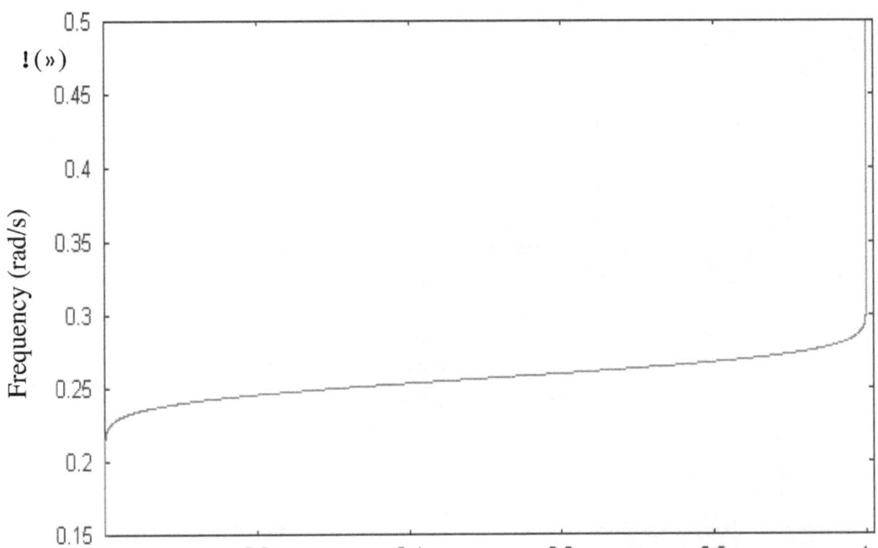

Figure 6

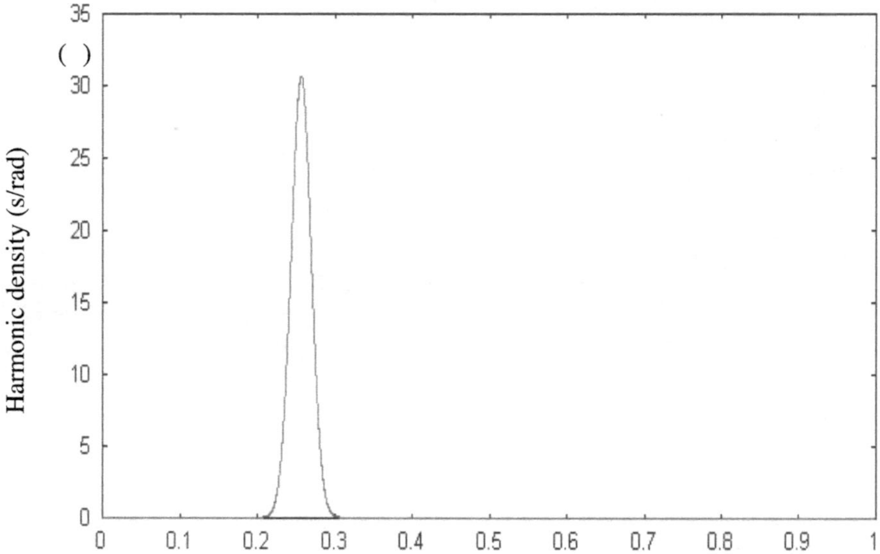

Figure 7

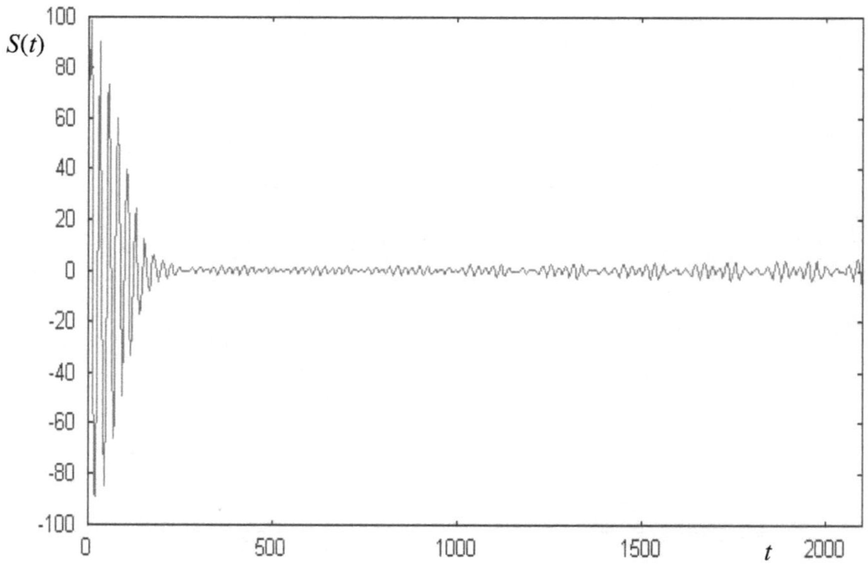

Figure 8

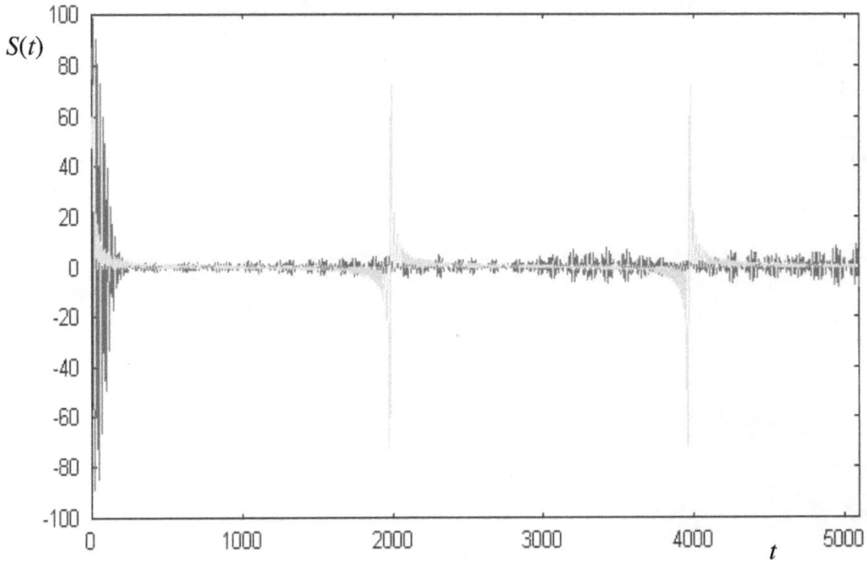

Figure 9

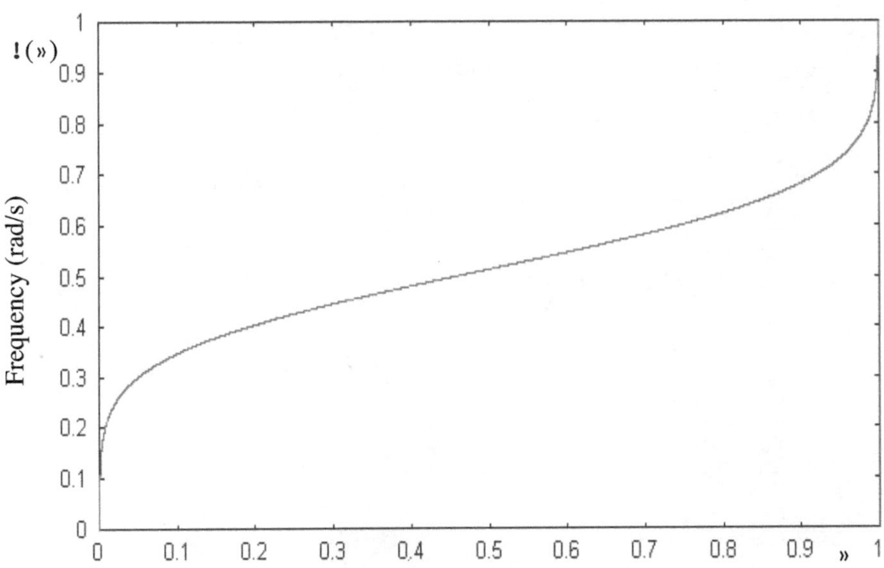

Figure 10

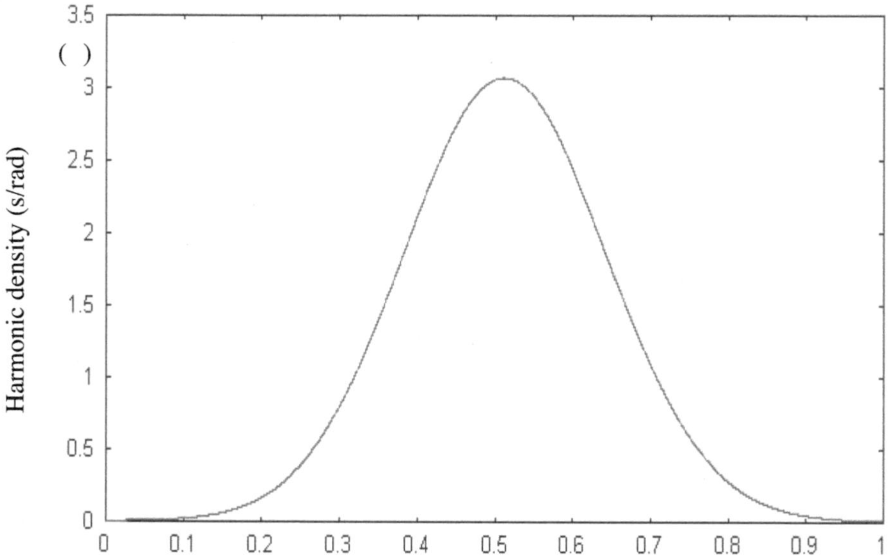

Figure 11

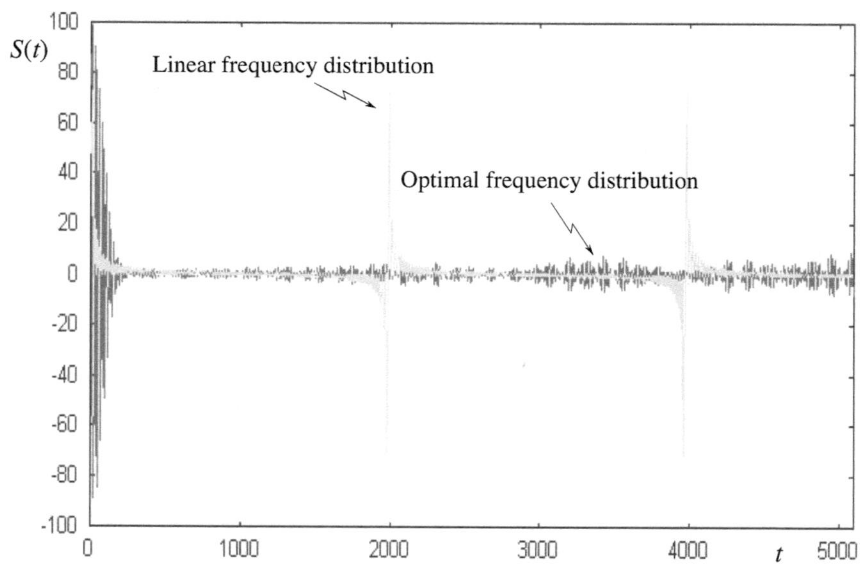

Figure 12

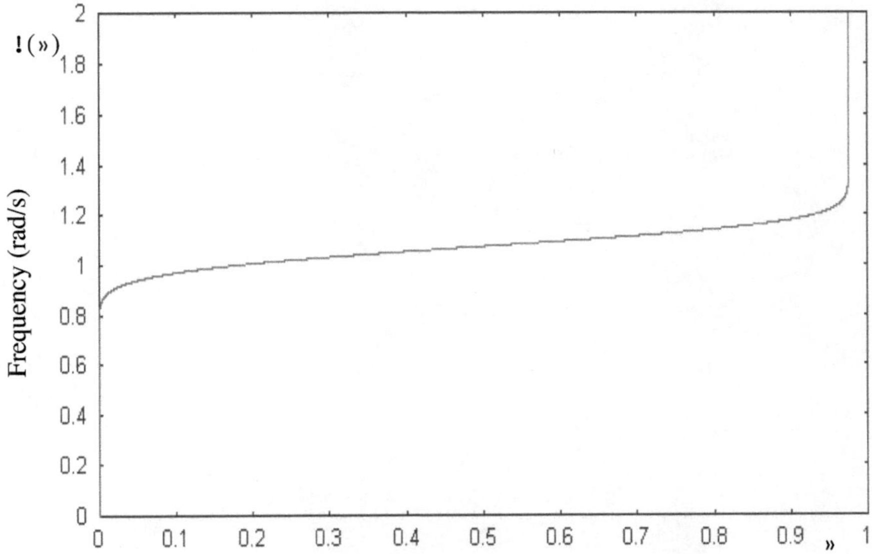

Figure 13

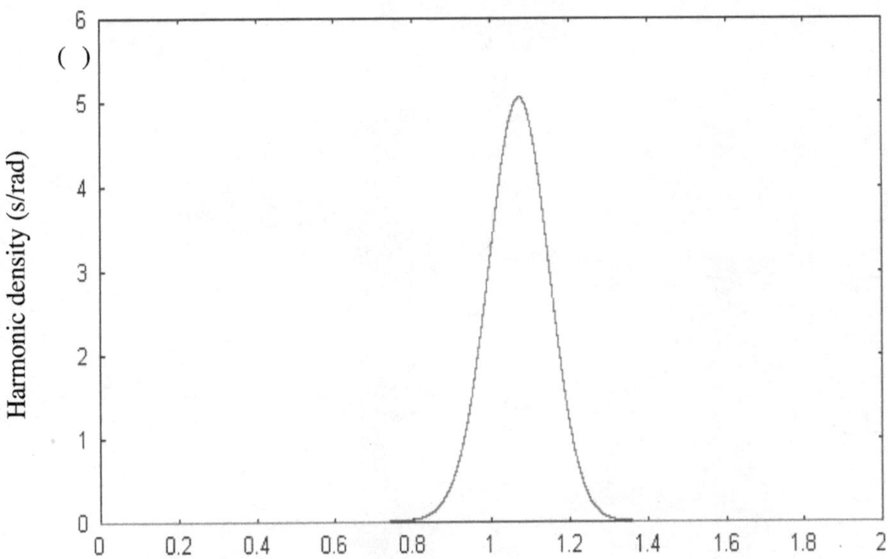

Figure 14

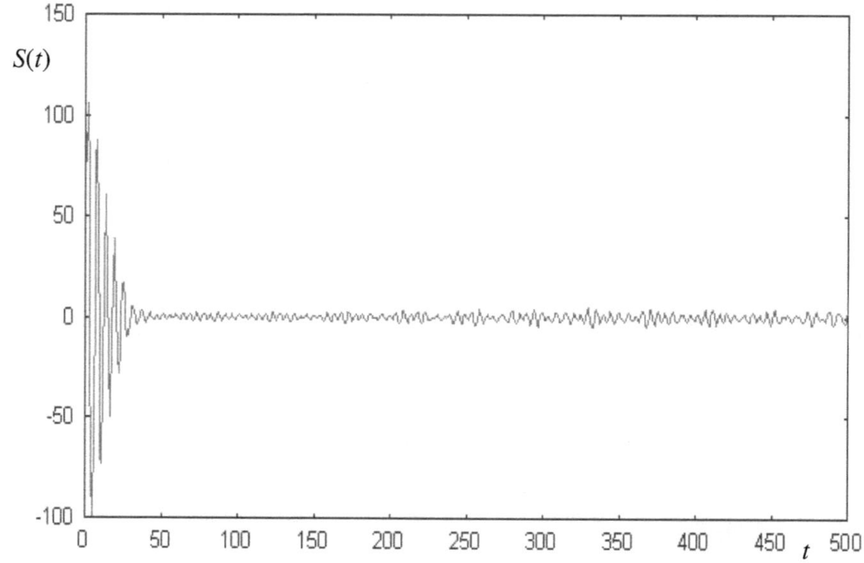

Figure 15

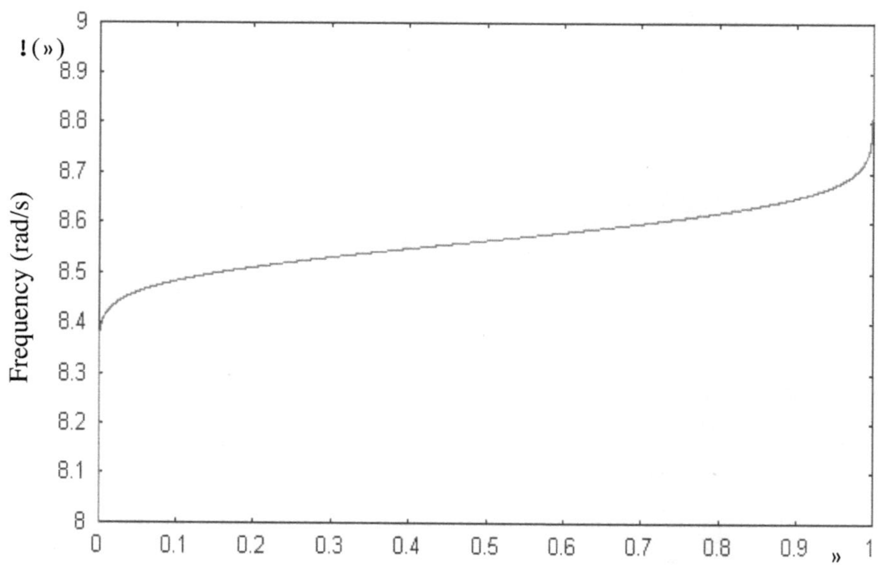

Figure 16

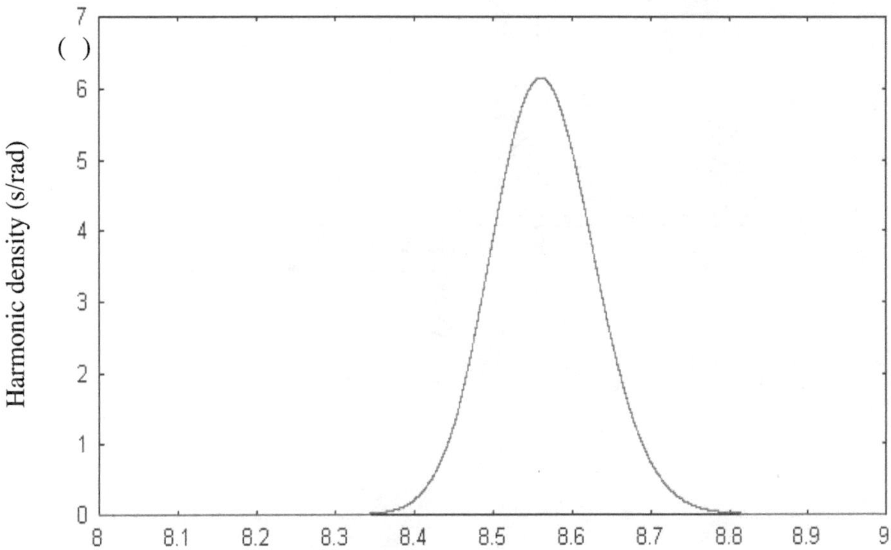

Figure 17

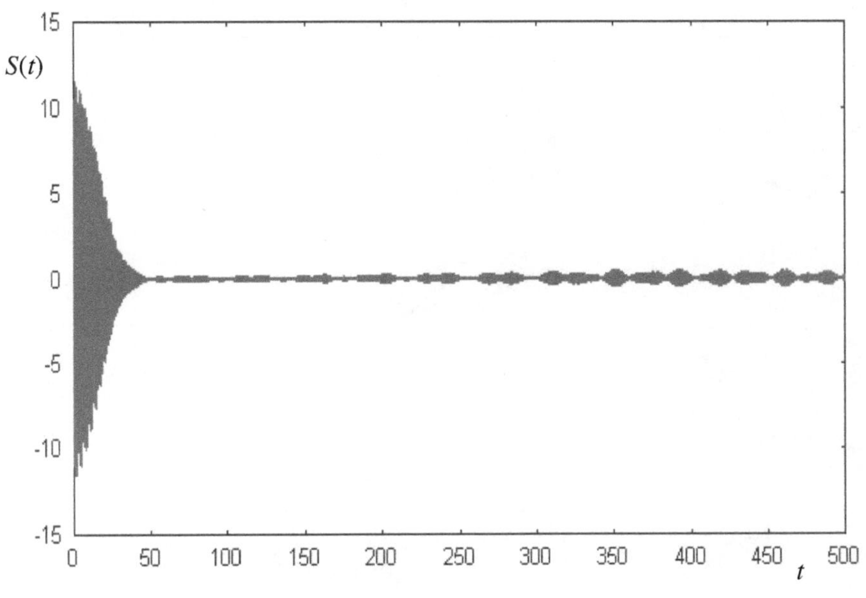

Figure 18

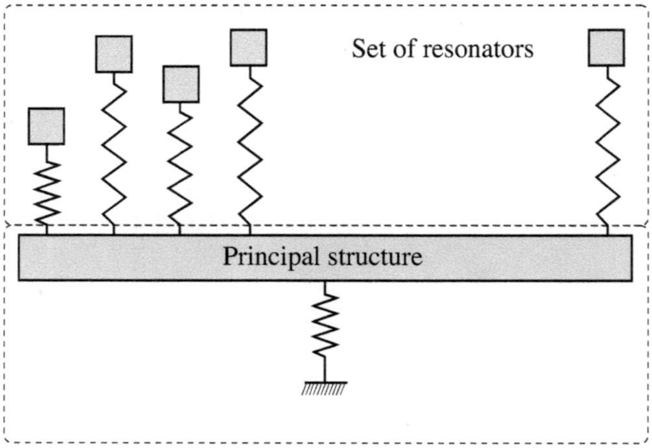

Figure 19

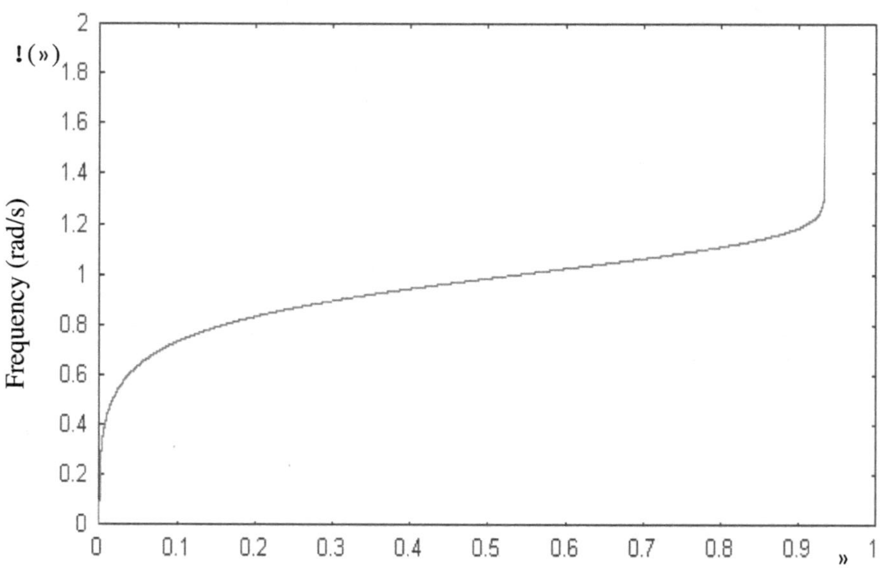

Figure 20

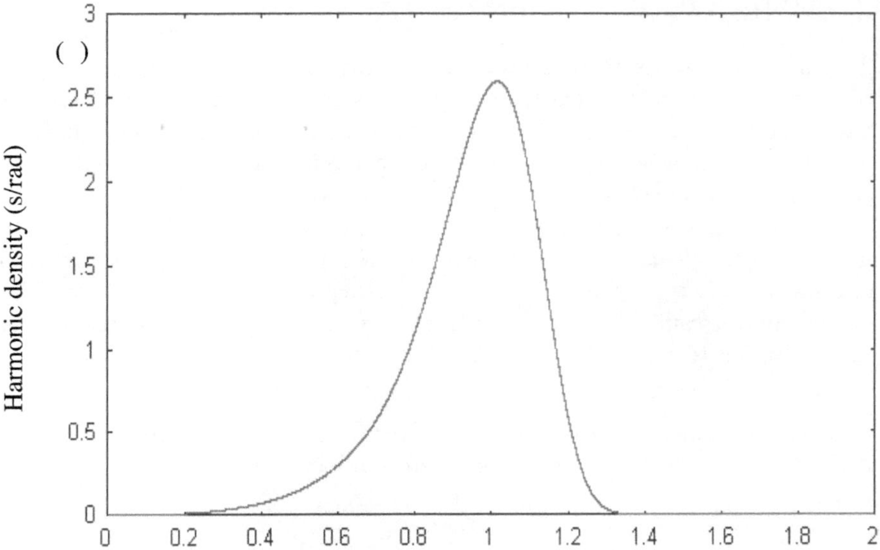

Figure 21

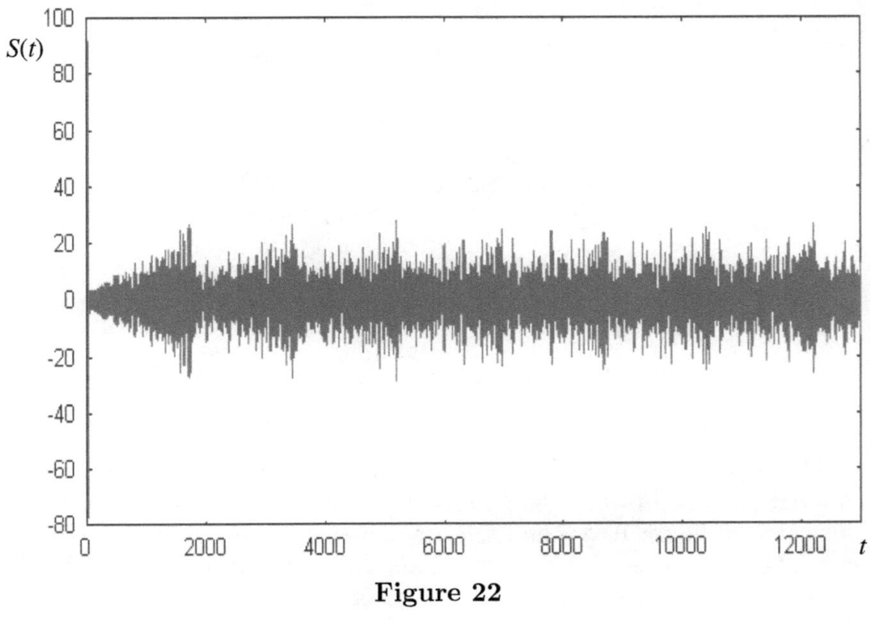

Figure 22

3 UNISAT: An Engineering Application

This section presents an application of a vibration damper based on a cluster of beams, based on the theory developed in section B. The theory is applied to a cluster of continuous beams structures attached to a continuous master, outlining an analysis for predicting the expected performances of the damper with particular emphasis to an optimal design of the device.

A built-up system is applied to a satellite aerospace structure to be launched next year, capable of absorbing the vibration energy at the lift-off, accordingly with the findings of the theory outlined in the paper. The experimental results illustrate the feasibility and the attractiveness of this new damping technique.

a. Pseudo-damping effect induced by a cluster of beams
Aim of the section is to derive the master-cluster coupled equations of motion.

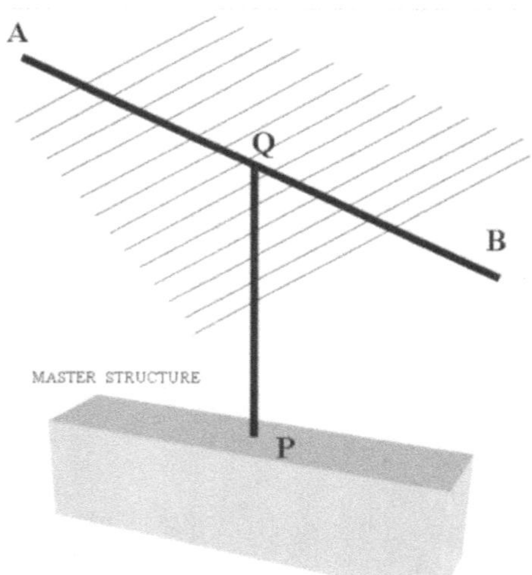

Figure 1. Sketch of the mastercluster coupling.

The cluster consists of a set of N beams, all clamped at the same support AB, attached to the main structure (master) by the connector QP as illustrated in figure 1. AB and QP are assumed to be rigid (their stiffness much higher than those of the beams) and the attachment between QP and the master is assumed to be a point connection.

The coupled equations of motion The master, a continuous linear undamped structure, is described by the governing equation:

$$L\left[w(\mathbf{x},t)\right] + m'_M \frac{\partial^2 w(\mathbf{x},t)}{\partial t^2} = R\delta(x - x_P) \tag{34}$$

where δ is the Dirac's distribution, $L[\,]$, $w(\mathbf{x},t)$, m'_M are the structural operator, the structural displacement and the mass density. x is the space variable along the master and R is the reaction force exerted by the cluster on the master at $\mathbf{x}_P$.

The master is coupled to the beams of the cluster through R:

$$R = \sum_{i=1}^{N} T_i = \sum_{i=1}^{N} B_i \left.\frac{\partial^3 w_i}{\partial \xi_i^3}\right|_{\xi_i=0} \tag{35}$$

where T_i, B_i, w_i, ξ_i are: the shear force at the clamping support, the bending stiffness, the flexural displacement (in the same direction of w) and the abscissa along the *i-th* beam of the cluster, respectively.

The $N+1$ coupled equations of motion for the N beams and for the master are:

$$\begin{cases} L\left[w(\mathbf{x},t)\right] + m'_M \dfrac{\partial^2 w(\mathbf{x},t)}{\partial t^2} - \delta(\mathbf{x} - \mathbf{x}_P) \displaystyle\sum_{i=1}^{N} B_i \left.\frac{\partial^3 w_i}{\partial \xi_i^3}\right|_{\xi_i=0} = 0 \\ B_i \dfrac{\partial^4 w_i(\xi_i,t)}{\partial \xi_i^4} + m'_i \dfrac{\partial^2 w_i(\xi_i,t)}{\partial t^2} = m'_i \dfrac{\partial^2 w(\mathbf{x}_P,t)}{\partial t^2}, \qquad i = 1,2,\ldots N \end{cases} \tag{36}$$

where m'_i is the mass per unit length of the *i-th* beam, and the beams of the cluster are forced by the inertial term related to the motion of AB that moves as P.

$\Phi_{ih}(\xi_i)$ and $q_{ih}(t)$ are the *i-th* orthonormal mode of the *h-th* beam (clamped-free) and the associated principal co-ordinates of the master, respectively. The beam vibration fields within the cluster are described by:

$$w_i(\xi_i,t) = \sum_{i=1}^{\infty} \Phi_{ih}(\xi_i) q_{ih}(t) \tag{37}$$

Substitution for these expressions into the beams equations (the second of system (36)) and the use of the orthonormality conditions produces:

$$\ddot{q}_{ih}(t) + \omega_{ih}^2 q_{ih}(t) = L_{ih} \frac{\partial^2 w\left(\mathbf{x}_P,t\right)}{\partial t^2} \tag{38}$$

where ω_{ih} is the *i-th* natural frequency of the *h-th* beam and

$$L_{ih} = \int_0^{l_i} m_i' \, \Phi_{ih}(\xi_i) \, d\xi_i \tag{39}$$

For equations (38), the convolution-form solutions hold:

$$q_{ih}(t) = \left[\Gamma(t) \frac{L_{ih}}{\omega_{ih}} \sin \omega_{ih} t \right] * \frac{\partial^2 w\,(\mathbf{x}_P, t)}{\partial t^2} \tag{40}$$

where $\Gamma(t)$ is the Heaviside distribution.

Equations (37) and (40) permit, after simple mathematics, to rewrite the equation of the master (first of equations (36)) in the form:

$$L\left[w(\mathbf{x}, t)\right] + m_M' \frac{\partial^2 w(\mathbf{x}, t)}{\partial t^2} - \delta(\mathbf{x} - \mathbf{x}_P) \left[S(t)\Gamma(t)\right] * \frac{\partial^2 w\,(\mathbf{x}_P, t)}{\partial t^2} = 0 \tag{41}$$

where

$$S(t) = \sum_{h=1}^{\infty} \sum_{i=1}^{N} \frac{U_{ih}}{\omega_{ih}} \sin \omega_{ih} t, \qquad U_{ih} = B_i L_{ih} \left. \frac{\partial^3 \Phi_{ih}}{\partial \xi_i^3} \right|_{\xi_i = 0} \tag{42}$$

Equation (41) is integral-differential in terms of the master displacement w. This prototype equation has been studied in the previous section and remarkable properties of the kernel$S(t)H(t)$ have been established. Of particular interest, the possibility, under certain conditions, of replacing the summation through the set of beams (index i) appearing into $S(t)$ by an integral, enlightening in this way the damping effect the cluster produces in the master motion. Physically, this means that the beams of the cluster, having a discrete spectrum of natural frequencies within the set, are replaced by a continuous distribution of frequencies with a spacing tending to zero. Introduction of the dummy variable χ, varying between 0 and 1 through the set of resonators with $\Delta\chi = 1/N$, allows summation (42) to be written as an integral $I(t)$:

$$\begin{aligned} S(t) &= \sum_{h=1}^{\infty} N \sum_{i=1}^{N} \frac{U_h\,(\chi_i)}{\omega_h(\chi_i)} \sin \omega_h(\chi_i) t \frac{1}{N} \approx \\ \approx I(t) &= \sum_{h=1}^{\infty} N \int_0^1 \frac{U_h(\chi)}{\omega_h(\chi)} \sin \omega_h(\chi) t \, d\chi \end{aligned} \tag{43}$$

where$\omega_h(\chi)$ represents the continuous distribution of the *h-th* natural frequency through the set of the beams that depends on the section and lengths of the beams within the cluster. This integral can be fruitfully manipulated replacing the integration variable χ byω_h. Let $\frac{d\omega_h(\chi)}{d\chi} = f(\omega_h)$ and change the integration limits:

$$I(t) = \sum_{h=1}^{\infty} N \int_{\omega_h(0)}^{\omega_h(1)} \frac{U_h(\chi)}{\omega_h f(\omega_h)} \sin \omega_h t \, d\omega_h \tag{44}$$

$f(\omega_h)$ is the frequency density of the *h-th* mode of the beams. The final step consists in expressing $U_h(\chi)$ in terms of ω_h. As shown in Appendix A, for a set of beams of different lengths and cross section area:

$$U_h(\chi) = \frac{4B(\chi)\beta_h^2}{l^3(\chi)} \left(\frac{\sinh \beta_h - \sin \beta_h}{\cos \beta_h - \cosh \beta_h} \right)^2 \tag{45}$$

where $\beta_1 = 1.875$, $\beta_2 = 4.694$, $\beta_3 = 7.855 \ldots$ etc.

The distribution of natural frequencies $\omega_h(\chi)$ depends on the bending stiffness $B(\chi)$ and the length $l(\chi)$ distributions through $\omega_h(\chi) = \frac{\beta_h^2}{l^2(\chi)}\sqrt{\frac{B(\chi)}{m'(\chi)}}$, where $m'(\chi)$ is the mass per unit length across the beams set. Substitution for this last expression into equation (45) gives:

$$U_h(\chi) = \mu_h \omega_h^2(\chi) \, m(\chi), \qquad \mu_h = \frac{4}{\beta_h^2} \left(\frac{\sinh \beta_h - \sin \beta_h}{\cos \beta_h - \cosh \beta_h} \right)^2$$

where $m(\chi) = m'(\chi) l(\chi)$ is the beams mass distribution within the set; $m(\chi)$ can be replaced by $m(\omega_h)$, since, through the distribution $\omega_h(\chi)$, a frequencyω_h remains associated at anyχ.

Thus, S can be conveniently approximated by $I(t)$:

$$S(t) \approx I(t) = \sum_{h=1}^{\infty} N \int_{\omega_h(0)}^{\omega_h(1)} \mu_h \frac{\omega_h m(\omega_h)}{f(\omega_h)} \sin \omega_h t \, d\omega_h \tag{46}$$

and the final equation for the master motion is obtained by substitution of (46) into (41):

$$L[w(\mathbf{x},t)] + m'_M \frac{\partial^2 w(\mathbf{x},t)}{\partial t^2} + \\ - \delta(\mathbf{x} - \mathbf{x}_P) \left[\sum_{h=1}^{\infty} N\Gamma(t) \int_{\omega_h(0)}^{\omega_h(1)} \mu_h \frac{\omega_h m(\omega_h)}{f(\omega_h)} \sin \omega_h t \, d\omega_h \right] * \frac{\partial^2 w(x_P,t)}{\partial t^2} = 0 \tag{47}$$

The third addend represents the global action of the cluster on the master.

Equivalent damping of the cluster The crucial point of the present theory relies in showing how the frequency representation of (47), through Fourier transform F,

$$L\left[W(\mathbf{x},\Omega)\right]-m'_M\Omega^2 W(\mathbf{x},\Omega)+ \\ -\delta(\mathbf{x}-\mathbf{x}_P)F\left\{\left[\sum_{h=1}^{\infty}N\Gamma(t)\int_{\omega_h(0)}^{\omega_h(1)}\mu_h\frac{\omega_h m(\omega_h)}{f(\omega_h)}\sin\omega_h t\,d\omega_h\right]*\frac{\partial^2 w\left(x_P,t\right)}{\partial t^2}\right\}=0 \tag{48}$$

produces for the third term a nonzero imaginary part, implying the cluster effect amounts to an equivalent dissipation.

The third addend can be indeed written as:

$$W\left(\mathbf{x}_P,\Omega\right)\sum_{h=1}^{\infty}-\Omega^2 NF\left\{\Gamma(t)\right\}*F\left\{\int_{\omega_h(0)}^{\omega_h(1)}\mu_h\frac{\omega_h m(\omega_h)}{f(\omega_h)}\sin\omega_h t\,d\omega_h\right\} \tag{49}$$

where:

$$F\left\{\Gamma\right\}=\frac{1}{2}\delta\left(\Omega\right)+\frac{1}{j\Omega} \tag{50}$$

and

$$\begin{aligned} F&\left\{\int_{\omega_h(0)}^{\omega_h(1)}\mu_h\frac{\omega_h m(\omega_h)}{f(\omega_h)}\sin\omega_h t\,d\omega_h\right\}= \\ &=\int_{\omega_h(0)}^{\omega_h(1)}\mu_h\frac{\omega_h m(\omega_h)}{f(\omega_h)}\int_{-\infty}^{+\infty}e^{-j\Omega t}\sin\omega_h t\,dt\,d\omega_h= \\ &=-j\pi\int_{\omega_h(0)}^{\omega_h(1)}\mu_h\frac{\omega_h m(\omega_h)}{f(\omega_h)}\left[\delta\left(\Omega+\omega_h\right)-\delta\left(\Omega-\omega_h\right)\right]d\omega_h= \\ &=-j\pi\left[\mu_h\frac{\Omega m(\Omega)}{f\left(\Omega\right)}\right]\Pi_h\left(\Omega\right) \end{aligned} \tag{51}$$

with

$$\Pi_h=\begin{cases}1 & \text{for } \Omega\in\left[\omega_h(0),\omega_h(1)\right]\\ 0 & \text{elsewhere}\end{cases}$$

Therefore:

$$F\left\{\Gamma(t)\right\} * F\left\{\int\limits_{\omega_h(0)}^{\omega_h(1)} \mu_h \frac{\omega_h m(\omega_h)}{f(\omega_h)} \sin\omega_h t \, d\omega_h\right\} =$$

$$= -j\Omega\left[\frac{\pi}{2}\mu_h \frac{m(\Omega)}{f(\Omega)}\right]\Pi_h(\Omega) - \int\limits_{-\infty}^{+\infty} \pi\mu_h \frac{\Pi_h(\zeta)\,\zeta m(\zeta)}{f(\zeta)(\zeta-\Omega)} d\zeta$$

In the integral on the rhs, the kernel of the Hilbert transform H is recognized:

$$\int\limits_{-\infty}^{+\infty} \pi\mu_h \frac{\Pi_h(\zeta)\,\zeta m(\zeta)}{f(\zeta)(\zeta-\Omega)} d\zeta = \pi^2\mu_h H\left\{\frac{\Pi_h(\Omega)\,\Omega m(\Omega)}{f(\Omega)}\right\}$$

Thus, equation (48) takes the final equivalent form:

$$L\left[(\mathbf{x},\Omega)\right] - m'_M\Omega^2 W(\mathbf{x},\Omega) +$$

$$+\,\delta(\mathbf{x}-\mathbf{x}_P)W(\mathbf{x},\Omega)N\sum_{h=1}^{\infty} -j\Omega\left[\frac{\pi}{2}\mu_h\frac{\Omega^2 m(\Omega)}{f(\Omega)}\right]\Pi_h(\Omega) +$$

$$-\,\pi^2\mu_h\Omega^2 H\left\{\frac{\Pi_h(\Omega)\,\Omega m(\Omega)}{f(\Omega)}\right\} = 0$$

Neglecting the real part of the series (small compared to $L\left[W(\mathbf{x},\Omega)\right] - m'_M\Omega^2 W(\mathbf{x},\Omega)$), the equation of the master finally reads:

$$L\left[W(\mathbf{x},\Omega)\right] - m'_M\Omega^2 W(\mathbf{x},\Omega) +$$

$$+\,j\Omega\left[\frac{\pi}{2}N\frac{\Omega^2 m(\Omega)}{f(\Omega)}\sum_{h=1}^{\infty}\mu_h\Pi_h(\Omega)\right]W(\mathbf{x},\Omega)\delta(\mathbf{x}-\mathbf{x}_P) = 0$$

This expression provides the frequency domain counterpart of the master equation and apparently suggests that the set of oscillators introduces a frequency dependent equivalent damping

$$C_{eq}(\Omega) = \frac{\pi}{2}N\frac{\Omega^2 m(\Omega)}{f(\Omega)}\sum_{h=1}^{\infty}\mu_h\Pi_h(\Omega) \qquad (52)$$

even in the absence of any dissipation effect.

An alternative and still significant expression of the apparent damping is obtained. In fact $dm = m(\Omega)N\,d\chi = m(\Omega)\,dN$ is the mass of the number dN of beams having natural frequency within the bandwidth $[\Omega, \Omega + d\Omega]$. Thus, equation (52) becomes:

$$C_{eq}(\Omega) = \frac{\pi}{2}\Omega^2 \frac{dm}{d\Omega} \sum_{h=1}^{\infty} \mu_h \Pi_h(\Omega) \tag{53}$$

b. Physical considerations and properties of C_{eq}

Physical meaning of C_{eq}: frequency and time domain The presence of a damping term in the master equation, in the absence of any energy dissipation within the whole system, amounts to an effect of energy transfer from the master to the beams, equivalent for the master to a loss of energy. In this view the cluster works as a vibration or shock absorber.

However, deriving C_{eq}, in equation (43) an approximation is made replacing the summation $S(t)$, through the set of beams, by the integral $I(t)$, and the simple frequency domain expressions (52) or (53) are produced. These hold only if suitable conditions for this replacement occur, and the question can be more conveniently examined in time domain. In general S and $I(t)$ differ indeed for a remainder term $\mathfrak{R} = I - S$. As shown in (3), $\mathfrak{R}$ is small at early times and, within a suitable time interval $[0, t^*]$, $\mathfrak{R} \approx 0$. Thus, for t smaller than t^*, $S \approx I$, $|\mathfrak{R}| \ll |S| \approx |I|$, the replacement of S by I is legitimated and equations (52) and (53) hold.

More in detail, applying the argument based on the *wave analogy* illustrated in (3), for the present case, t^* can be estimated by (see Appendix B):

$$t^* \approx \frac{2\pi N}{f(\Omega_{tun})} \tag{54}$$

where Ω_{tun} is the frequency in correspondence of which the set of beams is tuned. It follows that for $t \in [0, t^*]$, C_{eq} is the actual apparent damping of the master, while for later times the cluster effect is not anymore correctly represented by it. Physically, after t^*, the energy temporarily released to the set of resonators is indeed suddenly returned back to the master.

In section 2 it is shown that, in some remarkable cases, this energy return phenomenon can be prevented. In fact, $\mathfrak{R}$ depends on the functions $\omega_h(\chi)$ and a special class of them, able to minimize a suitable average of $\mathfrak{R}$, is found, permitting the replacement of the summation by an integral with a minimized error. In a sense, these 'optimal' frequency distributions makes t^* the largest possible, and the energy return to the master is not practically observed. In fact, a large t^*, in real structures, is substantially equivalent to an infinite t^*. Indeed even small dissipation transforms the most of vibration energy trapped in the cluster into heat and the energy returned back to the master, after t^*, is only a negligible fraction of that trapped initially.

Finally, as shown in the previous section 2, the return time and apparent damping are related: the faster the energy transfer from the master to the cluster, i.e. the higher C_{eq}, the longer t^*, as it also appears by comparison of equations (53) and (54), from which results $C_{eq}(\Omega_M) \propto t^*$.

Thus, as detailed in the next section, an effective design of the damper gains an important benefit in using these optimal frequency distributions, having better performances both in terms of C_{eq} and t^*.

c. Optimal frequency distributions and properties of $C_{eq}(\Omega)$

Theoretically, the best frequency distribution $\omega_h(\chi)$ is found in found in the previous section by solving for a functional minimization of the remainder square average. The found solutions belong to exponential families. Among them, that proposed ahead is rather simple and effective:

$$\omega_{\text{opt}}(\chi) = \frac{W}{2}\left[\frac{2\chi - 1}{|2\chi - 1|}\frac{e^{\alpha|2\chi-1|} - 1}{e^{\alpha} - 1} + 1\right] + \omega_{\min}$$

where $\chi \in [0, 1]$, $W = \omega_{\max} - \omega_{\min}$ and α is a parameter that modifies the shape of the frequency distribution (its optimal value for best energy absorption is 2.5). Assuming $\omega_1(\chi) = \omega_{\text{opt}}(\chi)$, *i.e.* the optimal distribution across the set for the first mode, the set of the beams lengths within the cluster follows as $l(\chi) = \frac{\beta_1}{\sqrt{\omega_{\text{opt}}(\chi)}}\sqrt[4]{\frac{B}{m'}}$, assumed that all the beams of the cluster have the same bending stiffness B and same cross section area (thickness h, width b, mass per unit length m') differing only for their lengths.

Thus, the frequency distributions for the *h-th* mode across the beams set:

$$\omega_h(\chi) = \left(\frac{\beta_h}{\beta_1}\right)^2 \omega_{\text{opt}}(\chi) \tag{55}$$

All these distributions present a typical trend for$\chi \in [0, 1]$: a rapid growth close to$\chi = 0$, an almost flat branch around$\chi_{tun} = 0.5$, and again a rapid growth close to $\chi = 1$, as shown in figure 2.

Consequently, $\frac{1}{f} = \frac{1}{d\omega_h/d\chi}$ is a Gaussian-like function with sharp peak aroundχ_{tun}, to which corresponds the tuning frequency

$$\omega_{tun\,h} = \left(\frac{\beta_h}{\beta_1}\right)^2 \frac{\omega_{\max} + \omega_{\min}}{2},$$

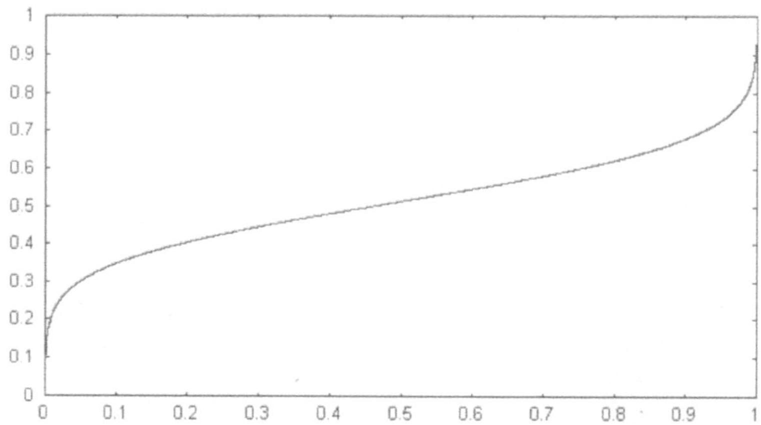

Figure 2. Typical optimal frequency distribution within the cluster.

the values around which the natural frequencies of the distribution $\omega_h(\chi)$ condense (see figure 3). An intuitive understanding of the optimal characteristics in terms of t^* and C_{eq} of these distributions is seen from expressions (52) and (54) that contain the factor $\frac{1}{f}$, implying they confer a large t^* and a large C_{eq} around the tuning frequency $\omega_{tun\,h}$.

Besides these typical peaks of the equivalent damping related to $\frac{1}{f}$, expressions (52) and (53) reveal other remarkable properties of C_{eq} discussed ahead.

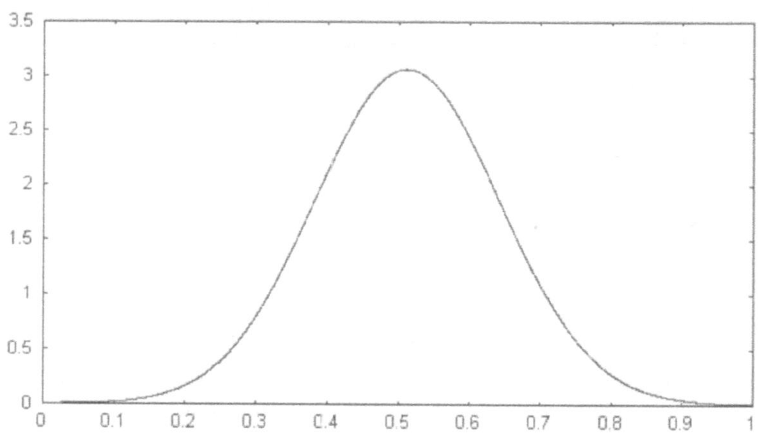

Figure 3. Gauss-like frequency density distribution within the cluster.

From equation (54) the frequency average damping can be estimate. The frequency intervals $[\omega_h(0), \omega_h(1)]$ can be partially overlapped or not. In the last case (as in the built-up device here presented) $\omega_h(1) < \omega_{h+1}(0)$ for any h, meaning $\frac{\beta_{h+1}}{\beta_h} \frac{l_{\min}}{l_{\max}} > 1$, where $l_{\min}$, $l_{\max}$ are the minimum and maximum beams lengths through the set. Thus, in each frequency window Π_h, the damping expression simplifies as $C_{eq\,h}(\Omega) = \frac{\pi}{2}\Omega^2 \frac{dm}{d\Omega}\mu_h$. An estimate of its average over the frequency bandwidth $\Omega \in [\omega_h(0), \omega_h(1)]$ is:

$$\bar{C}_{eq\,h} = \frac{1}{W_h}\int_{\omega_h(0)}^{\omega_h(1)} \frac{\pi}{2}\Omega^2 \frac{dm}{d\Omega}\mu_h \, d\Omega \approx \frac{\pi\omega_{tun\,h}^2}{2W_h}\mu_h M_{cl} \tag{56}$$

where $W_h = \omega_h(1) - \omega_h(0)$ and M_{cl} are the *h-th* device tuning bandwidth and the total mass of the cluster, respectively.

Finally a relationship between M_{cl} and N holds:

$$\begin{aligned} M_{cl} &= m'\sum_{i=1}^{N} l(\chi_i) = m'N\sum_{i=1}^{N} l(\chi_i)\,\Delta\chi \approx m'N\int_0^1 l(\chi)\,d\chi = \\ &= m'N\beta_1\sqrt[4]{\frac{B}{m'}}\int_0^1 \frac{1}{\sqrt{\omega_{\text{opt}}(\chi)}}\,d\chi \end{aligned} \tag{57}$$

Equations (52), (54), (55), (56) and (57) provide the basis for the cluster design.

d. Cluster design and performances of the built-up device

A device based on the previous theory has been designed to be used on board of UNISAT, that stands for UNIversity SATellite, a permanent space project developed at the University of Rome La Sapienza by the Gauss Group. It is a small scientific satellite (14 kg- 20 kg depending on the flying payload) launched four times in orbit (2000-2006) and next launch, equipped with the presented vibration suppressor, is planned in 2008. Severe vibrations occur to the electronic equipment of the satellite during the lift-off operations of the carrier and the present device is aimed at reducing shock and vibration on the plate carrying the electronic package.

The material used for the damper is still (namely $\rho = 7780$ kg/m^3, $E = 187.5$ GPa,) and the cluster of beams are realized by milling machines from a still sheet of thickness $h = 0.6$mm. The maximum allowed room on board for the device is 90 mm × 90 mm × 40 mm, with a maximum allowed mass agreed by the satellite designers equal to 150 g.

The design procedure follows the steps ahead:

- a) Frequency bandwidth and tuning frequency.

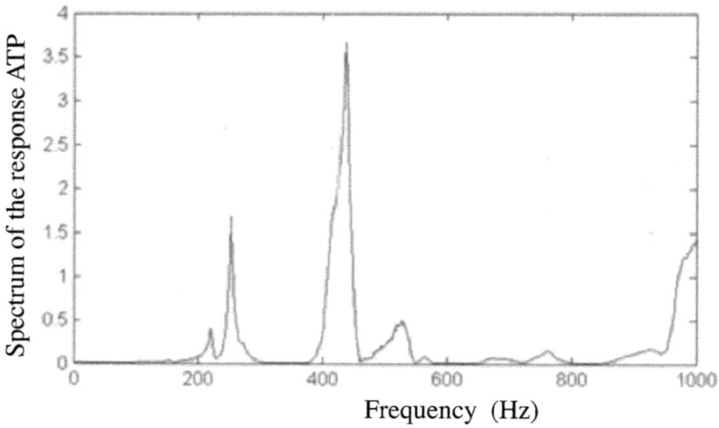

Figure 4. Experimental response at the selected attachment point.

On the basis of the experimental drive point frequency response at the attachment point P for $\Omega \in [0, 1000]$ Hz (see figure 4), the frequency bandwidth W and its central frequency ω_{tun} are selected. Looking at the highest peak, the choice is $\omega_{tun} = 440$ Hz, $W = 140$ Hz.

• b) Optimal frequency distribution.
With $\omega_{tun} = 440$ Hz, $W = 140$ Hz, $\omega_{\max}, \omega_{\min}$ are determined, and they completely define the optimal frequency distributions given by equation (55) (see figure 5) and the related values $\omega_{tun\,h}$, W_h.

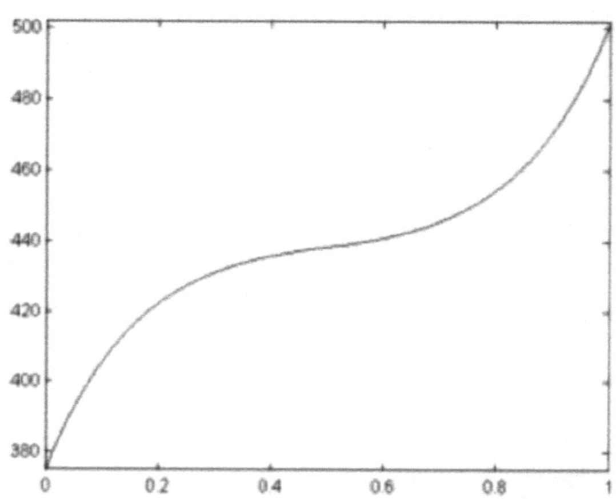

Figure 5. Optimal frequency distribution.

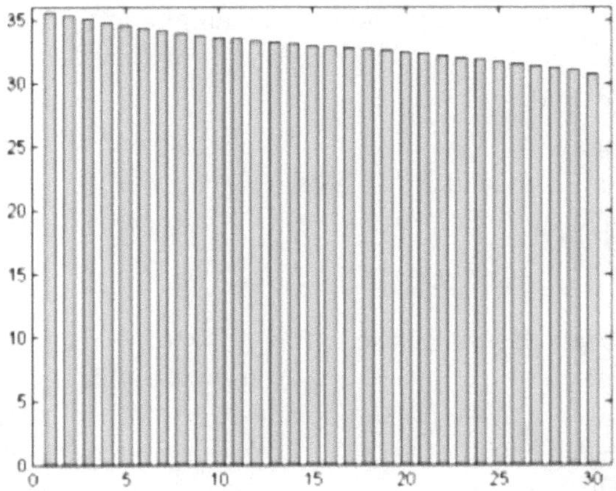

Figure 6. Optimal length distribution within the set of beams of the device.

• c) Lengths distribution.

The length distribution (see figure 6) is determined on the basis of $\omega_{\text{opt}}(\chi)$ as $l(\chi) = \frac{\beta_1\sqrt{h}}{\sqrt{\omega_{\text{opt}}(\chi)}}\sqrt[4]{\frac{E}{12\rho}}$, that yields minimum and maximum lengths of 30.7 and 35.6 mm, respectively. Considering that the device is made of symmetric beams (see figure 6), the maximum dimension along the beams axes is twice the maximum beam length, *i.e.* 71.2 mm, to which the clamped part of width of the beams, 10 mm, should be added leading to a maximum dimension equal to 81.2 mm that satisfies the design constraint (max 90 mm).

• d) Number of beams.

The number N of the beams is to guarantee a return time large enough to prevent energy comes back to the master. As it appears indeed form equation (56), once given the frequency bandwidth and the tuning frequency, only the total mass of the cluster has an effect on C_{eq}, but not the number of beams over which the mass is spread.

t^* is the time the energy is stored within the beams. It is desired that only a small fraction r of the initially trapped energy should be returned to the master. If η is the hysteretic damping factor of a beam, roughly energy decays as $e^{-\eta\omega t}$ for a harmonic motion at frequency ω. Lower frequencies have a slower decay, and the lowest decay is with the lowest frequency ω_{min} within the cluster. Thus, we can ask the factor $e^{-\eta\omega_{\text{min}}t^*}$ be equal to r:

$$e^{-\eta\omega_{\text{min}}t^*} = r \quad \Longrightarrow \quad t^* = -\frac{\ln r}{\eta\omega_{\text{min}}}$$

and from (54):

$$N = -\frac{f(\omega_{tun})\ln r}{\eta\omega_{\text{min}}} \tag{58}$$

that provides the order of magnitude for the number of beams to be included within the cluster. Since $f(\omega_{tun}) \approx 500$ rad/s, $\omega_{\min} \approx 2350$ rad/s, $\eta \approx 0.03$ and assuming $r = 0.01$, $N \approx 30$.

• e) Width and gap.
The maximum allowed width of the device is $D = 90$ mm, g is the gap between two adjacent beams, and b the width, then:

$$D = Nb + (N-1)g \tag{59}$$

Since, as it appears from equation (56), it is convenient to increase the mass of the cluster, *i.e.* b must be the largest, compatibly with equation (59). This implies g must be the smallest. Actually, it depends on the smallest thickness of the milling cutter disk that generates the gap between the beams, in this case is $g = 1$ mm. Thus, form the previous equation follows $b = 2$ mm.

• f) Maximization of the cluster mass (multiple layer).
The previous analysis completely defines the optimal cluster satisfying the prescribed design requirements. However, since the maximum allowed height of the whole device is up to 40 mm, a multiple layered structure can be hosted, with the advantage of increasing the total mass of the cluster, as suggested by equation (56). Thus, three equal sets of beams are in column, separated by two aluminum spacers with height 5mm that guarantee each beam does not hit the upper or the lower beam when undergoing the maximum accelerations at lift-off provided by the satellite builder.

The final mass M_{DEV} of the complete assembled device (represented in figure 7) is 130 g.

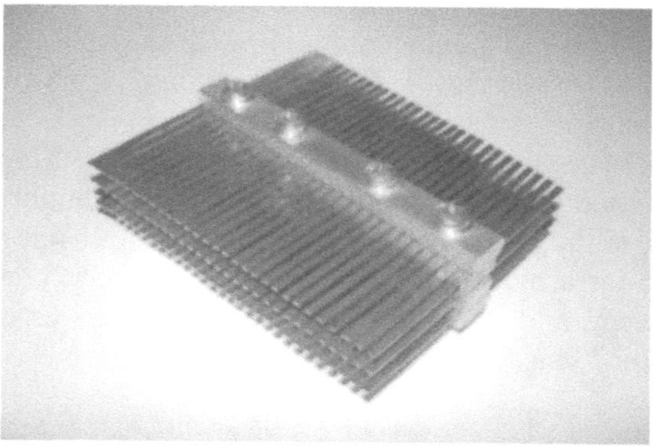

Figure 7. View of the build up device tuning frequency 440 Hz, total weight 130 g.

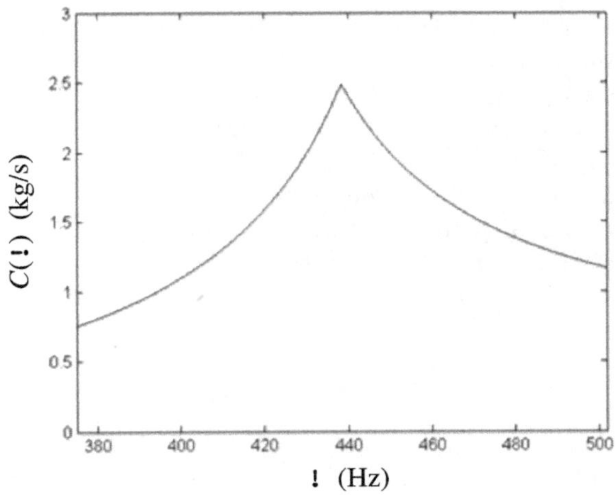

Figure 8. Apparent damping coefficient *vs* frequency of the device theoretically determined with an infinite number of resonators

With the previous data, the theoretical equivalent damping is determined through equation (52) and represented in figure 8.

Finally, the validation experiments are performed following the lines explained below.

As a first step, the best location for attachment point P is identified (fig. 9). An electro-dynamic shaker excites the structure with a spectrum

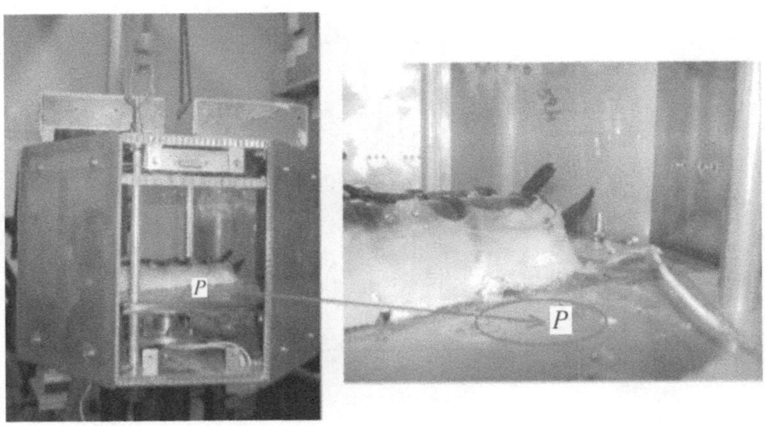

Figure 9. Selection of the test "point P" on board at which the suppression vibration device is applied.

similar to that meet in operating conditions and the maximum displacement point is identified as P. At the same time the drive point FRF at P is determined and a peak frequency identified (see figure 4, frequency peak at 440 Hz) used for determining bandwidth and tuning frequency.

A first comparison is made between the theoretical $C_{eq}(\Omega)$ and the experimentally identified apparent damping of the device. The FRF at the device attachment point is determined and the experimental apparent damping $C_{eq}^{EXP}(\Omega)$ is identified by fitting the obtained response by using a theoretical FRF of the form:

$$FRF_{TH}(\Omega) = \frac{1}{-\Omega^2 M_{DEV} + j\Omega C_{eq}(\Omega)}$$

where M_{DEV} is known. The plot of $C_{eq}^{EXP}(\Omega)$ is shown in figure 10, and the comparison with figure 8 shows a good agreement. The smoothness of the theoretical curve is because the developed theory uses an integral instead of the discrete summation through the set, while the experimental peaks corresponds to the separate resonance frequencies of the single beams of the cluster.

The built-up device is then installed on the satellite plate. Comparison of the new attenuated drive point FRF and the old one is shown in figure 12

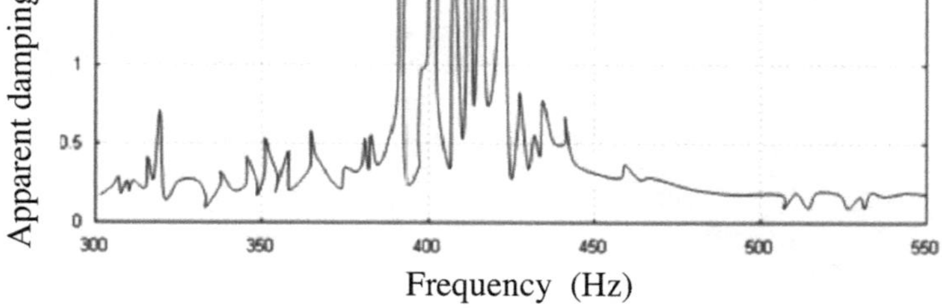

Figure 10. Experimental measurement of the apparent damping of the device (tuning frequency 438 Hz).

Figure 11. View of the final installation of the vibration suppression device on board of UNISAT.

in the frequency band covered by the first mode of the cluster of beams. In figure 13, the second modes of the device are indeed within the frequency bandwidth 2200-3000 Hz. It appears how also in this higher frequency band the damper effectively reduces the amplitude of vibration.

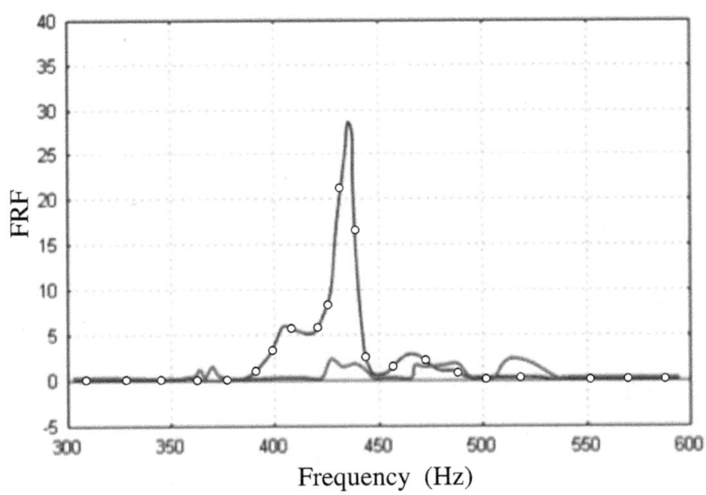

Figure 12. Comparison of the FRFs at "point P" with (black curve) and without (circled curve) the vibration suppression device.

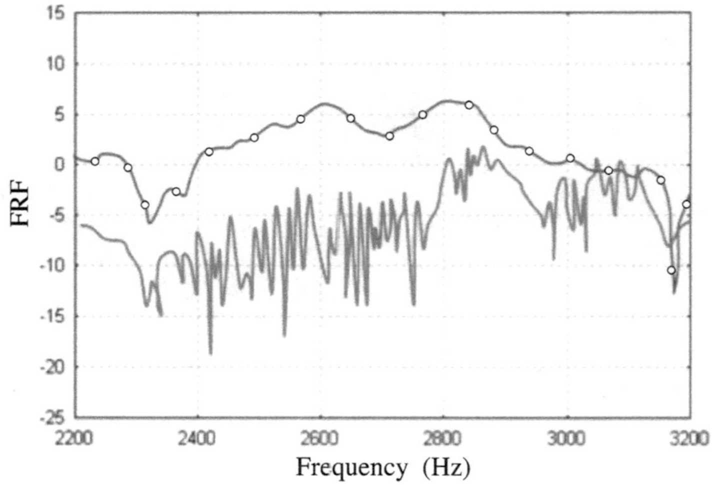

Figure 13. Experimental evidence of the second modes effect on the suppression of vibration.

Bibliography

[1] A.D. Pierce, V.W. Sparrow and D.A. Russel. Fundamental structural-acoustic idealization for structure with fuzzy internals. *Journal of Vibration and Acoustics*, 117:339–348, 1995.

[2] M. Strasberg, D. Feit. Vibration damping of large structures induced by attached small resonant structures. *Journal of Acoustical Society of America*, 99:335–344, 1996.

[3] G. Maidanik. Induced damping by a nearly continuous distribution of a nearly undamped oscillators: linear analysis. *Journal of Sound and Vibration*, 240:717–731, 2001.

[4] R.L. Weaver. The effect of an undamped finite degree of freedom 'fuzzy' substructure: numerical solution and theoretical discussion. *Journal of Acoustical Society of America*, 101:3159–3164, 1996.

[5] R.J. Nagem, I. Veljkovic, G. Sandri. Vibration damping by a continuous distribution of undamped oscillators. *Journal of Sound and Vibration*, 207:429–434, 1997.

[6] C.E. Celik, A. Akay. Dissipation in solids: thermal oscillations of atoms. *Journal of Acoustical Society of America*, 108:184–191, 2000.

[7] R.L. Weaver. Equipartition and mean square response in large undamped structures. *Journal of Acoustical Society of America*, 110:894–903, 2001.

[8] A. Carcaterra, A. Akay. Transient energy exchange between a primary structure and a set of oscillators: return time and apparent damping. *Journal of Acoustical Society of America*, 115:683–696, 2004.

[9] I. Murat Koç, A. Carcaterra, Zhaoshun Xu, A. Akay. Energy sinks: vibration absorption by an optimal set of undamped oscillators. *Journal of Acoustical Society of America*, 118:3031–3042, 2005.

[10] A. Carcaterra, A. Akay, I.M. Koc. Near-Irreversibility and damped response of a conservative linear structure with singularity points in its modal density. *Journal of Acoustical Society of America*, 119:2124–, 2006.

[11] A. Akay, Zhaoshun Xu, A. Carcaterra, I. Murat Koc. Experiments on vibration absorption using energy sinks. *Journal of Acoustical Society of America*, 118: 3043–3049, 2005.

[12] A. Carcaterra. An entropy formulation for the analysis of energy flow between mechanical resonators. *Mechanical Systems and Signal Processing,* 16:905-920, 2002.

[13] A. Carcaterra. Ensemble energy average and energy flow relationships for nonstationary vibrating systems. *Journal of Sound and Vibration*, Special Issue *Uncertainty in structural dynamics*, 288:751-790, 2005.

[14] T.Y. Petrosky. Chaos and irreversibility in a conservative nonlinear dynamical system with a few degrees of freedom. *Physical Review A,* 29:2078-2091, 1984.

[15] P.K. Datta, K.Kundu. Energy transport in one-dimensional harmonic chains. *Physical Review B,* 51:6287-6295, 1995.

[16] R.Livi, M. Pettini, S. Ruffo, M. Sparpaglione, A. Vulpiani. Equipartition threshold in nonlinear large Hamiltonian systems: the Fermi-Pasta-Ulam model. *Physical Review A,* 31:1039-1045, 1985.

[17] R.R. Nigmatullin, A. Le Mehaute. To the nature of irreversibility in linear systems. *Magnetic Resonance in Solids,* 6:165-179, 2004.

[18] M.G. Kendall, A. Stuart. *The advanced theory of statistics.* Charles Griffin & Company Limited, London,, 1961.

[19] E.J.G. Pitman. Sufficient statistics and intrinsic accuracy. *Proc. Cambridge Philosophical Society*, 32:576, 1936.

[20] M. V. Drexel, J.H. Ginsberg, Modal overlap and dissipation effects of a cantilever beam with multiple attached oscillators, *Journal of Vibration and Acoustics*, vol.123, 181–187, 2001.

[21] A. Carcaterra, A. Akay, Theoretical foundation of apparent damping and energy irreversible energy exchange in linear conservative dynamical systems, *Journal of Acoustical Society of America*, ISSN: 0001-4966 121, 1971–1982, 2007.

[22] A. Carcaterra, A. Akay, F. Lenti, Pseudo-damping in undamped plates and shells, *Journal of Acoustical Society of America*, ISSN: 0001-4966, vol. 122, 804–813, 2007.

[23] A. Carcaterra Minimum-variance response and irreversible energy confinement, invited lecture, *IUTAM Symposium on "The Vibration Analysis of Structures with Uncertainties"*, book edited by R.S. Langley. and A. Belyaev, Saint Petersburg, Russia, 2009.

[24] F. dell'Isola, C. Maurini, M. Porfiri, Passive damping of beam vibrations through distributed electric networks and piezoelectric transducers: prototype design and experimental validation, Smart Material and Structures, 13(2), 2004.

[25] F. dell'Isola, C. Maurini, S. Vidoli, Piezo-ElectroMechanical (PEM) structures: passive vibration control using distributed piezoelectric transducers, Compte Rendus Mecanique, 331(1), 2003.

[26] F.Magionesi, A. Carcaterra, Insights into the energy equipartition principle in undamped engineering structures, *Journal of Sound and Vibration*, ISSN: 0022-460X, 322(4-5), 851–869, 2008.

[27] N. Roveri, A. Carcaterra, A. Akay, Vibration absorption using non-dissipative complex attachments with impacts and parametric stiffness, *Journal of the Acoustical Society of America*, 126(5), 2306–2314, 2009.

[28] N. Roveri, A. Carcaterra, A. Akay, Energy equipartition and frequency distribution in complex attachments, *Journal of the Acoustical Society of America* 126(1), 122–128, 2009.

[29] A. Carcaterra, A. Sestieri, Energy Density Equations and Power Flow in Structures, *Journal of Sound and Vibration*, Academic Press , ISSN 0022-460X, vol. 188(2), 269–282, 1995.

[30] A. Carcaterra, Wavelength scale effect in energy propagation in structures, 13–24, in *Statistical Energy Analysis,* della collana *Solid Mechanics and Its Applications* vol. 67, IUTAM, Kluwer Academic Publishers, Dordrecht, ISBN 0-7923-5457-5, 1999.

[31] A. Carcaterra, L. Adamo, Thermal Analogy in Wave Energy Transfer: Theoretical and Experimental Analysis, *Journal of Sound and Vibration*, ISSN 0022-460X, Academic Press, vol. 226(2), 253–284, 1999.

[32] I. Prigogine, G. Nicolis, *Self-Organization in Nonequilibrium Systems*, John Wiley & Sons, 1977.

[33] H. Mori, Y. Kuramoto, *Dissipative Structures and Chaos*, Springer, 1998.

[34] I. Prigogine, I. Stengers, *La Nouvelle Alliance*, Gallimard, Paris, 1979.

[35] O. Costa de Beauregard, *Le second principe de la science du temps*, Editions du Seuil, Paris, 1963.

[36] C. Cercignani, *Ludwig Boltzmann-The Man Who Trusted Atoms*, Oxford University Press, 1998.

[37] H.R. Brown, W. Myrvold, J. Uffink, Boltzmann *H-theorem*, its discontents, and the birth of statistical mechanics, *Studies in History and Philosophy of Modern Physics*, 40, 174–191, 2009.

[38] F. dell'Isola and S. Vidoli, Continuum modelling of piezoelectromechanical truss beams: An application to vibration damping, *Archive of Applied Mechanics*, 68:1–19, 1998

[39] S. Vidoli and F. dell'Isola, Modal coupling in one-dimensional electromechanical structured continua,*Acta Mechanica*, 141:37–50, 2000.

[40] U. Andreaus, F. dell'Isola, and M. Porfiri, Piezoelectric Passive Distributed Controllers for Beam Flexural Vibrations,*Journal of Vibration and Control*, 10:625, 2004.

[41] C. Maurini, F. dell'Isola, and D. Del Vescovo, Comparison of piezoelectronic networks acting as distributed vibration absorbers,*Mechanical Systems and Signal Processing*, 18:1243-1271, 2004.

[42] F. dell'Isola and L. Rosa, Almansi-type boundary conditions for electric potential inducing flexure in linear piezoelectric beams,*Continuum Mechanics and Thermodynamics*, 9:115–125, 1997.

[43] F. dell'Isola and S. Vidoli, Damping of bending waves in truss beams by electrical transmission lines with PZT actuators,*Archive of Applied Mechanics*, 68:626–636, 1998.

[44] S. Vidoli and F. dell'Isola, Vibration control in plates by uniformly distributed PZT actuators interconnected via electric networks,*European Journal of Mechanics, A/Solids*, 20:435–456, 2001.

[45] S. Alessandroni, F. dell'Isola, and F. Frezza, Optimal piezo-electromechanical coupling to control plate vibrations,International Journal of Applied Electromagnetics and Mechanics, 13:113–120, 2001.

[46] F. dell'Isola, E.G. Henneke, and M. Porfiri, Synthesis of electrical networks interconnecting PZT actuators to damp mechanical vibrations,*International Journal of Applied Electromagnetics and Mechanics*, 14:417–424, 2001.

[47] S. Alessandroni, F. dell'Isola, and M. Porfiri, A revival of electric analogs for vibrating mechanical systems aimed to their efficient control by PZT actuators,*International Journal of Solids and Structures*, 39:5295–5324, 2002.

Fluid Shock Wave Generation at Solid-Material Discontinuity Surfaces in Porous Media

Francesco dell'Isola [*‡] and Pierre Seppecher [†‡] and Angela Madeo [+‡]

[*] Dip. DISG, Università di Roma "La Sapienza", via Eudossiana 18, 00184, Rome, Italy and Laboratorio di Strutture e Materiali Intelligenti, Fondazione Tullio Levi-Civita, Cisterna di Latina, Italy

[†] Inst. de Mathematique, Université de Toulon, Av. de l'Université, BP 132, 83957 La Garde Cedex, France.

[+] Laboratoire de Génie Civil et Ingénierie Environnementale, Université de Lyon–INSA, Bâtiment Coulomb, 69621 Villeurbanne Cedex, France

[‡] International Research Centre on "Mathematics & Mechanics of Complex Systems" M& MOCS

Abstract A general set of boundary conditions at the interface between dissimilar fluid-filled porous matrices is established starting from an extended Hamilton-Rayleigh principle. These conditions do include inertial effects. Once linearized, they encompass boundary conditions relative to volume Darcy-Brinkman and to surface Saffman-Beavers-Joseph-Deresiewicz dissipation effects.

1 Introduction

Many interesting mechanical phenomena occur in porous media when the saturating fluid flows under the action of pressure and the solid matrix is deformable. Modeling these phenomena represents an important challenge for engineering sciences.

The aim of this paper is to use the principle of virtual work –when dissipative and inertial effects cannot be neglected– for deducing a set of evolution equations and coherent boundary conditions valid at a fluid-permeable interface between dissimilar fluid-filled porous matrices. The spirit of the approach adopted here for modeling porous systems is very similar to the one used to develop models for two fluid mixtures by Gavrilyuk et al. (1997), Gavrilyuk et al. (1998), Gavrilyuk and Perepechko (1998), Gouin and Gavrilyuk (1998), Gavrilyuk and Gouin (1999), Gouin (1990). It also has some similarities with the treatment used to describe fluid saturated porous media by Dormieux and Stolz (1992), Dormieux et al. (1991) and

Coussy and Bourbie (1984). To frame the results presented here and to compare them with those available in the literature it is necessary to detail some of the features of the model we develop. Slightly modifying the conceptual scheme used by e.g. Dormieux et al. (1991) we conceive a kinematical description which seems suitable to model porous systems which are open with respect to the fluid constituent: i.e. systems in which the fluid can freely leave or enter the porous solid matrix through which it flows. Indeed, while we still use as a basic kinematical descriptor the solid matrix macroscopic placement field χ_s we replace the fluid macroscopic placement field χ_f with a macroscopic field ϕ_s defined in the homogenized macroscopic solid reference configuration. The field ϕ_s maps any solid material particle $\mathbf{X}_s$ into a precise particle $\mathbf{X}_f$ in the fluid reference configuration: it is the fluid material particle which occupies, at the given instant, the same spatial position as $\mathbf{X}_s$. The adjective "macroscopic" in the previous sentences is intended to remind to the reader that the model which is used here does not attempt to describe in a detailed way how the complex geometrical structure of the pores varies in the deformation process (see infra for a discussion of this point). Obviously the placement field χ_f can be easily recovered as it equals $\left(\chi_s \circ \phi_s^{-1}\right)$.

The introduced kinematical description is adapted to describe the evolution of porous systems in which the solid matrix is open to fluid filtration (as it happens in consolidation problems, see e.g. Mandel (1953), Cryer (1963), Madeo (2008)). For such open systems, it is necessary to follow the placement of a solid matrix when an unknown amount of fluid, taken from a given reservoir and free to enter or leave the solid matrix, is saturating its pores and flowing through its interconnection canals.

It has to be remarked explicitly that our approach is "purely macroscopic" and "variational". The approach is purely "macroscopic" because the kinematical descriptors for the solid matrix and saturating fluid can be regarded as "averaged" placement fields obtained from corresponding "microscopic" ones (see infra). Moreover, in the scheme we use, all "microscopic" descriptors which may be relevant are assumed to be given by constitutive equations depending on the two previously introduced basic "macroscopic" placements χ_s and ϕ_s. Here, differently to what done for instance in Dormieux and Stolz (1992), we do not attempt to deduce any "macroscopic" constitutive equations from those valid at "microscopic" level. The approach is "variational" because the evolution equations for the kinematical fields are deduced by paralleling the Hamilton-Rayleigh approach. The variational approach has been successfully adapted to continuous systems in different contexts: see e.g. Seliger and Witham (1968), Germain (1973), Houlsbya, Puzrin (2002), Sonnet et al. (2004) for (dissipative or non-dissipative) Cauchy continua, Bedford and Drumheller (1979) for porous media, Bedford and Drumheller (1978) for immiscible mixtures, Bedford and Drumheller (1983) for structured mixtures and in Mobbs (1982) for viscous fluids.

In this paper we construct an action functional accounting for all conservative phenomena occurring in the system and a Rayleigh dissipation function (i.e. a dissipation given in terms of a quadratic pseudo-potential). When formulating the principle of virtual work, we assume that the work done by inertial and internal conservative forces can be expressed as the first variation of the action functional and that the work done by dissipative actions can be expressed in terms of the Rayleigh dissipation function. The action-based postulation scheme is well posed as the introduced kinematical fields χ_s and ϕ_s are both functions defined on the solid matrix reference configuration. To be able to deal with systems in which surface solid material discontinuities are present we allow χ_s and ϕ_s to present gradient discontinuities concentrated on surfaces. In our analysis we generalize some results found in Batra et al. (1986).

We recover the bulk evolution equations, already available in the literature, which are valid in the regularity points of the kinematical fields. We obtain the boundary conditions valid at solid material discontinuity surfaces which are open to fluid flow. These boundary conditions may be interpreted as a "surface balance of force" and a "surface continuity of chemical potential". An Eulerian form of the first of these conditions (including inertial terms) has been obtained in Dormieux et al. (1991) where the principle of virtual work was applied to multiphase systems.

Several authors (see e.g. Ochoa-Tapia and Whitaker (1995), Jager and Mikelic (2000), Ochoa-Tapia (1995b), Hassanizadeh and Gray (1989)) formulated different boundary conditions to be used at solid material interfaces separating porous media and pure fluid. The main part of their efforts was directed to the justification and discussion of the boundary conditions originally proposed by Beavers and Joseph (1967), and Saffman (1971) for describing dissipation phenomena at the external interface of a porous systems. Some authors also focused on the deduction of such conditions by means of a micro-macro identification method: see e.g. Buridge and Keller (1981), Prat (1988), Chateau and Dormieux (1998), Marle (1982), Chandesris and Jamet (2006) , Chandesris and Jamet (2007), Ochoa-Tapia and Withaker (1995) or Valdes-Paradaa et al. (2006) and references there cited. On the other hand, in Deresiewicz (1963) a set of boundary conditions valid at interfaces between dissimilar fluid-filled porous media are proposed which assure uniqueness of the solution of field equations proposed by Biot (1956).

The boundary condition stating the "surface continuity of chemical potential" obtained in the present paper include an inertial term which, to our knowledge, is not deduced in the literature. It generalizes the condition found e.g. in Baek and Srinivasa (2004) (the spirit of which is very similar to the one adopted here). Several versions of "surface balance of force" or "kinematical boundary conditions" can be found in Deresiewicz (1963), Ciezko and Kubik (1998a-b), Debergue et al. (1999), Goyeau et al. (2003), Haber and Mauri (1983), Kuznetsov (1997), Le Bars and Grae Worster (2006), Levy and Sanchez-Palencia (1975), Rajagopal and Tao (1995), Sharma (2008), Ochoa-Tapia and Whitaker (1998).

Our boundary conditions seem suitable to describe (macroscopically) the behavior of solid material interfaces open to fluid flow. They are deduced without introducing any "small perturbation" assumption, so that they seem suitable also when the assumptions of small deformations of the solid matrix and linearized Stokes fluid flow cannot be accepted. However our results are subject to all the limitations implicit in any Rayleigh-like description of dissipative phenomena based on the introduction of a pseudo-potential.

The newly found boundary conditions are effectively Galilean invariant. To check this statement we revisited Reynolds Transport Theorem and Hadamard Conditions to derive some kinematical formulas implicitly used already by Gavrilyuk et al. (1997-1998). In these papers these formulas were needed to show that some evolution equations and boundary conditions assume the form of conservation laws.

We explicit warn the reader that: i) we decided to introduce an explicit notation for distinguish fields defined on the solid-reference configuration from those defined on the spatial or fluid-reference configuration, ii) we found more convenient to deduce all kinematical formulas using a space-time (Galilean) four-dimensional formalism, iii) we did separate all kinematical deductions and properties from those which are more physical in nature.

Concerning point i) we start remarking that when studying one constituent continua it is possible to proceed in presence of an abuse of notation in which fields defined in different configurations (and therefore corresponding to different mathematical functions) are denoted with the same symbol. This does not seem careful enough when multicomponent continua are considered. Indeed such an abuse of notation is, in this case, even more risky than usual, as one is dealing with models where it is necessary to introduce many different placement fields and where discontinuity surfaces for at least one of these fields may be present. In fluid-saturated porous media at least three configurations, and therefore domains of definitions for all considered fields, need to be considered. Therefore we use a notation which is more precise than the usual one, as it allows us to specify clearly for every considered tensor field in which spatial or material domain it is defined. Should the reader be disturbed by the notation which we introduced he is invited to recover the standard one simply ignoring all the circled superscripts.

Concerning point ii) we remark that it is simpler and more convenient to consider (as done for instance by Gavrilyuk and Gouin (1999)) the four-dimensional Galilean space-time as domain for all handled kinematical fields. In this way: a) any moving bi-dimensional surfaces in the physical space becomes a fixed co-dimension one surface in four dimensional space-time, b) piecewise regular spatial fields depending on time when regarded as fields with domain in four-dimensional space-time suffer discontinuities across fixed surfaces, c) space and time differentiation, space gradients and time derivatives, deformation gradients and velocities are dealt with in a more compact and unified manner, d) Hadamard jump conditions and Reynolds transport theorems assume a very simple form, e) as a consequence, some useful –but involved– kinematical relationships are easily seen to stem from elementary differential geometric ones. Indeed, the four-dimensional Galilean space-time is the suitable setting to be used in order to deduce from some well-known results in differential geometry many properties of piecewise differentiable tensor fields. Even if it seems possible to consider weaker regularity conditions (see e.g. Savar and Tomarelli (1998)) we try to render the presentation the simplest possible still choosing the admissible kinematical fields to be general enough to describe the phenomena we have in mind.

Concerning point iii): in our deduction it was necessary to deal with some important topics in differential geometry, concerning the mathematical properties of tensor fields which can be expressed as gradients of other tensor fields. In the four-dimensional setting we have chosen, this is equivalent to study kinematical properties of multicomponent continua. Sometimes this kinematical study is presented together with topics the nature of which is more specifically mechanical, i.e. related to the postulation scheme -based on phenomenological considerations- which is assumed in a specific modeling situation. We have chosen to keep separate all kinematical considerations. The abuse of notation mentioned at point i) is even more misleading when kinematical assumptions for placement fields are mixted with the phenomenological ones characterizing either the solid or the fluid constituent behavior. The conjunction of all these confusing choices may loose the reader in an indistinct list of properties the origin of which is unclear.

Referring to de Boer (1996-2000-2005), Rajagopal and Tao (1995), Dormieux et al. (2006) for an exhaustive and clear review of the development of porous media theory we limit ourselves to recall the pioneering works of Fillunger

(1936) (which were made available to the engineers community by Terzaghi (1943) and Biot (1941)).

The stream of research efforts which were thus originated produced several different families of mathematical models differing in the detail in which they aim to describe the reference and current configurations of solid and fluid constituents.

Indeed, the solid matrix, when displacing from its reference configuration, occupies a different spatial region which delineates a different empty pore region left to the fluid constituent. Such a region can have a very complex time-variable shape: therefore the complete description of its evolution is correspondingly very difficult. Depending on the detail which is required in such a description one can introduce a macroscopic or a microscopic model.

In the context of the theory of porous solids a "*purely microscopic model*" is one in which the kinematical description allows for the complete characterization of the shapes of all matrix internal pores and of fluid density and velocity at any point inside these pores. In the present paper, instead, we consider a mathematical model for the description of saturating-fluid flow in a porous matrix (having enough interconnected pores so to allow such flow) which is *purely macroscopic* in nature. In purely macroscopic models the "internal" shape of the porous solid matrix, i.e. the shape of its internal pores, is not described by any kinematical field and therefore the solid matrix kinematical description is limited to the introduction of a "homogenized" or "macroscopic" placement field χ_s. This field is defined on a "homogenized" reference configuration for the solid matrix in which a solid material particle represents a cluster of pores together with that part of solid matrix which is delineating them (for a discussion of the mentioned homogenization procedure see e.g. Marle (1982), De Buhan et al. (1998a-b), Hornung (1997)). The placement of such a macroscopic particle represents the spatial region occupied by the quoted cluster of pores: clearly the Eulerian mass density related to it is related to the solid mass effectively placed in the given Eulerian volume. Thus an "apparent" solid mass density, differing from the mass density of the material constituting the solid matrix, is associated to the introduced macroscopic solid placement field. Similarly the description of the kinematics of the fluid constituent flowing through the pores, delineated by the solid matrix, is obtained in a purely macroscopic model by means of the "homogenized" placement function χ_f defined on a "homogenized" fluid reference configuration. The velocity and apparent mass density related to such a macroscopic placement field do not account for the variations of the "microscopic" fluid velocity and mass density fields which occur inside the pores. Recall that in the present paper we prefer to consider the field ϕ_s instead of χ_f: this is more convenient as ϕ_s is defined in the same domain as χ_s.

One particular aspect of *purely macroscopic models* has been sometimes regarded as their main conceptual weakness. It concerns the physical interpretation which has to be associated to their basic kinematical descriptors, i.e. the "homogenized" placement fields: indeed a "homogenized" solid particle occupies, at a given instant, the same place as a "homogenized" solid particle. Nevertheless this circumstance is not surprising if one carefully considers the conceptual modeling assumptions underlying purely macroscopic models. The intuitive interpretation we just came to give to purely macroscopic theories for fluid saturated porous solids gains merit once grounded from a mathematical point of view by the so called "Theory of Homogenization" i.e. the mathematical theory aiming to rigorously deduce macroscopic models from microscopic ones Hornung (1997) with the references there cited, Chateau and Dormieux (1995)).

The importance of the Theory of Homogenization cannot be denied. However it is always possible and very useful to formulate "directly" a macroscopic theory without being forced to deduce it from a "purely microscopic" one. Moreover up to now very few results are available about the rigorous deduction of the macroscopic theory of Darcy flow through a deformable porous medium.

Indeed, it is always possible (and often suitable) to develop a macroscopic model independently from any microscopic one. Recall that Cauchy continuum mechanical models for one constituent bodies are formulated in a "direct" way without any reference to "atomistic" or "molecular" models and that very few practical models are rigorously justified by means of homogenization procedures. In general a mathematically coherent macroscopic model can be always formulated and supplies a useful guidance to the deduction procedure which starts from microscopic models. These procedures are often used to supply effective macroscopic constitutive equations in terms of the relevant microscopic properties of considered systems Allaire (1989-1991), for deduction of rigorous results concerning Darcy flow and to Pan and Horne (2001) Lee (2004), Kaasschieter and Frijns (2003) and references there cited for those concerning deformable matrices).

One can call "microscopic models" those intermediate models in which the macroscopic kinematical description is refined enough to describe in a more or less detailed way the shapes of the spatial regions separately occupied by solid and fluid constituents and some aspects of the motion of the material occupying these regions. The more detailed is the description of the shape of the solid porous matrix, the more "microscopic" is the formulated model. It is clear that different microscopic features of the pore shapes may be retained in the kinematical description : in some descriptions one could decide to account only for the ratio of volumes of the regions occupied respectively by solid and fluid constituents (thus introducing the solid volume fraction kinematical field) or for the shape of the canals interconnecting the pore (thus introducing a tortuosity tensor field) or for some geometrical features of the pores (thus introducing, for instance, the ratio between two characteristic lengths of the pore). Adding more and more kinematical descriptors one can more and more precisely approximate the purely microscopic theory. The choice of an "approximating" or "intermediate" microscopic theory results from a compromise between the need of a precise description of complex phenomena and the (computational or analytical) difficulties encountered.

We conclude remarking that the aforementioned modeling efforts respond to a strong demand from applications. Innumerable engineering problems require the design and the control of complex systems in which the flow of a fluid occur in a region partially occupied by a deformable solid matrix, the pores of which are interconnected.

Soil mechanics, geotechnical engineering and geology must supply the theoretical tools for controlling consolidation and subsidence phenomena, which are often influenced by related fluid filtration or flow phenomena (see e.g. Mandel (1953), Terzaghi (1943)) or the phenomena involved in earthquakes (see e.g. Yang (1999) or in the bradyseism and in the related micro-earthquakes in the Phlegraean Fields - Campi Flegrei region (South Italy) (see dell'Isola a et al (1998) and references there cited, Casertano et al (1976), Orsi et al. (1999)).

In biomechanics some phenomena related to the flow of fluids in a deformable porous matrix are also of interest: bone tissues are porous and several different fluids, with different properties, filtrate or flow through those pores which are interconnected. Indeed it seems now evident that bone tissue growth is regulated by a feed-back control system in which the effect of tissue deformation on fluid flow plays a central role (see e.g. Cowin (2001)).

Underground engineering (e.g. when designing or maintaining underground cavities for stocking nuclear wastes or gas) also has to face relevant problems involving phenomena of fluid filtration and flow in a porous matrix coupled to cracks growth and related increase of pore volume fraction and cracks interconnection (see e.g. dell'Isola et al. (2000), dell'Isola et al. (2003) and references there quoted).

2 Properties of the Gradients of C^2 Vector Fields

In this section the properties of smooth tensor fields which are the gradient of some vector field are investigated.

It is usual practice in mathematical physics, when transporting tensor fields by means of changes of variables, to use the same notation for the different tensor fields. This leads to some difficulties which are overcome when dealing with a one-constituent medium by introducing the adapted notations of material and Eulerian space-time derivatives. Such abuse of notation can be very misleading in the case of multi-constituents media where several diffeomorphisms are present. This is why we introduce a more precise notation which, although burdening, seems to be unavoidable in this context.

Notation 1. *Let $\mathbb{B}_a$ and $\mathbb{B}_b$ be two regular open subdomains of $\mathbb{R}^n$, and let $\boldsymbol{\mathcal{X}}$ be an homeomorphism from $\mathbb{B}_a$ onto $\mathbb{B}_b$. Given two tensor fields $\mathbb{t}$ and $\mathbb{z}$ defined on $\mathbb{B}_a$ and $\mathbb{B}_b$ respectively, we denote*

$$\mathbb{t}^{ⓑ} := \mathbb{t} \circ \boldsymbol{\mathcal{X}}^{-1}, \quad \mathbb{z}^{@} := \mathbb{z} \circ \boldsymbol{\mathcal{X}}. \tag{1}$$

It is trivial that $\left(\mathbb{t}^{ⓑ}\right)^{@} = \mathbb{t}$ and $\left(\mathbb{z}^{@}\right)^{ⓑ} = \mathbb{z}$.

Notation 2. *For any differentiable k-th order tensor field $\mathbb{t} = \mathbb{t}_{i_1 \dots i_k}$ we denote by $\mathbb{V}\,\mathbb{t}$ its gradient and, when $k \geq 1$ and i_k varies in $\{1, 2, ..., n\}$, by $\mathbb{DIV}\,\mathbb{t}$ its divergence.*

The components ofe these tensors are given by[1]

$$\left(\mathbb{V}\,\mathbb{t}\right)_{i_1 i_2 \dots i_{k+1}} := \partial_{i_{k+1}} \left(\mathbb{t}_{i_1 i_2 \dots i_k}\right),$$

$$\left(\mathbb{DIV}\,\mathbb{t}\right)_{i_1 i_2 \dots i_{k-1}} := \sum_{i_k=1}^{n} \partial_{i_k} \left(\mathbb{t}_{i_1 i_2 \dots i_k}\right).$$

Notation 3. *In this section we assume that $\boldsymbol{\mathcal{X}}$ is a C^2-diffeomorphism and, in order to lighten notations, we simply denote $\mathbb{F} := \mathbb{V}\boldsymbol{\mathcal{X}}$ and $\mathbb{J} := \det \mathbb{F}$.*

Using the chain rule, it is easy to recognize that $\mathbb{V}\boldsymbol{\mathcal{X}}^{-1} = \left(\mathbb{F}^{ⓑ}\right)^{-1} = \left(\mathbb{F}^{-1}\right)^{ⓑ}$ and so $\det\left(\mathbb{V}\boldsymbol{\mathcal{X}}^{-1}\right) = \left(\mathbb{J}^{ⓑ}\right)^{-1} = \left(\mathbb{J}^{-1}\right)^{ⓑ}$. Moreover, if $\mathbb{t}$ and $\mathbb{z}$ are C^1 tensor

[1] Here and from now on the symbol ∂_j indicate the partial derivative of a function with respect to the j-th component of its argument. Moreover in order to lighten notations, we adopt the Einstein summation convention on repeated indexes dropping the summation symbol.

fields defined on $\mathbb{B}_a$ and $\mathbb{B}_b$ respectively: [2]

$$\mathbb{\nabla}\mathbb{z}^{@} = (\mathbb{\nabla}\mathbb{z})^{@} \cdot \mathbb{F} \quad \text{and} \quad \mathbb{\nabla}\mathbb{t}^{ⓑ} = (\mathbb{\nabla}\mathbb{t})^{ⓑ} \cdot \left(\mathbb{F}^{-1}\right)^{ⓑ}. \tag{2}$$

Using these notations, the well-known change of variables formula read

$$\int_{\mathbb{B}_b} \mathbb{z}\, d\mathbb{B}_b = \int_{\mathbb{B}_a} \mathbb{z}^{@}\, \mathbb{J}\, d\mathbb{B}_a, \quad \text{or}$$

$$\int_{\mathbb{B}_a} \mathbb{t}\, d\mathbb{B}_a = \int_{\mathbb{B}_b} \mathbb{t}^{ⓑ} \left(\mathbb{J}^{-1}\right)^{ⓑ} d\mathbb{B}_b, \tag{3}$$

where $d\mathbb{B}_a$ and $d\mathbb{B}_b$ denote the volume measures on $\mathbb{B}_a$ and $\mathbb{B}_b$ respectively and will be omitted in the sequel as no confusion can arise.

Next Proposition gives a transport formula for the divergence operator and states an important property for the gradient of a diffeomorphism.

Proposition 1. *Let $\boldsymbol{\chi}$ be a C^2-diffeomorphism between the domains $\mathbb{B}_a$ and $\mathbb{B}_b$. For any differentiable tensor field $\mathbb{z}$ (of order ≥ 1) defined on $\mathbb{B}_b$ the following equation holds*

$$\mathbb{DIV}\left(\mathbb{J}\, \mathbb{z}^{@} \cdot \mathbb{F}^{-T}\right) = \mathbb{J}\, (\mathbb{DIV}\, \mathbb{z})^{@}. \tag{4}$$

In particular

$$\mathbb{DIV}\left(\mathbb{J}\mathbb{F}^{-T}\right) = 0. \tag{5}$$

Proof. Let us consider a differentiable scalar field ψ with compact support included in $\mathbb{B}_a$. Owing to the regularity assumptions on $\boldsymbol{\chi}$, the corresponding scalar field $\psi^{ⓑ}$ on $\mathbb{B}_b$ has compact support and is differentiable on $\mathbb{B}_b$. A simple change of variables gives

$$\int_{\mathbb{B}_b} \psi^{ⓑ}\, \mathbb{DIV}\, \mathbb{z} = \int_{\mathbb{B}_a} \psi\, (\mathbb{DIV}\, \mathbb{z})^{@}\, \mathbb{J}. \tag{6}$$

On the other hand, using the divergence theorem and recalling that $\psi^{ⓑ}$ has compact support

$$\int_{\mathbb{B}_b} \psi^{ⓑ}\, \mathbb{DIV}\, \mathbb{z} = -\int_{\mathbb{B}_b} \mathbb{z} \,.\, \mathbb{\nabla}\psi^{ⓑ}. \tag{7}$$

[2] Given two tensors T and S of order k and h the components of which are $T_{i_1 \ldots i_k}$ and $S_{j_1 \ldots j_h}$ the tensor $T \cdot S$ is the $(k+h) - 2$ order tensor with components $(T \cdot S)_{i_1 \ldots i_{k-1}\, j_2 \ldots j_h} = \sum_m T_{i_1 \ldots i_{k-1} m} S_{m j_2 \ldots j_h}$.

Starting from Eq. (7), using successively formula (2) for $\psi^{ⓑ}$, a change of variables, the divergence theorem and the fact that ψ has compact support we get the following equalities

$$\int_{\mathbb{B}_b} \psi^{ⓑ} \mathbb{DIV}\, \mathbb{z} = -\int_{\mathbb{B}_b} \mathbb{z} \cdot \left(\mathbb{\nabla}\, \psi \,\cdot \mathbb{F}^{-1}\right)^{ⓑ} = -\int_{\mathbb{B}_a} \mathbb{z}^{@} \cdot \mathbb{F}^{-T} \cdot \mathbb{\nabla} \psi \mathbb{J}$$
$$= \int_{\mathbb{B}_a} \psi \mathbb{DIV} \left(\mathbb{J}\; \mathbb{z}^{@} \cdot \mathbb{F}^{-T}\right). \tag{8}$$

The comparison between Eq. (6) and the last term of Eq.(8) gives

$$\int_{\mathbb{B}_a} \psi \left(\mathbb{DIV}\, \mathbb{z}\,\right)^{@} \mathbb{J} = \int_{\mathbb{B}_a} \psi \; \mathbb{DIV} \left(\mathbb{J}\; \mathbb{z}^{@} \cdot \mathbb{F}^{-T}\right).$$

The fact that this last equality is satisfied for any ψ with compact support included in $\mathbb{B}_a$ proves (4). It is enough to apply (4) choosing for $\mathbb{z}$ the identity tensor to get (5). □

Note that the previous proposition can be applied to $\boldsymbol{\mathcal{X}}^{-1}$ so that, for any differentiable tensor field $\mathbb{t}$ (of order ≥ 1) defined on $\mathbb{B}_a$ one gets

$$\mathbb{DIV} \left(\mathbb{J}^{-1}\; \mathbb{t} \cdot \mathbb{F}^{T}\right)^{ⓑ} = \left(\mathbb{J}^{-1} \mathbb{DIV}\, \mathbb{t}\,\right)^{ⓑ} \quad \text{and}$$
$$\mathbb{DIV} \left(\mathbb{J}^{-1}\; \mathbb{F}^{T}\right)^{ⓑ} = 0. \tag{9}$$

Let $\mathbb{U}$ be a C^1 vector field defined in $\mathbb{B}_b$ and let $\mathbb{U}^{@}$ be its corresponding vector field on $\mathbb{B}_a$. Let $\mathbb{N}_a$ and $\mathbb{N}_b$ be the outward unit normal vectors to $\partial\mathbb{B}_a$ and $\partial\mathbb{B}_b$ respectively, then

$$\int_{\partial\mathbb{B}_b} \mathbb{U} \cdot \mathbb{N}_b = \int_{\partial\mathbb{B}_a} \mathbb{U}^{@} \cdot \left(\mathbb{J}\; \mathbb{F}^{-T} \cdot \mathbb{N}_a\right), \tag{10}$$

Proof. Recalling Eq. (4), one gets

$$\int_{\partial\mathbb{B}_b} \mathbb{U} \cdot \mathbb{N}_b = \int_{\mathbb{B}_b} \mathbb{DIV} \left(\mathbb{U}\right) = \int_{\mathbb{B}_a} \left(\mathbb{DIV} \left(\mathbb{U}\right)\right)^{@} \mathbb{J}$$
$$= \int_{\mathbb{B}_a} \mathbb{DIV}(\mathbb{J}\, \mathbb{U}^{@} \cdot \mathbb{F}^{-T}) = \int_{\partial\mathbb{B}_a} \mathbb{U}^{@} \cdot \left(\mathbb{J}\, \mathbb{F}^{-T} \cdot \mathbb{N}_a\right).$$

□

Note that the last corollary applied to $\boldsymbol{\mathcal{X}}^{-1}$ reads

$$\int_{\partial\mathbb{B}_a} \mathbb{U}^{@} \cdot \mathbb{N}_a = \int_{\partial\mathbb{B}_o} \mathbb{U} \cdot \left(\left(\mathbb{J}^{-1}\mathbb{F}^T\right)^{ⓑ} \cdot \mathbb{N}_b \right).$$

3 Properties of the Gradients of Piecewise C^1 Vector Fields

In this section we define piecewise differentiable vector fields and investigate the properties of their gradients.

Let $\mathbb{S}_a$ be a smooth codimension-one hyper-surface in $\mathbb{B}_a$; this means that, at least locally, there exists a parametric representation of $\mathbb{S}_a$, i.e. an open subset Ω of $\mathbb{R}^{n-1}$ and a smooth embedding $\varphi \in C^1(\Omega, \mathbb{B}_a \subset \mathbb{R}^n)$ such that $\mathbb{S}_a := \varphi(\Omega)$. By definition of an embedding, for any $\mathbf{x} = \varphi(\mathbf{s}) \in \mathbb{S}_a$ the vectors

$$\mathbb{T}_i(\mathbf{x}) := \partial_i \varphi|_{\mathbf{s}=\varphi^{-1}(\mathbf{x})}, \quad i = 1, 2, ..., n-1$$

make a basis spanning the tangent space $\mathcal{T}_{\mathbf{x}}(\mathbb{S}_a)$ of $\mathbb{S}_a$ at $\mathbf{x}$. The orthogonal space to $\mathcal{T}_{\mathbf{x}}(\mathbb{S}_a)$ is one-dimensional: there exists a unique unit vector $\mathbb{N}_a(\mathbf{x})$ in this space which completes $\{\mathbb{T}_i(\mathbf{x})\}$ in a direct basis of $\mathbb{R}^n$. This vector $\mathbb{N}_a$ locally provides an orientation for $\mathbb{S}_a$ and we call it the *normal* to $\mathbb{S}_a$.

Notation 4. *Let $\mathbb{t}$ be a tensor field defined on $\mathbb{B}_a$ (and consequently on $\mathbb{S}_a$). We say that $\mathbb{t}$ is* differentiable on $\mathbb{S}_a$ *if $\mathbb{t} \circ \varphi \in C^1(\Omega)$. The* surface gradient $\mathbb{\nabla}^{\mathbb{S}_a}\mathbb{t}(\mathbf{x})$ *at point $\mathbf{x}$ is the linear operator which, to any tangent vector $\mathbb{T} = \sum_{i=1}^{n-1} v_i \mathbb{T}_i$, associates the derivative of $\mathbb{t}$ in the direction $\mathbb{T}$ defined by*

$$\mathbb{\nabla}^{\mathbb{S}_a}\mathbb{t}(\mathbf{x}) \cdot \mathbb{T} := \sum_{i=1}^{n-1} v_i \, \partial_i \left(\mathbb{t} \circ \varphi \right).$$

Recall that, even if the basis $\mathbb{T}_i$ depends on the choice of the parametrization φ, the surface gradient (regarded as a linear operator) does not.

Notation 5. *We say that a tensor field $\mathbb{t}$ defined on $\mathbb{B}_a$ is* piecewise continuous *(or briefly C^0_{pw}) if there exists a smooth codimension-one C^1 hypersurface $\mathbb{S}_a$ in $\mathbb{B}_a$ such that $\mathbb{t}$ belongs to $C^0(\mathbb{B}_a \backslash \mathbb{S}_a, \mathbb{R}^p)$ and admits continuous traces $\mathbb{t}^+$ and $\mathbb{t}^-$ on both sides of $\mathbb{S}_a$. The quantity*

$$[|\mathbb{t}(\mathbf{x})|] := \mathbb{t}^+(\mathbf{x}) - \mathbb{t}^-(\mathbf{x})$$

is called jump *of $\mathbb{t}$ through the surface $\mathbb{S}_a$ at point $\mathbf{x}$. Moreover, the surface $\mathbb{S}_a$ is said to be the* singularity surface *of the field $\mathbb{t}$. When $\mathbb{t}$ has vanishing* *jump across the singularity surface, we simply indicate by* $\mathbb{t}$ *the common value* $\mathbb{t}^+ = \mathbb{t}^-$.

Indeed, at least locally, the normal $\mathbb{N}_a$ to $\mathbb{S}_a$ defines the "upper" part $\mathbb{B}_a^+$ of $\mathbb{B}_a$ toward which the normal is pointing and the "lower" part $\mathbb{B}_a^-$ of $\mathbb{B}_a$ in the opposite direction. Then, for any $\mathbf{x} \in \mathbb{S}_a$

$$\mathbb{t}^+(\mathbf{x}) = \lim_{\mathbf{y} \to \mathbf{x}} \mathbb{t}(\mathbf{y}), \quad \mathbf{y} \in \mathbb{B}_a^+; \qquad \mathbb{t}^-(\mathbf{x}) = \lim_{\mathbf{y} \to \mathbf{x}} \mathbb{t}(\mathbf{y}), \quad \mathbf{y} \in \mathbb{B}_a^-.$$

Notation 6. *We say that a tensor field* $\mathbb{t}$ *defined on* $\mathbb{B}_a$ *is* piecewise differentiable *(or briefly* C^1_{pw}*) if it is continuous and if its gradient* $\mathbb{\nabla}\mathbb{t}$ *is* C^0_{pw}.

Property 1. *The well-known Hadamard property (see e.g. Kosinski (1986)) states that the jump of the gradient of a* C^1_{pw} *tensor field* $\mathbb{t}$ *is a rank-one matrix field in the form*

$$\begin{aligned} [|\mathbb{\nabla}\mathbb{t}\,|] &= [|\mathbb{\nabla}\mathbb{t}\,|] \cdot \mathbb{N}_a \otimes \mathbb{N}_a = \left((\nabla\mathbb{t})^+ \cdot \mathbb{N}_a - (\nabla\mathbb{t})^- \cdot \mathbb{N}_a\right) \otimes \mathbb{N}_a \\ &= \left[\left|\frac{\partial \mathbb{t}}{\partial \mathbb{N}_a}\right|\right] \otimes \mathbb{N}_a \end{aligned} \tag{11}$$

In other words, for any $\mathbb{T} \in \mathcal{T}_{\mathbf{x}}(\mathbb{S}_a)$

$$[|\mathbb{\nabla}\mathbb{t}(\mathbf{x}) \cdot \mathbb{T}\,|] = 0. \tag{12}$$

This property simply reflects the fact that $\mathbb{t}$, when restricted to $\mathbb{S}_a$, reduces to a differentiable field and

$$\mathbb{\nabla}^{\mathbb{S}_a}\, \mathbb{t} \cdot \mathbb{T} = (\mathbb{\nabla}\mathbb{t}\,)^+ \cdot \mathbb{T} = (\mathbb{\nabla}\mathbb{t}\,)^- \cdot \mathbb{T} \tag{13}$$

The following proposition states some important consequences of Property 1.

Let $\mathbb{B}_a, \mathbb{B}_b$ be two regular open subsets of $\mathbb{R}^n$ respectively, and let $\boldsymbol{\mathcal{X}} \in C^1_{pw}(\mathbb{B}_a, \mathbb{B}_b)$ with singularity surface $\mathbb{S}_a \subset \mathbb{B}_a$. Assume that $\mathbb{J} \neq 0$ everywhere on $\mathbb{S}_a$, then

(i) For any $\mathbb{T}$ tangent to $\mathbb{S}_a$, $[|\mathbb{F} \cdot \mathbb{T}|] = 0$ on $\mathbb{S}_a$.

(ii) The surface $\mathbb{S}_b := \boldsymbol{\mathcal{X}}(\mathbb{S}_a)$ is a smooth codimension-one C^1 surface in $\mathbb{B}_b$ with tangent vectors $(\mathbb{F} \cdot \mathbb{T}_i)^{ⓑ}$, $i \in \{1, 2, ..., n-1\}$.

(iii) For any normal vector field $\mathbb{M}_b$ to $\mathbb{S}_b$, the following jump condition holds on $\mathbb{S}_a$

$$[|\mathbb{J}^{-1}\mathbb{F}^T|] \cdot \mathbb{M}_b^{@} = 0, \tag{14}$$

(iv) Moreover, the quantity $\mathbb{J}^{-1}\mathbb{F}^T \cdot \mathbb{M}_b^{@}$ which is continuous through the surface is orthogonal to $\mathbb{S}_a$.

Proof. Point (i) is an immediate consequence of (12) if we recall that $\mathbb{F}$ denotes $\mathbb{\nabla}\boldsymbol{\mathcal{X}}$. To prove point (ii) we note that, as $\boldsymbol{\mathcal{X}}$ is C^1_{pw}, its restriction to $\mathbb{S}_a$ is differentiable and so is $\xi := \boldsymbol{\mathcal{X}} \circ \varphi$ which makes a parametrization for $\mathbb{S}_b := \boldsymbol{\mathcal{X}}(\mathbb{S}_a)$. Moreover, for any $\mathbf{y} = \xi(\mathbf{s}) \in \mathbb{S}_b$ and for any $i = 1, 2, ..., n-1$, the vectors

$$\partial_i \xi|_{\mathbf{s}=\xi^{-1}(\mathbf{y})} = \mathbb{F}^{\pm}|_{\boldsymbol{\mathcal{X}}^{-1}(\mathbf{y})} \cdot \partial_i \varphi|_{\varphi^{-1}(\boldsymbol{\mathcal{X}}^{-1}(\mathbf{y}))} = \left(\mathbb{F}^{\pm} \cdot \mathbb{T}_i\right)^{ⓑ} = \left(\mathbb{F} \cdot \mathbb{T}_i\right)^{ⓑ}$$

make a basis spanning the tangent space $\mathcal{T}_{\mathbf{y}}(\mathbb{S}_b)$ of $\mathbb{S}_b$ at $\mathbf{y}$.

To prove points (iii) and (iv) we consider, at any point $\mathbf{x}$ of $\mathbb{S}_a$, the three linear applications L_a and $L_b^{\pm}$ respectively defined on $\mathbb{R}^n$ by

$$L_a(\mathbb{U}) := \det(\mathbb{T}_1, \mathbb{T}_2, ..., \mathbb{T}_{n-1}, \mathbb{U}), \tag{15}$$

$$L_b^{\pm}(\mathbb{V}) := \det\left(\left(\mathbb{F}^{\pm} \cdot \mathbb{T}_1\right)^{ⓑ}, \left(\mathbb{F}^{\pm} \cdot \mathbb{T}_2\right)^{ⓑ}, ..., \left(\mathbb{F}^{\pm} \cdot \mathbb{T}_{n-1}\right)^{ⓑ}, \mathbb{V}\right). \tag{16}$$

Owing to point (ii) one easily gets $L_b^{+}(\mathbb{V}) = L_b^{-}(\mathbb{V})$. In virtue of the Rietz theorem there exist unique vectors $\tilde{\mathbb{M}}_a$ and $\tilde{\mathbb{M}}_b$ such that

$$L_a(\mathbb{U}) = \tilde{\mathbb{M}}_a \cdot \mathbb{U}, \quad \forall \mathbb{U} \in \mathbb{R}^n, \quad L_b(\mathbb{V}) = \tilde{\mathbb{M}}_b \cdot \mathbb{V} \quad \forall \mathbb{V} \in \mathbb{R}^n. \tag{17}$$

Hence

$$\tilde{\mathbb{M}}_a \cdot \mathbb{U} = \det(\mathbb{T}_1, \mathbb{T}_2, ..., \mathbb{T}_{n-1}, \mathbb{U}), \tag{18}$$

and

$$\begin{aligned} &\tilde{\mathbb{M}}_b \cdot \left(\mathbb{F}^{\pm} \cdot \mathbb{U}\right)^{ⓑ} \\ &= \det\left(\left(\mathbb{F}^{\pm} \cdot \mathbb{T}_1\right)^{ⓑ}, \left(\mathbb{F}^{\pm} \cdot \mathbb{T}_2\right)^{ⓑ}, ..., \left(\mathbb{F}^{\pm} \cdot \mathbb{T}_{n-1}\right)^{ⓑ}, \left(\mathbb{F}^{\pm} \cdot \mathbb{U}\right)^{ⓑ}\right) \\ &= \left(\mathbb{J}^{\pm}\right)^{ⓑ} \det\left(\left(\mathbb{T}_1\right)^{ⓑ}, \left(\mathbb{T}_2\right)^{ⓑ}, ..., \left(\mathbb{T}_{n-1}\right)^{ⓑ}, \mathbb{U}^{ⓑ}\right). \end{aligned} \tag{19}$$

From (18) and (19) we get

$$\tilde{\mathbb{M}}_b^{ⓐ} \cdot \left(\mathbb{F}^{\pm} \cdot \mathbb{U}\right) = \mathbb{J}^{\pm} \det(\mathbb{T}_1, \mathbb{T}_2, ..., \mathbb{T}_{n-1}, \mathbb{U})$$

and so

$$\tilde{\mathbb{M}}_b^{ⓐ} \cdot \left(\mathbb{F}^{\pm} \cdot \mathbb{U}\right) = \mathbb{J}^{\pm}\, \tilde{\mathbb{M}}_a \cdot \mathbb{U}$$

Since this last identity is satisfied for any $\mathbb{U} \in \mathbb{R}^n$ then

$$\tilde{\mathbb{M}}_a = \left(\mathbb{J}^{+}\right)^{-1} \left(\mathbb{F}^{+}\right)^{T} \cdot \tilde{\mathbb{M}}_b^{ⓐ} = \left(\mathbb{J}^{-}\right)^{-1} \left(\mathbb{F}^{-}\right)^{T} \cdot \tilde{\mathbb{M}}_b^{ⓐ}, \tag{20}$$

thus

$$[|\mathbb{J}^{-1}\mathbb{F}^{T}|] \cdot \tilde{\mathbb{M}}_b^{ⓐ} = 0. \tag{21}$$

From (18), (19) we also get that for any $i \in \{1, 2, ..., n-1\}$,

$$\tilde{\mathbb{M}}_a \cdot \mathbb{T}_i = 0, \quad \text{and} \quad \tilde{\mathbb{M}}_b \cdot \left(\mathbb{F}^{\pm} \cdot \mathbb{T}_i\right)^{ⓑ} = 0.$$

Hence, $\tilde{\mathbb{M}}_a$ and $\tilde{\mathbb{M}}_b$ belong to the one dimensional orthogonal spaces to $\mathbb{S}_a$ and $\mathbb{S}_b$ respectively. As they clearly are non vanishing, Eq. (14) remains valid for any $\mathbb{M}_b$ normal to $\mathbb{S}_b$. □

We call *piecewise diffeomorphism* a C^1_{pw} homeomorphism $\boldsymbol{\mathcal{X}}$ from $\mathbb{B}_a$ onto $\mathbb{B}_b$ such that $\boldsymbol{\mathcal{X}}^{-1} \in C^1_{pw}(\mathbb{B}_b, \mathbb{B}_a)$.

Note that if $\boldsymbol{\mathcal{X}}$ is a piecewise diffeomorphism with singularity surface $\mathbb{S}_a$ then the previous proposition can be applied to both $\boldsymbol{\mathcal{X}}$ and $\boldsymbol{\mathcal{X}}^{-1}$. Thus we also have, for any $\mathbb{M}_a \perp \mathbb{S}_a$ the following jump condition on $\mathbb{S}_b = \boldsymbol{\mathcal{X}}(\mathbb{S}_a)$

$$\left[\left|\left(\mathbb{J}\mathbb{F}^{-T}\right)^{ⓑ}\right|\right] \cdot \mathbb{M}_a^{ⓑ} = 0, \tag{22}$$

and the quantity $\left(\mathbb{J}\mathbb{F}^{-T} \cdot \mathbb{M}_a\right)^{ⓑ}$ is orthogonal to $\mathbb{S}_b$. If $\mathbb{t}$ is a differentiable tensor field defined on $\mathbb{B}_a$ then its corresponding tensor field $\mathbb{t}^{ⓑ}$ on $\mathbb{B}_b$ may be not differentiable on $\mathbb{S}_b$. However, $\mathbb{t}^{ⓑ}$ is C^1_{pw} and on $\mathbb{S}_b$ we have

$$\left(\mathbb{\nabla}\mathbb{t}^{ⓑ}\right)^+ = \left(\mathbb{\nabla}\mathbb{t} \cdot \left(\mathbb{F}^+\right)^{-1}\right)^{ⓑ}; \quad \left(\mathbb{\nabla}\mathbb{t}^{ⓑ}\right)^- = \left(\mathbb{\nabla}\mathbb{t} \cdot \left(\mathbb{F}^-\right)^{-1}\right)^{ⓑ}. \tag{23}$$

4 Bulk Kinematical Identities and Hadamard Conditions at Moving Boundaries

Let $\chi : \mathbb{B}_a := B_a \times (0, T) \to \mathbb{R}^3$ be the placement map of a three dimensional continuum; the fixed domain $B_a \subset \mathbb{R}^3$ is usually referred to as the reference configuration and the moving volume $B_b(t) := \chi(B_a, t)$ as the current configuration. We assume that, at any instant t, $\chi(., t)$ is a C^1_{pw} diffeomorphism.

From now on, we denote $\mathbf{F} := \nabla\chi$ and $\mathbf{v} := \partial\chi/\partial t$, the usual 3D space gradient of the map χ in the domain B_a and the usual velocity field in B_a.

The singularity surface $S_a(t)$ is a moving surface which can be parametrized, at least locally, by a function $\varphi : (\Omega \times (0, T)) \to B_a$ where Ω is an open subset of $\mathbb{R}^2$.

It is well known that, as only the global position of the set $S_a(t)$ is relevant, the physical quantities attached to $S_a(t)$ should not depend on the choice of its parametrization.

From now on, we assume that χ is C^1_{pw} on the space-time domain $\mathbb{B}_a$ and thus that the function φ is C^1. Therefore it is possible to introduce, for any point $\mathbf{x}(t) = \varphi(\mathbf{s},t) \in S_a(t)$, the vectors

$$\tau_i(\mathbf{x}(t)) := \partial_i \varphi|_{\varphi^{-1}(\mathbf{x}(t),t)}, \qquad i = 1,2$$

which span a basis for the tangent plane to $S_a(t)$ and the unit vector $\mathbf{N}_a(\mathbf{x}(t))$ orthogonal to the tangent plane to S_a which completes $\{\tau_i\}$ in a direct basis of $\mathbb{R}^3$. For any $\mathbf{x}$, we introduce the "*velocity of the surface* $S_a(t)$" as:

$$\mathbf{w}(\mathbf{x}(t)) := \left.\frac{\partial \varphi}{\partial t}\right|_{\varphi^{-1}(\mathbf{x}(t),t)}.$$

It is well known, (see e.g.Kosinski (1986)), that even if this velocity depends on the choice of the parametrization φ, its normal part $c_a := \mathbf{w} \cdot \mathbf{N}_a$, commonly called the *celerity* of the surface, does not. Analogously we introduce $\mathbf{N}_b$ and c_b the normal and celerity of $S_b = \chi(S_a)$.

In this section we show that kinematical conditions which are usually derived in different (and sometimes intricate) ways are simple consequences of Hadamard's property. We emphasize that they are purely kinematical constraints and that neither physical assumptions nor balance principles should be used, as it is sometimes done, to derive them.

Proposition 2. *For any tensor fields f and $\mathbf{f}$ defined on B_b ($\mathbf{f}$ of order $k \geq 1$), we have at any regular point of χ:*

$$\operatorname{div}\left(J\,\mathbf{f}^{@} \cdot \mathbf{F}^{-T}\right) = J(\operatorname{div}\mathbf{f})^{@} \quad \textit{and}$$

$$\frac{\partial\left(J\,f^{@}\right)}{\partial t} - \operatorname{div}\left(J f^{@} \otimes \mathbf{F}^{-1} \cdot \mathbf{v}\right) = J\left(\frac{\partial f}{\partial t}\right)^{@}, \tag{24}$$

and in particular,

$$\operatorname{div}\left(J\mathbf{F}^{-T}\right) = 0, \qquad \frac{\partial J}{\partial t} = \operatorname{div}\left(J\mathbf{F}^{-1} \cdot \mathbf{v}\right), \tag{25}$$

$$(\operatorname{div}\mathbf{f})^{@} = \nabla \mathbf{f}^{@} : \mathbf{F}^{-1}, \qquad \left(\frac{\partial f}{\partial t}\right)^{@} = \frac{\partial f^{@}}{\partial t} - \nabla f^{@} \cdot \mathbf{F}^{-1} \cdot \mathbf{v}. \tag{26}$$

Moreover,at any point of the singularity surface S_a:

$$[|\mathbf{F}|] \cdot \tau_1 = 0, \quad [|\mathbf{F}|] \cdot \tau_2 = 0, \quad [|\mathbf{F}|] \cdot c_a \mathbf{N}_a + [|\mathbf{v}|] = 0, \tag{27}$$

$$\left[\left|J^{-1}\mathbf{F}^T \cdot \mathbf{N}_b^{@}\right|\right] = 0, \qquad \left[\left|J^{-1}\left(\mathbf{v} \cdot \mathbf{N}_b^{@} - c_b^{@}\right)\right|\right] = 0 \tag{28}$$

and

$$[|\mathbf{v}|] \otimes \left(J^{-1}\mathbf{F}^T \cdot \mathbf{N}_b^{@}\right) + J^{-1}\left(c_b^{@} - \mathbf{v} \cdot \mathbf{N}_b^{@}\right)[|\mathbf{F}|] = 0. \tag{29}$$

Finally, the normals and celerities of S_a and S_b are related by

$$\mathbf{N}_b = \frac{\left(J\mathbf{F}^{-T} \cdot \mathbf{N}_a\right)^{ⓑ}}{\left\|\left(J\mathbf{F}^{-T} \cdot \mathbf{N}_a\right)^{ⓑ}\right\|} \quad and \quad c_b = \frac{\left(Jc_a + J\left(\mathbf{F}^{-1} \cdot \mathbf{v}\right) \cdot \mathbf{N}_a\right)^{ⓑ}}{\left\|\left(J\mathbf{F}^{-T} \cdot \mathbf{N}_a\right)^{ⓑ}\right\|}. \tag{30}$$

Notice that considering $\boldsymbol{\mathcal{X}}^{-1}$ instead of $\boldsymbol{\mathcal{X}}$ gives equations symmetrical to (24)-(30)

Proof. To prove this proposition, let us extend χ in order to consider it as a fourth-dimensional piecewise diffeomorphism. We consider the C^1_{pw}-diffeomorphism $\boldsymbol{\mathcal{X}}$ defined on $\mathbb{B}_a := B_a \times (0,T)$ by $\boldsymbol{\mathcal{X}}(a,t) := (\chi(a,t),t)$ and let us denote by $\mathbb{B}_b$ its image. The map $\boldsymbol{\mathcal{X}}$ resumes all needed information about the motion of the considered continuum. In particular, $\mathbb{B}_b := \cup_{t\in(0,T)} B_b(t) \times \{t\}$. The singularity surface $\mathbb{S}_a$ of $\boldsymbol{\mathcal{X}}$ is related to the motion of the singularity surface $S_a(t)$ by $\mathbb{S}_a := \cup_{t\in(0,T)} S_a(t) \times \{t\}$.

Notation 7. *If $\mathbb{f}$ is a tensor field of order $k \geq 1$ the components of which are $\mathbb{f}_{i_1,i_2,\ldots,i_k}$, with $i_k \in \{1,2,3,4\}$, we decompose $\mathbb{f}$ writing $\mathbb{f} = (\mathbf{f}\ ,\ f)$. Here $\mathbf{f}$ and f are the tensors of order k and $k-1$ defined by*

$$\mathbf{f}_{i_1 i_2 \ldots i_{k-1}\, j} = \mathbb{f}_{i_1 i_2 \ldots i_{k-1}\, j}, \quad j = 1,2,3, \qquad f_{i_1,i_2,\ldots,i_{k-1}} = \mathbb{f}_{i_1 i_2 \ldots i_{k-1}\, 4}.$$

Using this decomposition for $\mathbb{f}$ and $\mathbb{U}$ we write

$$\mathbb{f} \cdot \mathbb{U} = \mathbf{f} \cdot \mathbf{u} + fu. \tag{31}$$

Notation 8. *The 4D space-time gradient and divergence of a tensor $\mathbb{f}$ defined in the space-time are related to its 3D gradient and time derivative by*

$$\mathbb{\nabla}\mathbb{f} = (\nabla\mathbb{f}, \partial\mathbb{f}/\partial t) \quad and \quad \mathbb{DIV}\mathbb{f} = \operatorname{div}\mathbf{f} + \partial f/\partial t. \tag{32}$$

In particular, if $\mathbb{f}$ is a vector field, the 4×4 matrix $\mathbb{\nabla}\mathbb{f}$ admits the block decomposition[3]:

$$\mathbb{\nabla}\mathbb{f} = \begin{pmatrix} \nabla\mathbf{f} & (\partial\mathbf{f}/\partial t)^T \\ \nabla f & \partial f/\partial t \end{pmatrix}. \tag{33}$$

Applying this block decomposition to the gradient of $\boldsymbol{\mathcal{X}}$ gives

$$\mathbb{F} := \mathbb{\nabla}\,\boldsymbol{\mathcal{X}} = \begin{pmatrix} \mathbf{F} & \mathbf{v} \\ \mathbf{0} & 1 \end{pmatrix}, \tag{34}$$

[3] When defining matrices, we identify any vector with the corresponding row matrix.

and we can remark that $\mathbb{J} := \det \mathbb{F}$ coincides with $J := \det \mathbf{F}$.

Applying the chain rule to the identity $\chi^{-1} \circ \chi = Id$ we get $\partial \chi^{-1}/\partial t = -\mathbf{F}^{-1} \cdot \mathbf{v}$ and consequently

$$\mathbb{F}^{-1} = \begin{pmatrix} \mathbf{F}^{-1} & (-\mathbf{F}^{-1} \cdot \mathbf{v})^T \\ \mathbf{0} & 1 \end{pmatrix}, \qquad \mathbb{F}^{-T} = \begin{pmatrix} \mathbf{F}^{-T} & \mathbf{0} \\ -\mathbf{F}^{-1} \cdot \mathbf{v} & 1 \end{pmatrix}. \tag{35}$$

Let $\mathbb{f} = (\mathbf{f}, f)$ be of order $k \geq 1$. Equation (4) reads

$$\begin{aligned} \operatorname{div}\left(J\, \mathbf{f}^{\,@} \cdot \mathbf{F}^{-T} - J f^{\,@} \otimes \mathbf{F}^{-1} \cdot \mathbf{v}\right) + \frac{\partial \left(J\, f^{\,@}\right)}{\partial t} \\ = J \left(\operatorname{div} \mathbf{f} + \frac{\partial f}{\partial t} \right)^{@} \end{aligned} \tag{36}$$

which implies (24). Similarly, Eq. (5) becomes (25). It is then easy to prove that using (25) in (24) gives (26). We notice that equations (26) encompass the classical relationships between material and Eulerian derivatives.

The singularity surface $\mathbb{S}_a$ admits the parametrization Φ defined by $\Phi(\mathbf{s}, t) = (\varphi(\mathbf{s}, t), t)$ on the set $\Omega \times (0, T)$. A 4D vector $(\mathbf{m}, m)$ is orthogonal to $\mathbb{S}_a$ if it is orthogonal to the three tangent vectors $\partial_1 \Phi = (\tau_1, 0)$, $\partial_2 \Phi = (\tau_2, 0)$ and $\partial \Phi / \partial t = (\mathbf{w}, 1)$; that is if

$$\tau_1 \cdot \mathbf{m} = 0, \quad \tau_2 \cdot \mathbf{m} = 0, \quad \mathbf{w} \cdot \mathbf{m} + m = 0. \tag{37}$$

From the two first equalities we deduce that $\mathbf{m}$ is proportional to the normal $\mathbf{N}_a$ to S_a. Hence the vectors orthogonal to $\mathbb{S}_a$ are those proportional to $\mathbb{M}_a = (\mathbf{N}_a, -c_a)$.

Noticing that the three vectors

$$\mathbb{T}_1 = (\tau_1, 0), \quad \mathbb{T}_2 = (\tau_2, 0), \quad \mathbb{T}_3 = (c_a \mathbf{N}_a, 1) \tag{38}$$

span the tangent hyper-plane to $\mathbb{S}_a$ and applying point (i) of Proposition 3 to these vectors gives (27).

We notice that $\mathbb{M}_b := (\mathbf{N}_b, -c_b)$ is orthogonal to $\mathbb{S}_b$ so that point (iii) of Proposition 3

$$[|\mathbb{J}^{-1}\mathbb{F}^T|] \cdot \mathbb{M}_b^{@} = \left[\left|\left(J^{-1}\mathbf{F}^T \cdot \mathbf{N}_b^{@},\ J^{-1}(\mathbf{v} \cdot \mathbf{N}_b^{@} - c_b^{@})\right)\right|\right] = 0, \tag{39}$$

which implies (28).

Point (iv) of Proposition 3 states that

$\mathbb{J}^{-1}\mathbb{F}^T \cdot \mathbb{M}_b^@ = \left(J^{-1}\mathbf{F}^T \cdot \mathbf{N}_b^@ \, , \, J^{-1}(\mathbf{v} \cdot \mathbf{N}_b^@ - c_b^@)\right)$ is orthogonal to $\mathbb{S}_a$ and so co-linear to $(\mathbf{N}_a \, , \, -c_a)$. This implies

$$\mathbf{N}_a = \frac{J^{-1}\mathbf{F}^T \cdot \mathbf{N}_b^@}{\left\| J^{-1}\mathbf{F}^T \cdot \mathbf{N}_b^@ \right\|} \quad \text{and} \quad c_a = \frac{J^{-1}\left(c_b^@ - \mathbf{v} \cdot \mathbf{N}_b^@\right)}{\left\| J^{-1}\mathbf{F}^T \cdot \mathbf{N}_b^@ \right\|}. \tag{40}$$

Equations (30) are the symmetrical relationships obtained by considering $\boldsymbol{\mathcal{X}}^{-1}$.

Finally let us apply the rank-one property (11) for the jump of the gradient of the map $\boldsymbol{\mathcal{X}}$. It states the existence of a 4D vector $\mathbb{U} = (\mathbf{u} \, , \, u)$, such that

$$[|\mathbb{F}|] = \begin{pmatrix} [|\mathbf{F}|] & [|\mathbf{v}^T|] \\ \mathbf{0} & 0 \end{pmatrix} = \mathbb{U} \otimes \mathbb{M}_a,$$

the space-time decomposition of which gives

$$[|\mathbf{F}|] = \mathbf{u} \otimes \mathbf{N}_a \qquad \text{and} \qquad [|\mathbf{v}|] = -\mathbf{u}\, c_a.$$

Eliminating $\mathbf{u}$ gives the following jump condition on $S_a(t)$:

$$c_a \, [|\mathbf{F}|] = - \, [|\mathbf{v}|] \otimes \mathbf{N}_a. \tag{41}$$

Replacing in this equation $\mathbf{N}_a$ and c_a by formulas (40) gives (29). □

5 Balance Equations and Corresponding Jump Conditions in the Space-Time

Any balance equation for a quantity f on $\mathbb{B}_b$ is of the type $\operatorname{div} \mathbf{f} + \partial f / \partial t = r$, where $\mathbf{f}$ is the corresponding flux and r is a source term. This equation is written in the time-space (see Eq. (32)), introducing the 4D field defined on $\mathbb{B}_b$ by $\mathbb{f} = (\mathbf{f} \, , \, f)$, in the simple form

$$\mathbb{DIV}\, \mathbb{f} = r, \tag{42}$$

Owing to proposition 1 this balance equation is easily transported on $\mathbb{B}_a$ in the form $\mathbb{DIV}\left(\mathbb{J}\, \mathbb{f}^@ \cdot \mathbb{F}^{-T}\right) = \mathbb{J} r^@$ which reads

$$\operatorname{div}\left(J\, \mathbf{f}^@ \cdot \mathbf{F}^{-T} - J f^@ \otimes \mathbf{F}^{-1} \cdot \mathbf{v}\right) + \frac{\partial \left(J\, f^@\right)}{\partial t} = J r^@. \tag{43}$$

The jump condition on $\mathbb{S}_b$ associated to this balance equation is easily recovered by considering equation (42) in the sense of distributions. If we do not consider any surface source term, this jump condition reads $[|\mathbb{f} \cdot \mathbb{M}_b|] = 0$ which, recalling that $\mathbb{M}_b = (\mathbf{N}_b, -c_b)$, reduces to the more usual equation

$$[|\mathbf{f} \cdot \mathbf{N}_b - f c_b|] = 0. \tag{44}$$

This jump condition is easily transported on S_a: it takes the form $\left[\left|\mathbf{f}^{@} \cdot \mathbf{N}_b^{@} - f^{@} c_b^{@}\right|\right] = 0$ which recalling (30) also reads

$$[|J\mathbf{f}^{@} \cdot (\mathbf{F}^{-T} \cdot \mathbf{N}_a) - J f^{@} \otimes (\mathbf{F}^{-1} \cdot \mathbf{v}) \cdot \mathbf{N}_a - J c_a f^{@}|] = 0. \tag{45}$$

6 Porous Medium with a Solid-Material Surface Singularity

6.1 Kinematics

As we intend to give a macroscopic description of a porous medium, we consider a continuum made by the superposition of two continuous phases: a fluid one and a solid one.

Let us introduce the domains $B_s \subset \mathbb{R}^3$ and $B_f \subset \mathbb{R}^3$ (usually referred to as the Lagrangian configurations of the two constituents) and the maps

$$\chi_s : \mathbb{B}_s := B_s \times (0, T) \to \mathbb{R}^3, \quad \text{and} \quad \chi_f : \mathbb{B}_f := B_f \times (0, T) \to \mathbb{R}^3$$

which represent the placement of the solid and fluid constituents. The motion of the fluid inside the solid matrix is described by the function $\phi_s : \mathbb{B}_s \to B_f$ which, at any time t, associates to each solid particle $\mathbf{X}_s$ that particular fluid material particle $\mathbf{X}_f = \phi_s(\mathbf{X}_s, t)$ occupying the same physical position as $\mathbf{X}_s$. The three introduced maps are related by $\chi_s = \chi_f \circ \phi_s$. We can assume, extending B_s and ϕ_s if necessary, that ϕ_s is an homeomorphism from B_s to B_f. This extension and the resulting extension of χ_s have no physical sense, but make easier the description of open porous media. It will be mandatory to check that our final equations do not depend on the choice of this extension. Therefore $\chi_s(B_s, t) = \chi_f(B_f, t)$ and we denote $B_e(t)$ this time-varying 3D domain referred to as the Eulerian configuration. In the sequel, in order to apply our previous results, we assume that the 4D-counterparts of χ_s, χ_f and φ_s are piecewise C^1-diffeomorphisms.

We still adopt the superscript notation ⓢ (respectively ⓕ and ⓔ) to denote the transport of a tensor field from the configuration where it is defined to B_s (resp. B_f, B_e). For instance, if a tensor $\mathbf{t}$ is defined on B_f, then $\mathbf{t}^{ⓢ} := \mathbf{t} \circ \phi_s$, while if it is an Eulerian field defined on B_e, then $\mathbf{t}^{ⓢ} := \mathbf{t} \circ \chi_s$.

We denote the space gradient of the three placements by

$$\mathbf{F}_s := \nabla \chi_s, \qquad \mathbf{F}_f := \nabla \chi_f, \qquad \mathbf{G}_s := \nabla \phi_s, \tag{46}$$

and its determinant by

$$J_s := \det \mathbf{F}_s, \qquad J_f := \det \mathbf{F}_f, \qquad I_s := \det \mathbf{G}_s$$

It is immediate to check, that the chain rule gives $\mathbf{F}_s = \mathbf{F}_f^{\textcircled{s}} \cdot \mathbf{G}_s$ and $J_s = J_f^{\textcircled{s}} I_s$.

We define now the classical Lagrangian velocity fields $\mathbf{v}_s$ and $\mathbf{v}_f$, associated to the motion of the solid and of the fluid constituent, on B_s and B_f and, on B_s, the time derivative $\mathbf{u}_s$ of the map ϕ_s, which is not a velocity in the classical sense, but plays a central role in further calculations:

$$\mathbf{v}_s := \frac{\partial \chi_s}{\partial t}, \qquad \mathbf{v}_f := \frac{\partial \chi_f}{\partial t}, \qquad \mathbf{u}_s := \frac{\partial \phi_s}{\partial t}. \tag{47}$$

By the chain rule we get

$$0 = \frac{\partial}{\partial t}\left(\phi_s\left(\phi_s^{-1}\left(\mathbf{X}_f, t\right), t\right)\right) = \mathbf{G}_s^{\textcircled{f}} \cdot \frac{\partial \phi_s^{-1}}{\partial t} + \mathbf{u}_s^{\textcircled{f}}. \tag{48}$$

This relationship allows us to link $\mathbf{v}_s$, $\mathbf{v}_f$ and $\mathbf{u}_s$:

$$\begin{aligned} \mathbf{v}_f &= \frac{\partial \chi_f}{\partial t} = \frac{\partial}{\partial t}\left(\chi_s\left(\phi_s^{-1}\left(\mathbf{X}_f, t\right), t\right)\right) = \mathbf{F}_s^{\textcircled{f}} \cdot \frac{\partial \phi_s^{-1}}{\partial t} + \mathbf{v}_s^{\textcircled{f}} \\ &= -\mathbf{F}_s^{\textcircled{f}} \cdot (\mathbf{G}_s^{\textcircled{f}})^{-1} \cdot \mathbf{u}_s^{\textcircled{f}} + \mathbf{v}_s^{\textcircled{f}}. \end{aligned} \tag{49}$$

Transporting this relationship on B_s gives

$$\mathbf{v}_f^{\textcircled{s}} = \mathbf{v}_s - \mathbf{F}_s \, . \, \mathbf{G}_s^{-1} . \, \mathbf{u}_s. \tag{50}$$

Let us define the acceleration fields γ_s, γ_f and $\mathbf{a}_s$ as the time derivatives of $\mathbf{v}_s$, $\mathbf{v}_f$ and $\mathbf{u}_s$ respectively. Using Eq. (26) for the diffeomorphism ϕ_s, it is straightforward that

$$\frac{\partial}{\partial t} \mathbf{v}_f^{\textcircled{s}} = \gamma_f^{\textcircled{s}} + \nabla \mathbf{v}_f^{\textcircled{s}} \cdot \mathbf{G}_s^{-1} \cdot \mathbf{u}_s \tag{51}$$

Finally, since it is needed for further calculations, we compute the time derivative of the tensor $\mathbf{F}_s \cdot \mathbf{G}_s^{-1}$; using (26) for the map ϕ_s it is straightforward to recover that

$$\begin{aligned} \frac{\partial}{\partial t}\left(\mathbf{F}_s \cdot \mathbf{G}_s^{-1}\right) &= \frac{\partial}{\partial t} \mathbf{F}_f^{\textcircled{s}} = \nabla \mathbf{F}_f^{\textcircled{s}} \cdot \mathbf{G}_s^{-1} \cdot \mathbf{u}_s + (\nabla \mathbf{v}_f)^{\textcircled{s}} \\ &= \nabla\left(\mathbf{F}_s \cdot \mathbf{G}_s^{-1}\right) \cdot \mathbf{G}_s^{-1} \cdot \mathbf{u}_s + \nabla \mathbf{v}_f^{\textcircled{s}} \cdot \mathbf{G}_s^{-1}. \end{aligned} \tag{52}$$

In the sequel we focus on a surface $S_s(t)$ which may be a surface of singularity for χ_s and/or ϕ_s and on the image surfaces $S_f(t) = \phi_s(S_s(t), t)$ and $S_e(t) = \chi_s(S_s(t), t) = \chi_f(S_f(t), t)$ which are moving surfaces in B_f and $B_e(t)$. We apply to these surfaces the notations and formulas stated in the previous sections. In particular, we introduce the celerities c_s, c_f and c_e of $S_s(t)$, $S_f(t)$ and $S_e(t)$ respectively.

Actually, in this paper we only consider the case in which the surface S_s is a *solid-material surface discontinuity* which means that it does not depend on time. From now on we assume that S_s is parametrized by a function φ which does not depend on time. Consequently, the celerity c_s of the surface S_s is vanishing. This particular case of solid-material surface has many applications. It models all those phenomena in which S_s divides the solid skeleton in two parts with different mechanical properties (e.g. different porosities, rigidities, etc.) It also models, as a limit case, the boundary of a fluid-filled porous matrix in contact with a pure fluid.

The hypothesis that the surface is solid-material ($c_s = 0$), applying (41) to both χ_s and ϕ_s, implies

$$[|\mathbf{v}_s|] = 0 \quad \text{and} \quad [|\mathbf{u}_s|] = 0 \quad \text{on } S_s. \tag{53}$$

We underline that these equations do not imply $[|\mathbf{v}_f|] = 0$.

We finally remark that if $(\mathbf{v}_s^{ⓔ} - \mathbf{v}_f^{ⓔ}) \cdot \mathbf{N}_e = 0$, or equivalently by (50) and (30) $\mathbf{u}_s \cdot \mathbf{N}_f^{ⓢ} = 0$, then from (27)

$$\left[\left|\mathbf{F}_f^{ⓢ} \cdot \mathbf{u}_s\right|\right] = \left[\left|\mathbf{v}_s - \mathbf{v}_f^{ⓢ}\right|\right] = 0. \tag{54}$$

6.2 Balance of Masses

The masses $\mathcal{M}_s(B)$ and $\mathcal{M}_f(B)$ of solid skeleton and fluid contained in a part $B \subset B_e$ of the physical space at time t are represented by means of two Eulerian densities ρ_s and ρ_f respectively in the form

$$\mathcal{M}_s(B) = \int_B \rho_s, \qquad \mathcal{M}_f(B) = \int_B \rho_f.$$

These densities are usually called "*apparent densities*" and they do not coincide with the mass densities of the materials which constitute the solid skeleton or the fluid. Simple changes of variables give

$$\mathcal{M}_s(B) = \int_{\chi_s^{-1}(B)} \rho_s^{ⓢ} J_s \quad \text{and}$$
$$\mathcal{M}_f(B) = \int_{\chi_s^{-1}(B)} \rho_f^{ⓢ} J_s = \int_{\chi_f^{-1}(B)} \rho_f^{ⓕ} J_f, .$$

which leads us to introduce the "*solid-Lagrangian apparent densities*" η_s, m_f for the solid and the fluid constituent defined on B_s by $\eta_s := J_s \rho_s^{ⓢ}$, $m_f := J_s \rho_f^{ⓢ}$ and the "*fluid-Lagrangian apparent density*" η_f of the fluid constituent on B_f by $\eta_f := J_f \rho_f^{ⓕ}$. The densities m_f and η_f are related by $m_f = I_s \eta_f^{ⓢ}$.

As we do not intend to model melting, dissolution or erosion phenomena, we assume conservation of mass for each constituent. Mass conservation for the solid skeleton and the fluid take the form of the balance laws

$$\frac{\partial}{\partial t}\eta_s = 0, \quad \text{and} \quad \frac{\partial}{\partial t}\eta_f = 0. \tag{55}$$

which are of the type studied in section 5. The results of section 5 give the associated jump conditions $[|\eta_s|]\, c_s = 0$ and $[|\eta_f|]\, c_f = 0$ on S_s and S_f respectively. As $c_s = 0$ the first equation is trivially satisfied. So is is the second one if one assumes (which, as it is well known, can be done for a fluid without loss of generality) that η_f is constant in space and time. However, the pull-back on B_s of the fluid balance and jump equations, using the transport formulas (43) and (45) together with equation (50) gives the non-trivial equations

$$\begin{aligned} \dot{m}_f + \operatorname{div} \mathbf{D} = \mathbf{0} \quad & \text{on } B_s, \\ [|\mathbf{D}|] \cdot \mathbf{N}_s = 0 \quad & \text{on } S_s, \end{aligned} \tag{56}$$

where $\mathbf{D} := -m_f \mathbf{G}_s^{-1} \cdot \mathbf{u}_s$ and $\dot{m}_f := \partial m_f / \partial t$. The vector $\mathbf{D}$ is interpreted as the mass fluid flux through the porous medium in the Lagrangian configuration of the skeleton. The quantity $\mathbf{D} \cdot \mathbf{N}_s$, which is well defined at the interface S_s, is the flux (per unit area of S_s) of fluid flowing through the interface. We introduce

$$d := \left(\frac{\mathbf{D} \cdot \mathbf{N}_s}{\| J_s \mathbf{F}_s \cdot \mathbf{N}_s \|} \right)^{ⓔ} = \rho_f (\mathbf{v}_f^{ⓔ} - \mathbf{v}_s^{ⓔ}) \cdot \mathbf{N}_e$$

$$= \eta_f^{ⓔ} \| (J_f^{-1} \mathbf{F}_f^T)^{ⓔ} \cdot \mathbf{N}_e \| (\mathbf{N}_f \cdot \mathbf{u}_s)^{ⓔ} \tag{57}$$

which is well defined at the interface S_e and corresponds to the flux (per unit area of S_e) of fluid flowing through the interface.

7 Evolution Equations and Associated Jump Conditions in Presence of Dissipation

7.1 Action and Rayleigh Functionals

Recall that the kinematics of the considered porous medium is described by means of the fields χ_s and ϕ_s defined on B_s.

We introduce the kinetic energy

$$\frac{1}{2}\int_{B_e}\left(\rho_s(\mathbf{v}_s^{\textcircled{e}})^2+\rho_f(\mathbf{v}_f^{\textcircled{e}})^2\right)=\int_{B_s}\Lambda\left(\eta_s,m_f,\mathbf{v}_s,\mathbf{v}_f^{\textcircled{s}}\right)$$

where $\Lambda\left(\eta_s,m_f,\mathbf{v}_s,\mathbf{v}_f^{\textcircled{s}}\right)=1/2\left(\eta_s(\mathbf{v}_s)^2+m_f(\mathbf{v}_f^{\textcircled{s}})^2\right)$ is the solid-Lagrangian pull-back of the kinetic energy density.

We now assume that the potential energy of the porous medium is characterized by a local density Ψ on B_s which depends on the kinematic descriptors χ_s and ϕ_s through the placement χ_s, the strain tensor $\varepsilon := 1/2(\mathbf{F}_s^T\cdot\mathbf{F}_s-\mathbf{I})$ and the quantity of fluid contained in the porous medium $m_f = I_s\,\eta_f$. For instance Ψ can be the sum of a non-homogeneous deformation energy potential $\Psi_i(\varepsilon,m_f,\mathbf{X}_s)$ and a potential accounting for external body forces $\Psi_g = (\eta_s+m_f)\,E_p(\chi_s(\mathbf{X}_s))$. As we do not intend to model surface tension phenomena, we do not consider any concentration of energy on the singularity surface S_s. Neither do we consider any dependence of Ψ on higher gradients of the kinematical fields as done for instance in Sciarra et al. (2008).

Setting $\mathbb{B}_s := B_s\times(0,T)$, we define the action functional $\mathcal{A}$ for the porous system as

$$\mathcal{A}:=\int_{\mathbb{B}_s}(\Lambda-\Psi)\,. \tag{58}$$

It is well known that, in absence of dissipation, imposing the stationarity of the action implies that the kinematic descriptors satisfy the virtual power principle i.e. a weak form of the balance of momentum. As we want to account for dissipation phenomena, we introduce a generalized Rayleigh dissipation pseudo-potential $\mathcal{R}$ on the Eulerian configuration. In linear thermodynamics the dissipation $2\mathcal{R}$ is a quadratic form of the velocity fields

$$\begin{aligned}2\mathcal{R}=&\int_{B_e\backslash S_e}\mathcal{D}\left(\mathbf{v}_s^{\textcircled{e}}-\mathbf{v}_f^{\textcircled{e}},\nabla\left(\mathbf{v}_s^{\textcircled{e}}\right),\nabla(\mathbf{v}_f^{\textcircled{e}})\right)\\&+\int_{S_e}\mathcal{D}_S\left((\mathbf{v}_s^{\textcircled{e}})^-,(\mathbf{v}_s^{\textcircled{e}})^+,(\mathbf{v}_f^{\textcircled{e}})^-,(\mathbf{v}_f^{\textcircled{e}})^+\right)\end{aligned} \tag{59}$$

where the volume density $\mathcal{D}$ is a positive quadratic form, the surface density $\mathcal{D}_S$ is a Galilean invariant quadratic form (the coefficients of these two forms may also depend on all the static parameters). In the sequel, we limit ourselves to the case

$$\mathcal{D} = (\mathbf{v}_s^{\textcircled{e}} - \mathbf{v}_f^{\textcircled{e}}) \cdot \mathbf{K} \cdot (\mathbf{v}_s^{\textcircled{e}} - \mathbf{v}_f^{\textcircled{e}}) + \nabla\left(\mathbf{v}_s^{\textcircled{e}} - \mathbf{v}_f^{\textcircled{e}}\right) : \mathbf{B} : \nabla\left(\mathbf{v}_s^{\textcircled{e}} - \mathbf{v}_f^{\textcircled{e}}\right) + \nabla \mathbf{v}_f^{\textcircled{e}} : \mathbf{M} : \nabla \mathbf{v}_f^{\textcircled{e}} \tag{60}$$

$$\mathcal{D}_S = \left[\left|\mathbf{v}_f^{\textcircled{e}}\right|\right] \cdot \mathbf{S} \cdot \left[\left|\mathbf{v}_f^{\textcircled{e}}\right|\right] \tag{61}$$

where $\mathbf{K}$ and $\mathbf{S}$ are second order symmetric, positive tensors, $\mathbf{M}$ and $\mathbf{B}$ are symmetric positive fourth order tensors, the symbol : stands for the double contraction product. The tensor $\mathbf{K}$ accounts for the Darcy dissipation; its inverse (if invertible) is called the Darcy permeability tensor. The tensor $\mathbf{B}$ accounts for Brinkman dissipation. Classical fluid viscous effects are described by $\mathbf{M}$, while $\mathbf{S}$ describes friction effects at the interface. We already noticed that, as the extension of χ_s is arbitrary in a pure fluid region, the model should not depend on $\mathbf{v}_s$ in this region. Hence, $\mathbf{K}$ and $\mathbf{B}$ have to vanish in any pure fluid region.

We introduce respectively, the Darcy friction force κ, the Brinkman stress tensor Π, the fluid viscous stress tensor Π_f and the friction surface force σ as

$$\kappa := \mathbf{K} \cdot (\mathbf{v}_s^{\textcircled{e}} - \mathbf{v}_f^{\textcircled{e}}), \qquad \Pi := \mathbf{B} : \nabla\left(\mathbf{v}_s^{\textcircled{e}} - \mathbf{v}_f^{\textcircled{e}}\right), \tag{62}$$

$$\Pi_f := \mathbf{M} : \nabla \mathbf{v}_f^{\textcircled{e}}, \qquad \sigma := \mathbf{S} \cdot \left[\left|\mathbf{v}_f^{\textcircled{e}}\right|\right]. \tag{63}$$

This Rayleigh dissipation is pulled back on B_s by simply changing the variables

$$2\mathcal{R} = \int_{B_s \setminus S_s} J_s \mathcal{D}^{\textcircled{s}} + \int_{S_s} \mathcal{D}_S^{\textcircled{s}} \left\| J_s \mathbf{F}_s^{-T} \cdot \mathbf{N}_s \right\|. \tag{64}$$

In the application of this formula, the change of variables in $\mathcal{D}_S$ and in the first term of $\mathcal{D}$ is straightforward. The terms involving gradients in $\mathcal{D}$ need to be transported according to formula (2).

7.2 Equations of Motion

Let us denote by $\mathbf{q} := (\chi_s, \phi_s)$ the kinematic descriptor of the medium (a field defined on $\mathbb{B}_s$). Hence the action $\mathcal{A}$ is a functional of $\mathbf{q}$. Moreover, let us denote by $\mathbf{q}_t$, $\dot{\mathbf{q}}_t$ the fields defined at any instant t on B_s by $\mathbf{q}_t(\mathbf{X}_s) := \mathbf{q}(\mathbf{X}_s, t)$ and $\dot{\mathbf{q}}_t(\mathbf{X}_s) := \partial \mathbf{q}/\partial t(\mathbf{X}_s, t)$. The Rayleigh potential $\mathcal{R}$ is at each instant t a functional of $(\mathbf{q}_t, \dot{\mathbf{q}}_t)$.

The physical principle which determines the motion of a system can be alternatively stated in the framework of second Newton's law (balance of momentum), of D'Alembert principle (weak formulation of momentum balance) or of Rayleigh-Hamilton principle. We adopt this last approach which reads

$$\frac{\partial \mathcal{A}}{\partial \mathbf{q}} \cdot \delta \mathbf{q} = \int_0^T \left(\frac{\partial \mathcal{R}}{\partial \dot{\mathbf{q}}_t} \mid \delta \mathbf{q}_t \right) dt. \tag{65}$$

Here, $\partial \mathcal{A}/\partial \mathbf{q}$ and $\partial \mathcal{R}/\partial \dot{\mathbf{q}}_t$ must be understood in the sense of Frechet differentials. Considering $\delta \mathbf{q} = (\delta \chi_s, \delta \phi_s)$ with compact support included in B_s, we get after a long but straightforward calculation (cfr. Appendix B):

$$\begin{aligned}
\delta \mathcal{A} &:= \frac{\partial \mathcal{A}}{\partial \mathbf{q}} \cdot \delta \mathbf{q} \\
&= \int_{\mathbb{B}_s \backslash \mathbb{S}_s} \left[-\left(\eta_s \gamma_s + m_f \gamma_f^{ⓢ} \right) + \operatorname{div} \left(\mathbf{F}_s \cdot \frac{\partial \Psi}{\partial \varepsilon} \right) - \frac{\partial \Psi}{\partial \chi_s} \right] \cdot \delta \chi_s + \\
& \int_{\mathbb{B}_s \backslash \mathbb{S}_s} \left[\mathbf{G}_s^{-T} \cdot m_f \left(\mathbf{F}_s^T \cdot \gamma_f^{ⓢ} + \nabla \left(\frac{\partial \Psi}{\partial m_f} \right) \right) \right] \cdot \delta \phi_s \\
& + \int_{\mathbb{S}_s} \left(\left[\left| \mathbf{v}_f^{ⓢ} \otimes \mathbf{D} - \mathbf{F}_s \cdot \frac{\partial \Psi}{\partial \varepsilon} \right|\right] \cdot \mathbf{N}_s \right) \cdot \delta \chi_s \\
& + \int_{\mathbb{S}_s} \left(\left[\left| \mathbf{G}_s^{-T} \cdot \left(m_f \left(\frac{1}{2} \left(\mathbf{v}_f^{ⓢ} \right)^2 - \frac{\partial \Psi}{\partial m_f} \right) \mathbf{I} \right.\right.\right. \\
& - \mathbf{F}_s^T \cdot \mathbf{v}_f^{ⓢ} \otimes \mathbf{D} \cdot \mathbf{N}_s \cdot \delta \phi_s .
\end{aligned} \tag{66}$$

On the other hand, computation of the right hand side of (65), reads (cfr. Appendix C for details)

$$\begin{aligned}
& \int_{\mathbb{B}_s \backslash \mathbb{S}_s} - \operatorname{div} \left(J_s \Pi_f^{ⓢ} \cdot \mathbf{F}_s^{-T} \right) \cdot \delta \chi_s + \int_{\mathbb{S}_s} \left[\left| J_s \Pi_f^{ⓢ} \cdot \mathbf{F}_s^{-T} \cdot \mathbf{N}_s \right|\right] \cdot \delta \chi_s + \\
& \int_{\mathbb{B}_s \backslash \mathbb{S}_s} \mathbf{G}_s^{-T} \cdot \mathbf{F}_s^T \cdot \left(J_s \kappa^{ⓢ} - \operatorname{div} \left(J_s (\Pi^{ⓢ} - \Pi_f^{ⓢ})^T \cdot \mathbf{F}_s^{-T} \right) \right) \cdot \delta \phi_s \\
& + \int_{\mathbb{S}_s} \left[\left| J_s \mathbf{G}_s^{-T} \cdot \mathbf{F}_s^T \cdot \left(\left((\Pi^{ⓢ} - \Pi_f^{ⓢ})^T \right) \cdot \mathbf{F}_s^{-T} \cdot \mathbf{N}_s \right.\right. \\
& - \left\| J_s \mathbf{F}_s^{-T} \cdot \mathbf{N}_s \right\| \sigma^{ⓢ} \cdot \delta \phi_s .
\end{aligned} \tag{67}$$

Considering arbitrary test functions $\delta \chi_s$ and $\delta \phi_f$ with compact support included in $\mathbb{B}_s \backslash \mathbb{S}_s$ the variational principle (65) implies the following system of equations valid in $\mathbb{B}_s \setminus \mathbb{S}_s$

$$
\begin{aligned}
&-\left(\eta_s \gamma_s + m_f \gamma_f^{ⓢ}\right) + \operatorname{div}\left(\mathbf{F}_s \cdot \frac{\partial \Psi}{\partial \varepsilon}\right) - \frac{\partial \Psi}{\partial \chi_s} \\
&= -\operatorname{div}\left(J_s \Pi_f^{ⓢ} \cdot \mathbf{F}_s^{-T}\right), \\
&m_f\left(\mathbf{F}_s^T \cdot \gamma_f^{ⓢ} + \nabla\left(\frac{\partial \Psi}{\partial m_f}\right)\right) \\
&= \mathbf{F}_s^T \cdot \left(J_s \kappa^{ⓢ} - \operatorname{div}\left(J_s (\Pi^{ⓢ} - \Pi_f^{ⓢ})^T \cdot \mathbf{F}_s^{-T}\right)\right).
\end{aligned}
\tag{68}
$$

Then, considering test functions with compact support which intersects $\mathbb{S}_s$, we get the jump conditions valid on $\mathbb{S}_s$:

$$
\left[\left|\mathbf{v}_f^{ⓢ} \otimes \mathbf{D} - \mathbf{F}_s \cdot \frac{\partial \Psi}{\partial \varepsilon}\right|\right] \cdot \mathbf{N}_s = \left[\left| J_s \Pi_f^{ⓢ} \cdot \mathbf{F}_s^{-T} \right|\right] \cdot \mathbf{N}_s, \tag{69}
$$

$$
\begin{aligned}
&\left[\left|\mathbf{G}_s^{-T} \cdot \left(m_f\left(\frac{1}{2}\left(\mathbf{v}_f^{ⓢ}\right)^2 - \frac{\partial \Psi}{\partial m_f}\right)\mathbf{I} - \mathbf{F}_s^T \cdot \mathbf{v}_f^{ⓢ} \otimes \mathbf{D}\right)\right|\right] \cdot \mathbf{N}_s \\
&= \left[\left|\mathbf{G}_s^{-T} \cdot \mathbf{F}_s^T \cdot \left(J_s \left((\Pi^{ⓢ} - \Pi_f^{ⓢ})^T\right) \cdot \mathbf{F}_s^{-T} \cdot \mathbf{N}_s \right.\right.\right. \\
&- \left\| J_s \mathbf{F}_s^{-T} \cdot \mathbf{N}_s \right\| \sigma^{ⓢ}.
\end{aligned}
\tag{70}
$$

We remark that equations (68) and (69) encompass the well known Lagrangian balance equation for the total stress and the corresponding jump condition; these equations only involve physical quantities. As for equations (70), they are not available in the literature. It is not obvious that in their present form they are Galilean invariant, that they do not depend on the choice of the reference configuration of the fluid and that, when considering pure fluid regions, they do not depend on the extension of χ_s in this region.

7.3 Galilean Invariance

In this section we check that the set of evolution equations (68), together with the associated jump conditions (69), (70), respect Galilean invariance. In order to do so we just rewrite them in a new Galilean reference frame which moves with a constant velocity $\mathbf{v}_0$ with respect to the original one and we check that the system of equations do not change. We start by noticing that the dissipative terms appearing on the right hand sides of (68)-(70) are Galilean invariant since they only involve relative velocities or gradients of velocities.

As for the equations of motion (68) they are Galilean invariant since the accelerations γ_s and $\gamma_f^{ⓢ}$ do not change when changing the Galilean reference frame, as the relative velocity $\mathbf{v}_0$ is constant with respect to time. The invariance of the jump condition (69) is immediately verified simply noticing that $\left[\left|\mathbf{v}_f^{\textcircled{s}}\right|\right] = \left[\left|\mathbf{v}_f^{\textcircled{s}} - \mathbf{v}_0\right|\right]$ and that $\mathbf{D}$, being a relative velocity, is also Galilean invariant.

Equation (70) does not have an evident Galilean invariant form. However, let us prove that it is equivalent to

$$\left[\left|\mathbf{G}_s^{-T} \cdot \left(m_f \left(\frac{1}{2}\left(\mathbf{v}_f^{\textcircled{s}} - \mathbf{v}_s\right)^2 - \frac{\partial \Psi}{\partial m_f}\right)\mathbf{I} \right.\right.\right.$$
$$\left. - \mathbf{F}_s^T \cdot \left(\mathbf{v}_f^{\textcircled{s}} - \mathbf{v}_s\right) \otimes \mathbf{D} \cdot \mathbf{N}_s \right.$$
$$= \left[\left|\mathbf{G}_s^{-T} \cdot \mathbf{F}_s^T \cdot \left(J_s \left((\Pi^{\textcircled{s}} - \Pi_f^{\textcircled{s}})^T\right) \cdot \mathbf{F}_s^{-T} \cdot \mathbf{N}_s - \left\| J_s \mathbf{F}_s^{-T} \cdot \mathbf{N}_s \right\| \sigma^{\textcircled{s}}\right)\right|\right], \quad (71)$$

in which only Galilean invariant quantities appear.
Indeed, the difference Q between the left hand sides of (71) and (70), using continuity of $\mathbf{v}_s$ and recalling Eq. (50), reads

$$Q = \frac{1}{2}\mathbf{v}_s^2 \left[\left| m_f \mathbf{G}_s^{-T} \right|\right] \cdot \mathbf{N}_s$$
$$+ \left[\left| -m_f \left(\mathbf{v}_f^{\textcircled{s}} \cdot \mathbf{v}_s\right) \mathbf{G}_s^{-T} + \mathbf{G}_s^{-T} \cdot \mathbf{F}_s^T \cdot \mathbf{v}_s \otimes \mathbf{D} \right|\right] \cdot \mathbf{N}_s.$$

We know that, applying formula (28) to the map ϕ_s^{-1} under the assumption $[|\eta_f|] = 0$, one gets

$$\left[\left| m_f \mathbf{G}_s^{-T} \right|\right] \cdot \mathbf{N}_s = 0. \quad (72)$$

Hence, using again the continuity of $\mathbf{v}_s$ it is easy to get

$$Q = \mathbf{v}_s \cdot \left[\left| -m_f \mathbf{v}_f^{\textcircled{s}} \otimes \mathbf{G}_s^{-T} + \mathbf{F}_s \cdot \mathbf{G}_s^{-1} \otimes \mathbf{D} \right|\right] \cdot \mathbf{N}_s,$$

which using the balance of mass (56) and again (72) is equivalent to

$$Q = \mathbf{v}_s \cdot \left\{ - \left[\left| \mathbf{v}_f^{\textcircled{s}} \right|\right] \otimes m_f \mathbf{G}_s^{-T} + \left[\left| \mathbf{F}_s \cdot \mathbf{G}_s^{-1} \right|\right] \otimes \mathbf{D} \right\} \cdot \mathbf{N}_s.$$

Moreover, by Eq. (30) applied to the map χ_s it is clear that

$$\mathbf{N}_s = \left(J_s^{-1} \mathbf{F}_s^T \cdot \mathbf{N}_e^{\textcircled{s}}\right) / \left\| J_s^{-1} \mathbf{F}_s^T \cdot \mathbf{N}_e^{\textcircled{s}} \right\|,$$

where $\mathbf{N}_e$ is the unit normal vector to the Eulerian surface S_e. Using this result and recalling the definitions of m_f and $\mathbf{D}$, it can be recognized that

$$Q = -\frac{\eta_f^{ⓢ} \mathbf{v}_s}{\left\| J_s^{-1} \mathbf{F}_s^T \cdot \mathbf{N}_e^{ⓢ} \right\|} \cdot \left[\left| \mathbf{v}_f^{ⓢ} \right|\right] I_s J_s^{-1} \left(\mathbf{F}_s \cdot \mathbf{G}_s^{-1} \right)^T$$

$$+ \frac{\eta_f^{ⓢ} \mathbf{v}_s}{\left\| J_s^{-1} \mathbf{F}_s^T \cdot \mathbf{N}_e^{ⓢ} \right\|} \cdot I_s J_s^{-1} \left[\left| \mathbf{F}_s \cdot \mathbf{G}_s^{-1} \right|\right] \otimes \left(\mathbf{v}_f^{ⓢ} - \mathbf{v}_s \right) \cdot \mathbf{N}_e^{ⓢ}.$$

Therefore, the quantity Q vanishes owing to the purely kinematical formula (29) applied to the diffeomorphism $\chi_s \circ \phi_s^{-1}$ (notice that $\mathbf{v}_s \cdot \mathbf{N}_e^{ⓢ} = c_e^{ⓢ}$). It follows that the Galilean invariance of the last jump condition remains proven.

7.4 The equations do not depend on the choice of the fluid reference configuration

Equation (71) seems to depend on the choice of the fluid reference configuration through the tensor $\mathbf{G}_s$ (all others equations clearly do not depend on this choice). Let us first project this equation on the tangent plane to S_f. Hence, if τ_e is an arbitrary tangent vector to S_e, multiplication of equation (71) by $\mathbf{G}_s \cdot \mathbf{F}_s^{-1} \cdot \tau_e^{ⓢ}$ gives

$$\tau_e^{ⓢ} \cdot \left[\left| \left(\mathbf{v}_f^{ⓢ} - \mathbf{v}_s \right) \right|\right] d^{ⓢ} + \tau_e^{ⓢ} \cdot \left[\left| \left(\Pi^{ⓢ} - \Pi_f^{ⓢ} \right)^T \right|\right] \cdot \mathbf{N}_e^{ⓢ}$$

$$- \left\| J_s \mathbf{F}_s^{-T} \cdot \mathbf{N}_s \right\| \tau_e^{ⓢ} \cdot \mathbf{S} \cdot \left[\left| \mathbf{v}_f^{ⓢ} - \mathbf{v}_s \right|\right] = 0 \qquad (73)$$

Now let us project equation (71) in a different direction, multiplying it by the continuous quantity $-\mathbf{u}_s = \mathbf{G}_s \cdot \mathbf{F}_s^{-1} \cdot (\mathbf{v}_f^{ⓢ} - \mathbf{v}_s)$: notice that, owing to (57) and (54), this projection gives additional information to (73) only if $d \neq 0$. We get

$$\left[\left| \left(-\frac{1}{2} \left(\mathbf{v}_f^{ⓢ} - \mathbf{v}_s \right)^2 - \frac{\partial \Psi}{\partial m_f} \right) \right|\right] d^{ⓢ}$$

$$= \left[\left| \left(\mathbf{v}_f^{ⓢ} - \mathbf{v}_s \right) \cdot \left(\left(\left(\Pi^{ⓢ} - \Pi_f^{ⓢ} \right)^T \right) \cdot \mathbf{N}_e^{ⓢ} - \sigma^{ⓢ} \right) \right|\right]. \qquad (74)$$

Notice that this equation does not depend on the choice of the reference configuration of the fluid.

8 The Case of a Deformable Porous Medium Surrounded by a Pure Fluid

We now assume that the surface discontinuity S_s separates a given porous medium (which occupies the volume B_s^+) from a pure fluid (which occupies the volume B_s^-). Proper evolution equations and boundary conditions are deduced from our previous results.

Clearly, the equations of motion of the porous medium in B_s^+ coincide with (68) where the potential Ψ is replaced by its restriction Ψ^+ in B_s^+.

When the fluid is pure, its Eulerian energy density, its chemical potential and its pressure are functions of its mass density only. These three real functions are denoted respectively by Ψ_f, μ_f and p_f. They are related by

$$\mu_f(y) = \Psi'_f(y) \qquad \text{and} \qquad p_f\,(y) = -\Psi_f(y) + y\,\Psi'_f(y), \tag{75}$$

Therefore, the restriction Ψ^- of the potential Ψ in B_s^- is that of a pure fluid: transporting the Eulerian density $\Psi_f(\rho_f)$ on B_s, we get

$$\Psi^-\,(\varepsilon, m_f) = J_s \Psi_f\left(\rho_f^{ⓢ}\right) = J_s \Psi_f\left(J_s^{-1} m_f\right), \tag{76}$$

note that J_s is a function of ε only, as we have $J_s = \det \mathbf{F}_s = \det\sqrt{2\varepsilon + \mathbf{I}} = \sqrt{\det(2\varepsilon + \mathbf{I})}$.

In the pure fluid region B_s^- we clearly have $\eta_s = 0$ and, as already noticed, $\mathbf{K}^{ⓢ} = 0$, $\mathbf{M}^{ⓢ} = 0$. We also neglect external body forces and viscous forces by setting $\partial\Psi^-/\partial\chi_s = 0$ and $\Pi = \Pi_f = 0$.

With these assumptions, the equation of motion (68) in B_s^- reads

$$\begin{aligned} m_f \gamma_f^{ⓢ} &= \operatorname{div}\left(\mathbf{F}_s \cdot \frac{\partial \Psi^-}{\partial \varepsilon}\right) \\ &= \operatorname{div}\left[\mathbf{F}_s \cdot \left(\Psi_f|_{\rho_f^{ⓢ}} \frac{\partial J_\varepsilon}{\partial \varepsilon} + J_s \frac{\partial}{\partial \varepsilon}\,\Psi_f|_{\rho_f^{ⓢ}}\right)\right] \\ &= \operatorname{div}\left[\mathbf{F}_s \cdot \left(\Psi_f|_{\rho_f^{ⓢ}} \frac{\partial J_s}{\partial \varepsilon} + J_s \left.\frac{\partial \Psi_f}{\partial \rho_f}\right|_{\rho_f^{ⓢ}} \frac{\partial}{\partial \varepsilon}\left(J_s^{-1} m_f\right)\right)\right]. \end{aligned} \tag{77}$$

Recalling that $\partial J_s/\partial\varepsilon = J_s \mathbf{F}_s^{-1} \cdot \mathbf{F}_s^{-T}$ and $\partial J_s^{-1}/\partial\varepsilon = -J_s^{-2}\partial J_s/\partial\varepsilon$, equation (77) can be simplified in

$$m_f \gamma_f^{ⓢ} = \operatorname{div}\left[\left(J_s\;\Psi_f|_{\rho_f^{ⓢ}} - m_f\;\mu_f|_{\rho_f^{ⓢ}}\right)\mathbf{F}_s^{-T}\right];$$

or equivalently, recalling the inverse of Eq. (28) and (75), in

$$m_f \gamma_f^{ⓢ} = -J_s \mathbf{F}_s^{-T} \cdot \nabla\left(p_f|_{\rho_f^{ⓢ}}\right). \tag{78}$$

This last equation, when transported on the Eulerian configuration, reads

$$\nabla p_f + \rho_f \gamma_f^{ⓔ} = 0; \tag{79}$$

we thus recover the usual equation for the motion of a non-viscous fluid fluid.

Let us now study the second of equations (68) also valid in B_s^-. Indeed, the existence of an supplementary equation may seem astonishing. Under the hypotheses we formulated, recalling (76) and (75), Eq. (68) can be rewritten

$$m_f \nabla \left(\mu_f|_{\rho_f^{ⓢ}} \right) + m_f \mathbf{F}_s^T \cdot \gamma_f^{ⓢ} = 0,$$

which, multiplied on the left by $\mathbf{F}_s^{-T}$, gives

$$J_s \rho_f^{ⓢ} \mathbf{F}_s^{-T} \cdot \nabla \left(\mu_f|_{\rho_f^{ⓢ}} \right) + m_f \gamma_f^{ⓢ} = 0,$$

This last equation is clearly equivalent to Eq. (78) as $\nabla(p_f)|_{\rho_f^{ⓢ}} = \rho_f^{ⓢ} \nabla(\mu_f)|_{\rho_f^{ⓢ}}$. Hence, as expected, the fluid is governed only by the usual equation for an non-viscous fluid.

Let us consider the jump condition given by Eq. (69) on the surface S_s which divides the porous medium region B_s^+ from the pure fluid region B_s^-. Considering the expression (76) for the potential of the pure fluid, it becomes

$$\left(-\mathbf{F}_s^+ \cdot \left(\frac{\partial \Psi}{\partial \varepsilon} \right)^+ + \left(\mathbf{v}_f^{ⓢ} \right)^+ \otimes \mathbf{D} \right) \cdot \mathbf{N}_s$$
$$= \left(J_s^+ \left(\mathbf{F}_s^+ \right)^{-T} p_f|_{\rho_f^{ⓢ}} + \left(\mathbf{v}_f^{ⓢ} \right)^- \otimes \mathbf{D} \right) \cdot \mathbf{N}_s, \tag{80}$$

where we used the fact that $J_s^- \mathbf{F}_s^- \cdot \mathbf{N}_s = J_s^+ \mathbf{F}_s^+ \cdot \mathbf{N}_s$ (here again the superscripts + and − indicate the traces on S_s of quantities defined on B_s^+ and B_s^- respectively).

As for the second jump condition (70), it becomes

$$\left(\mathbf{G}_s^+ \right)^{-T} \cdot \left[\left(-m_f^+ \frac{\partial \Psi^+}{\partial m_f^+} + \frac{1}{2} m_f^+ \left(\mathbf{v}_f^{ⓢ} \right)^{+\,2} \right) \mathbf{I} \right.$$
$$- \left(\mathbf{F}_s^+ \right)^T \cdot \left(\mathbf{v}_f^{ⓢ} \right)^+ \otimes \mathbf{D} \cdot \mathbf{N}_s = \tag{81}$$
$$\left(\mathbf{G}_s^- \right)^{-T} \cdot \left[\left(\left(-m_f^- \, \mu_f|_{\rho_f^{ⓢ}} + \frac{1}{2} m_f^- \left(\mathbf{v}_f^{ⓢ} \right)^{-\,2} \right) \mathbf{I} \right.$$
$$- \left(\mathbf{F}_s^- \right)^T \cdot \left(\mathbf{v}_f^{ⓢ} \right)^- \otimes \mathbf{D} \cdot \mathbf{N}_s.$$

In conclusion, the motion of the porous medium is driven by two independent equations of the type (68), while the motion of the pure fluid is driven by a simple equation in the form (78). The differential system is completed by two independent jump conditions given by (80) and (81). Equation (80) states that at a permeable interface between a porous medium and a pure fluid the jump of stress equals the negative of the jump of transported fluid momentum. Equation (81) states how at the same interface the jump of chemical potential plus fluid kinetic energy is related to transported fluid momentum.

9 Conclusions

To our knowledge, the inertia terms appearing in the jump condition $(69)_2$ are not found in the literature. Moreover all boundary conditions (69) are valid also when the solid matrix is suffering large deformations and when the Stokes fluid-flow condition is not applicable.
A deduction of the jump condition valid in the particular case of absence of inertia, of Darcy-Brinkman and Beavers-Joseph dissipation is presented in Baek and Srinivasa (2004).
Other authors (see e.g. Neale and Nader (1974), Vafai and Thiyagaraja (1987), Vafai and Kim (1990), Poulikakos and Kazmierczak (1997)) based themselves on the pioneering works of Beavers and Joseph (1967) and Saffman (1971) to justify the so-called slip boundary conditions at the interface between a porous matrix and an external viscous fluid. Beavers-Joseph-Saffman conditions include the continuity of the normal components of the relative velocity and of the gradient of the relative velocity at the interface between the porous medium and the external fluid: nevertheless, they only describe phenomena related to the viscosity of the outflowing fluid with no consideration of inertial effects and Darcy-Brinkman dissipation. Beavers-Joseph-Saffman conditions can be deduced from jump condition (69), once assuming that the solid matrix is suffering small deformation and when Stokes fluid-flow condition is verified.
The jump conditions deduced in Deresiewicz (1960-1964), are suitable to assure that the differential problem of Darcy-Fillunger-Terzaghi-Biot is well-posed (see Fillunger (1936), Terzaghi (1943), Biot (1941-1963)): these last jump conditions can be obtained as a particular case from Beavers-Joseph-Saffman conditions, once it is possible to neglect dissipative phenomena at the considered interface. In Albers (2006), Wilmanski (1999-2006), De La Cruz et al. (1992), Quiroga-Goode and Carcione (1997) the jump conditions proposed by ? are used to study wave propagation phenomena at discontinuity surfaces in porous media.

In (Coussy and Bourbie (1984), Coussy et al. (1998), Rasolofosaon and Coussy (1985-1986), a variational approach is used to study some wave propagation phenomena of interest in oil industry: the boundary conditions proposed by Deresiewicz (1963) are examined there with a view towards the applications.
In Kubik and Cieszko (2005) a dissipative Rayleigh surface potential is proposed which is suitable to produce a particular form of Beavers-Joseph-Saffman boundary conditions: many versions of them are compared in Alazmi and Vafai (2001).

The jump conditions (69) also allow for describing phenomena in which inertial effects are relevant. The inertial terms here newly introduced are quadratic in the velocity fields at the interface: when Stokes fluid-flow conditions hold (and when the solid matrix is subject to "small deformations") they may be negligible. Indeed, when the equations are linearized in the neighborhood of a state of rest (i.e. when all velocity fields and their gradients are vanishing) the aforementioned inertial terms do not produce, in the resulting boundary conditions, any term additional to those appearing in Beavers-Joseph-Saffman conditions. However, when the linearization procedure is performed in the neighborhood of a state in which some velocity fields are not vanishing then inertial terms cannot be neglected, even when Rayleigh modeling hypothesis applies. Finally one should remark that in Ochoa-Tapia and Whitaker (1998) some inertial effects at the interface are considered. However, they deduce no-slip conditions for tangential velocity and a normal-to-the-interface boundary condition by means of an averaging procedure involving "excess quantities". Their assumptions produce inertial terms in which only the tangential part of the fluid velocity appears.

Future investigations will be aimed to get a generalization of the newly found boundary conditions to the case of shock waves.

A Appendix: Some Preliminary Variations

We now compute how the different quantities appearing in $\mathcal{A}$ vary when the two independent kinematical variables χ_s and ϕ_s change. We denote $\delta\chi_s$ and $\delta\phi_s$ the infinitesimal changes of χ_s and ϕ_s respectively.
Recall that we have assumed that the singularity surface S_s (if any) of the fields χ_s and ϕ_s is fixed in B_s. Owing to this assumption, the fact of considering the fields $\delta\chi_s$ and $\delta\phi_s$ regular in B_s is sufficient to explore the whole space of considered functions χ_s and ϕ_s i.e. the set of the C^1_{pw} functions the singularity surface of which, if any, must be the surface S_s. This would not be the case if one wants to consider a moving singularity surface. Such a situation requires a more delicate treatment and will be

developed in a forthcoming paper.

As far as the fields ε, η_s and $\mathbf{v}_s$ attached to the solid constituent are concerned, their variation in terms of the independent kinematical fields are given by:[4]

$$\begin{aligned}\delta\varepsilon &= \frac{1}{2}\delta\left(\mathbf{F}_s^T\cdot\mathbf{F}_s-\mathbf{I}\right)\\ &=\frac{1}{2}\left(\delta\mathbf{F}_s^T\cdot\mathbf{F}_s+\mathbf{F}_s^T\cdot\delta\mathbf{F}_s\right)=\left(\mathbf{F}_s^T\cdot\nabla\left(\delta\chi_s\right)\right)^{sym},\\ \delta\eta_s &= 0, \qquad \delta\mathbf{v}_s=\delta\dot{\chi}_s\end{aligned}\tag{82}$$

We now compute the variation of the fields m_f and $\mathbf{v}_f^{Ⓢ}$ attached to the fluid constituent. In particular, if we want to compute δm_f the preliminary variation of the fields I_s and $\eta_f^{Ⓢ}$ must be established[5]:

$$\delta I_s=\delta\left(\det\left(\nabla\phi_s\right)\right)=I_s\mathbf{G}_s^{-T}\mid\nabla\left(\delta\phi_s\right);$$

$$\delta\eta_f^{Ⓢ}=\delta\left(\eta_f\circ\phi_s\right)=\left(\nabla\eta_f\right)^{Ⓢ}\cdot\delta\phi_s=\nabla\eta_f^{Ⓢ}\cdot\mathbf{G}_s^{-1}\cdot\delta\phi_s.$$

The variation δm_f can be now computed as follows:

$$\begin{aligned}\delta m_f &= \delta\left(I_s\eta_f^{Ⓢ}\right)=\eta_f^{Ⓢ}\delta I_s+I_s\,\delta\eta_f^{Ⓢ}\\ &= I_s\left(\eta_f^{Ⓢ}\mathbf{G}_s^{-T}\mid\nabla\left(\delta\phi_s\right)+\nabla\eta_f^{Ⓢ}\cdot\mathbf{G}_s^{-1}\cdot\delta\phi_s\right)\\ &= \operatorname{div}\left(\eta_f^{Ⓢ}I_s\,\mathbf{G}_s^{-1}\cdot\delta\phi_s\right)=\operatorname{div}\left(m_f\mathbf{G}_s^{-1}\cdot\delta\phi_s\right),\end{aligned}\tag{83}$$

where we used the fact that $\operatorname{div}\left(I_s\mathbf{G}_s^{-T}\right)=0$ proved in Eq. (5).

The variation of the solid Lagrangian fluid velocity $\mathbf{v}_f^{Ⓢ}$ is now computed. Recalling Eq. (50) it is easy to show that[6]

$$\begin{aligned}\delta\mathbf{v}_f^{Ⓢ} &= \delta\mathbf{v}_s-\delta\mathbf{F}_s\cdot\mathbf{G}_s^{-1}\cdot\mathbf{u}_s+\mathbf{F}_s\cdot\mathbf{G}_s^{-1}\cdot\delta\mathbf{G}_s\cdot\mathbf{G}_s^{-1}\cdot\mathbf{u}_s-\mathbf{F}_s\cdot\mathbf{G}_s^{-1}\cdot\delta\mathbf{u}_s\\ &= \delta\dot{\chi}_s-\nabla\left(\delta\chi_s\right)\cdot\mathbf{G}_s^{-1}\cdot\mathbf{u}_s+\mathbf{F}_s\cdot\mathbf{G}_s^{-1}.\nabla\left(\delta\phi_s\right)\cdot\mathbf{G}_s^{-1}\cdot\mathbf{u}_s\\ &\quad-\mathbf{F}_s\cdot\mathbf{G}_s^{-1}\cdot\delta\dot{\phi}_s.\end{aligned}\tag{84}$$

[4]Given a tensor field σ, $\sigma^{sym}:=\frac{1}{2}\left(\sigma+\sigma^T\right)$ denotes the symmetric part of σ. Moreover, we denote in the sequel the partial time derivatives of solid-Lagrangian fields by a superposed dot.

[5]We recall that, given a tensor field $\mathbf{M}$, the derivation rule for the determinant reads: $\delta\left(\det\mathbf{M}\right)=\det\mathbf{M}\left(\mathbf{M}^{-T}\mid\delta\mathbf{M}\right)$.

[6]Given a matrix $\mathbf{G}$, the differentiation formula for its inverse $\mathbf{G}^{-1}$ reads $\delta\left(\mathbf{G}^{-1}\right)=-\mathbf{G}^{-1}.\delta\mathbf{G}.\mathbf{G}^{-1}$. This result is easily recovered differentiating the equality $\mathbf{G}\cdot\mathbf{G}^{-1}=\mathbf{I}$.

B Appendix: Variation of the Action Functional

In order to suitably perform the solid-Lagrangian variation $\delta\mathcal{A}$ of the action functional, an arbitrary compact set $K \subset B_s$ is defined across the discontinuity S_s and the variation of the action functional is performed assuming that the test functions $\delta\chi_s$ and $\delta\phi_s$ have support included in the set K. This means that $\delta\chi_s$ and $\delta\phi_s$ are assumed to be suitably regular and to vanish on the boundary and outside the set K. In the sequel, the intersection of the surface S_s with the set K is denoted by Σ.

Recalling that we assumed Ψ to be a function of ε, m_f and χ_s and that $\Lambda = 1/2\left(\eta_s\left(\mathbf{v}_s\right)^2 + m_f(\mathbf{v}_f^{ⓢ})^2\right)$ it is easy to recognize that

$$\delta\Psi = \frac{\partial\Psi}{\partial\varepsilon} \mid \delta\varepsilon + \frac{\partial\Psi}{\partial\chi_s}\delta\chi_s + \frac{\partial\Psi}{\partial m_f}\delta m_f \tag{85}$$

and

$$\delta\Lambda = \eta_s\mathbf{v}_s\cdot\delta\mathbf{v}_s + m_f\mathbf{v}_f^{ⓢ}\cdot\delta\mathbf{v}_f^{ⓢ} + \frac{1}{2}\mathbf{v}_s^2\,\delta\eta_s + \frac{1}{2}\left(\mathbf{v}_f^{ⓢ}\right)^2\delta m_f. \tag{86}$$

Recalling Eqs. (82), (83) and (84) and the fact that $\partial\Psi/\partial\varepsilon$ is a symmetric tensor, these equations can be rewritten as

$$\delta\Psi = \frac{\partial\Psi}{\partial\varepsilon} \mid \left(\mathbf{F}_s^T\cdot\nabla\left(\delta\chi_s\right)\right) + \frac{\partial\Psi}{\partial\chi_s}\cdot\delta\chi_s + \frac{\partial\Psi}{\partial m_f}\operatorname{div}\left(m_f\mathbf{G}_s^{-1}\cdot\delta\phi_s\right)$$

and

$$\delta\Lambda = \delta\Lambda_s + \delta\Lambda_f \tag{87}$$

where

$$\delta\Lambda_s := \left(\eta_s\mathbf{v}_s + m_f\mathbf{v}_f^{ⓢ}\right)\cdot\delta\dot{\chi}_s - m_f\mathbf{v}_f^{ⓢ}\cdot\nabla\left(\delta\chi_s\right)\cdot\mathbf{G}_s^{-1}\cdot\mathbf{u}_s,$$

$$\delta\Lambda_f := \frac{1}{2}\left(\mathbf{v}_f^{ⓢ}\right)^2\operatorname{div}\left(m_f\mathbf{G}_s^{-1}\cdot\delta\phi_s\right)$$

$$+ m_f\mathbf{v}_f^{ⓢ}\cdot\mathbf{F}_s\cdot\mathbf{G}_s^{-1}\cdot\nabla\left(\delta\phi_s\right)\cdot\mathbf{G}_s^{-1}\cdot\mathbf{u}_s - m_f\mathbf{v}_f^{ⓢ}\cdot\mathbf{F}_s\cdot\mathbf{G}_s^{-1}\cdot\delta\dot{\phi}_s.$$

Some processes of integration by parts in space and in time are now needed in order to rewrite the variations $\delta\Psi$ and $\delta\Lambda$ in terms of the variations $\delta\chi_s$ and $\delta\phi_s$ of the primitive kinematical fields.

We first compute the variation of the potential energy: integrating by parts in space Eq. (87) it is easy to get

$$\int_{B_s\times(0,T)} \delta\Psi = \int_{K\times(0,T)} \left[\frac{\partial\Psi}{\partial\chi_s} - \operatorname{div}\left(\mathbf{F}_s\cdot\frac{\partial\Psi}{\partial\varepsilon}\right)\right]\cdot\delta\chi_s$$
$$-\int_{K\times(0,T)} \left[m_f\mathbf{G}_s^{-T}\cdot\nabla\left(\frac{\partial\Psi}{\partial m_f}\right)\right]\cdot\delta\phi_s + \tag{88}$$
$$\int_{\Sigma\times(0,T)} \left[\left|\mathbf{F}_s\cdot\frac{\partial\Psi}{\partial\varepsilon}\mathbf{N}_s\right|\right]\cdot\delta\chi_s + \int_{\Sigma\times(0,T)} \left[\left|m_f\mathbf{G}_s^{-T}\cdot\frac{\partial\Psi}{\partial m_f}\mathbf{N}_s\right|\right]\cdot\delta\phi_s.$$

In order to obtain simpler calculations, the variation of the kinetic energy is now computed evaluating separately the two terms appearing in Eq. (87). Integrating by parts in space and time the first term, recalling that η_s is constant in space and time and that $\mathbf{D} = -m_f\mathbf{G}_s^{-1}\cdot\mathbf{u}_s$ it is easy to recognize that [7]

$$\int_{B_s\times(0,T)} \delta\Lambda_s$$
$$= -\int_{K\times(0,T)} \left[\eta_s\gamma_s + m_f\dot{\mathbf{v}}_f^{\text{Ⓢ}} + \dot{m}_f\mathbf{v}_f^{\text{Ⓢ}} + \operatorname{div}\left(\mathbf{v}_f^{\text{Ⓢ}}\otimes\mathbf{D}\right)\right]\cdot\delta\chi_s$$
$$+\int_{\Sigma\times(0,T)} \left(\left[\left|\mathbf{v}_f^{\text{Ⓢ}}\otimes\mathbf{D}\right|\right]\cdot\mathbf{N}_s\right)\cdot\delta\chi_s.$$

Finally, using expression (51) for $\dot{\mathbf{v}}_f^{\text{Ⓢ}}$, the balance of mass (56) for $\dot{m}_f$ and simplifying, it is easy to recognize that

$$\int_{B_s\times(0,T)} \delta\Lambda_s = -\int_{K\times(0,T)} \left(\eta_s\gamma_s + m_f\gamma_f^{\text{Ⓢ}}\right)\cdot\delta\chi_s$$
$$+\int_{\Sigma\times(0,T)} \left(\left[\left|\mathbf{v}_f^{\text{Ⓢ}}\otimes\mathbf{D}\right|\right]\cdot\mathbf{N}_s\right)\cdot\delta\chi_s. \tag{89}$$

The second term appearing on the right hand side of Eq. (87) is now evaluated: integrating by parts in space and time and recalling Eqs. (50) and the definition of the mass flux vector $\mathbf{D}$, we get

[7] From now on, we assume that $\delta\chi_s\ (\cdot,0) = \delta\chi_s\ (\cdot,T)$ and $\delta\phi_s\ (\cdot,0) = \delta\phi_s\ (\cdot,T)$ so that no time boundary terms are present in our calculation.

$$\int_{B_s\times(0,T)} \delta\Lambda_f = \int_{K\times(0,T)} \left[-\frac{1}{2}m_f \nabla\left(\left(\mathbf{v}_f^{\circledS}\right)^2\right)\cdot \mathbf{G}_s^{-1}\right]$$
$$+\int_{K\times(0,T)} \left[\operatorname{div}\left(\mathbf{v}_f^{\circledS}\cdot \mathbf{F}_s\cdot \mathbf{G}_s^{-1}\otimes \mathbf{D}\right)+\frac{\partial}{\partial t}\left(m_f \mathbf{v}_f^{\circledS}\cdot \mathbf{F}_s\cdot \mathbf{G}_s^{-1}\right)\right]\cdot \delta\phi_s$$
$$+\int_{\Sigma\times(0,T)} \left[\frac{1}{2}\left[\left|m_f\left(\mathbf{v}_f^{\circledS}\right)^2 \mathbf{G}_s^{-T}\right|\right]\cdot \mathbf{N}_s\right.$$
$$\left. -\left[\left|\mathbf{v}_f^{\circledS}\cdot \mathbf{F}_s\cdot \mathbf{G}_s^{-1}\otimes \mathbf{D}\right|\right]\cdot \mathbf{N}_s\cdot\delta\phi_s.\right.$$

Using expressions (56), (51) and (52) for $\dot{m}_f$, $\dot{\mathbf{v}}_f^{\circledS}$ and $\partial\left(\mathbf{F}_s\cdot \mathbf{G}_s^{-1}\right)/\partial t$ respectively, recalling again the definition of $\mathbf{D}$ and noting that $\nabla\left(\left(\mathbf{v}_f^{\circledS}\right)^2\right) = 2\,\mathbf{v}_f^{\circledS}\cdot \nabla \mathbf{v}_f^{\circledS}$, this equation can be simplified in

$$\int_{B_s\times(0,T)} \delta\Lambda_f = \int_{K\times(0,T)} \left(m_f \gamma_f^{\circledS}\cdot \mathbf{F}_s\cdot \mathbf{G}_s^{-1}\right)\cdot\delta\phi_s \tag{90}$$
$$+\int_{\Sigma\times(0,T)} \left(\left[\left|\frac{1}{2}m_f \mathbf{G}_s^{-T}\left(\mathbf{v}_f^{\circledS}\right)^2 - \mathbf{G}_s^{-T}\cdot \mathbf{F}_s^{T}\cdot \mathbf{v}_f^{\circledS}\otimes \mathbf{D}\right|\right]\cdot \mathbf{N}_s\right)\cdot\delta\phi_s.$$

Merging Eqs. (88), (89) and (90), and recalling that they hold for any $\delta\chi_s$, $\delta\phi_s$ with compact support K included in B_s, expression (66)for the variation of the action functional is finally recovered.

C Appendix: Computation of the Rayleigh-Hamilton Dissipation

We recall that, owing to (60), (61) and (2), (50), the pull back of the dissipation densities reads

$$\begin{aligned}\mathcal{D}^{\circledS} &= \left(\mathbf{F}_s\cdot \mathbf{G}_s^{-1}\cdot \mathbf{u}_s\right)\cdot \mathbf{K}^{\circledS}\cdot\left(\mathbf{F}_s\cdot \mathbf{G}_s^{-1}\cdot \mathbf{u}_s\right)\\ &+\left(\nabla \mathbf{v}_f^{\circledS}\cdot \mathbf{F}_s^{-1}\right):\mathbf{M}^{\circledS}:\left(\nabla \mathbf{v}_f^{\circledS}\cdot \mathbf{F}_s^{-1}\right)\\ &+\left(\nabla\left(\mathbf{F}_s\cdot \mathbf{G}_s^{-1}\cdot \mathbf{u}_s\right)\cdot \mathbf{F}_s^{-1}\right):\mathbf{B}^{\circledS}:\left(\nabla\left(\mathbf{F}_s\cdot \mathbf{G}_s^{-1}\cdot \mathbf{u}_s\right)\cdot \mathbf{F}_s^{-1}\right),\end{aligned} \tag{91}$$

$$\mathcal{D}_S^{\circledS} = \left[\left|\mathbf{v}_f^{\circledS}\right|\right]\cdot \mathbf{S}^{\circledS}\cdot\left[\left|\mathbf{v}_f^{\circledS}\right|\right]. \tag{92}$$

We consider arbitrary test functions $\delta\mathbf{q} = (\delta\chi_s, \delta\phi_s)$ with compact support included in $\mathbb{B}_s$ We now compute the term appearing on the right hand side of Eq.(65); in particular we start by noticing that owing to Eq. (64) we get

$$\int_0^T \left(\frac{\partial \mathcal{R}}{\partial \dot{\mathbf{q}}_t} \mid \delta \mathbf{q}_t\right) dt = \int_{\mathbb{B}_s \setminus \mathbb{S}_s} J_s (\mathbf{F}_s \cdot \mathbf{G}_s^{-1} \cdot \delta\phi_s) \cdot \kappa^{\circledS}) +$$
$$\int_{\mathbb{B}_s \setminus \mathbb{S}_s} J_s \left(\nabla(\delta\chi_s - \mathbf{F}_s \cdot \mathbf{G}_s^{-1} \cdot \delta\phi_s) \cdot \mathbf{F}_s^{-1}\right) : \Pi_f^{\circledS}$$
$$+ \int_{\mathbb{B}_s \setminus \mathbb{S}_s} J_s \left(\nabla \left(\mathbf{F}_s \cdot \mathbf{G}_s^{-1} \cdot \delta\phi_s\right) \cdot \mathbf{F}_s^{-1}\right) : \Pi^{\circledS} +$$
$$\int_{\mathbb{S}_s} \left\| J_s \mathbf{F}_s^{-T} \cdot \mathbf{N}_s \right\| \left[\left| \delta\chi_s - \mathbf{F}_s \cdot \mathbf{G}_s^{-1} \cdot \delta\phi_s \right|\right] \cdot \sigma^{\circledS}. \tag{93}$$

Integrating by parts in space the right hand side of equation (93) gives

$$\int_{\mathbb{B}_s \setminus \mathbb{S}_s} - \operatorname{div}\left(J_s \Pi_f^{\circledS} \cdot \mathbf{F}_s^{-T}\right) \cdot \delta\chi_s$$
$$+ \int_{\mathbb{B}_s \setminus \mathbb{S}_s} \mathbf{G}_s^{-T} \cdot \mathbf{F}_s^{T} \cdot \left(J_s \kappa^{\circledS} - \operatorname{div}\left(J_s (\Pi^{\circledS} - \Pi_f^{\circledS})^T \cdot \mathbf{F}_s^{-T}\right)\right) \cdot \delta\phi_s$$
$$+ \int_{\mathbb{S}_s} \left[\left| J_s \Pi_f^{\circledS} \cdot \mathbf{F}_s^{-T} \cdot \mathbf{N}_s \right|\right] \cdot \delta\chi_s$$
$$+ \int_{\mathbb{S}_s} \left[\left| \mathbf{G}_s^{-T} \cdot \mathbf{F}_s^{T} \cdot \left(J_s \left((\Pi^{\circledS} - \Pi_f^{\circledS})^T\right) \cdot \mathbf{F}_s^{-T} \cdot \mathbf{N}_s \right.\right.\right.$$
$$- \left\| J_s \mathbf{F}_s^{-T} \cdot \mathbf{N}_s \right\| \sigma^{\circledS} \cdot \delta\phi_s.$$

D Bibliography

Alazmi B., Vafai K.: Analysis of fluid flow and heat transfer interfacial conditions between a porous medium and a fluid layer. Int. J. Heat Mass Transf. 44, 1735-1749 (2001)

Albers B.: Monochromatic surface waves at the interface between poroelastic and fluid half-spaces. Proc. R. Soc. Lond. A 462, 701723 (2006)

Allaire G.: Homogenization of the Stokes flow in a connected porous medium. Asymptotic Anal. 2, 3, 203-222 (1989).

Allaire G.: Homoge ne isation des e quations de Stokes dans un domaine perfore de petits trous re partis pe riodiquement. C. R. Acad. Sci. Paris, Se r. I, 309, 11, 741-746 (1989).

Allaire G.: Homogenization of the Navier-Stokes equations in open sets perforated with tiny holes. Arch. Rational Mech. Anal., 113, 209298 (1991).

Allaire G.: Homogenization of the Navier-Stokes equations with a slip boundary condition. Comm. Pure Appl. Math., XLIV, 605-641 (1991).

Allaire, G.: Continuity of the Darcys law in the low-volume fraction limit. Ann. Scuola Norm. Sup. Pisa Cl. Sci. (4) 18, 475-499 (1991)

Altay G., Dokmeci M.C.: On the equations governing the motion of an anisotropic poroelastic material. Proc. R. Soc. A 462, 2373-2396 (2006)

Baek S., Srinivasa A.R.: Diffusion of a fluid through an elastic solid undergoing large deformation. Int. J. Non-Linear Mech. 39, 201-218 (2004)

Batra G., Bedford A., Drumheller D.S.: Applications of Hamiltons principle to continua with singular surfaces. Arch. Rational Mech. Anal. 93, 223251 (1986)

Bedford A., Drumheller D.S.: A variational theory of immiscible mixtures. Arch. Rational Mech. Anal. 68, 37-51 (1978)

Bedford A., Drumheller D.S.: A variational theory of porous media. Int. J. Solids Struct. 15, 967-980 (1979) 20 CISM Course C-1006 Udine, July 12-16 2010

Bedford A., Drumheller D.S.: Recent advances: theories of immiscible and structured mixtures. Int. J. Eng. Sci. 21, 863-960 (1983)

Biot M.A.: General theory of three-dimensional consolidation. J. Appl. Phys. 12, 155-164 (1941)

Biot M.A.: General solutions of the equations of elasticity and consolidation for a porous material. J. Appl. Mech. 23, 91-96 (1956)

Biot M.A.: Theory of propagation of elastic waves in fluid-saturated porous solid. J. Acoust. Soc. Am. 28, 168-191 (1956)

Biot M.A., Willis D G.: The elastic coefficients of the theory of consolidation. J. Appl. Mech. 24 594-601 (1957)

Biot M.A.: Mechanics of deformation and acoustic propagation in porous media. J. Appl. Phys. 33, 1482-1498 (1962)

Biot M.A.: Variational principles for acoustic gravity waves. Phys. Fluids 6, 772-778 (1963)

Beavers G.S., Joseph D.D.: Boundary conditions at a naturally permeable wall. J. Fluid Mech. 30, 197- 207 (1967)

Burridge R., Keller J.B.: Poroelasticity equations derived from microstructure. J. Acoust. Soc. Am. 70:4, 1140-1146 (1981)

Casertano L., Oliveri del Castillo A., Quagliariello M.T.: Hydrodynamics and geodynamics in Campi Flegrei Area of Italy. Nature 264, 161-164 (1976)

Caviglia G.,MorroA.:Harmonic waves in thermo-visco-elastic solids.Int. J. Eng. Sci.43, 1323-1336 (2005)

Chandesris M., Jamet D.: Boundary conditions at a planar fluidporous interface for a Poiseuille flow. Int. J. Heat and Mass Transf. 49, 2137- 2150

(2006).

Chandesris M., Jamet D.: Boundary conditions at a fluidporous interface: an a priori estimation of the stress jump coefficients. Int. J. Heat and Mass Transf. 50, 3422-3436 (2007)

Chateau X., Dormieux L.: Homoge ne isation dun milieu poreux non sature : Lemme de Hill et applica- tions. C.R. Acad. Sci. Paris 320, Se rie IIb, 627-634 (1995)

Chateau X., Dormieux L.: Approche microme canique du comportement dun milieu poreux non sature . C. R. Acad. Sci. Paris 326, Se rie II, 533-538 (1998)

Cieszko M., Kubik J.: Interaction of elastic waves with a fluid-saturated porous solid boundary. J. of Theor. Appl. Mech. 36, 561580 (1998)

Cieszko M. Kubik J.: Derivation of matching conditions at the contact surface between fluid-saturated porous solid and bulk fluid. Transp. Porous Med. 34, 319336 (1999)

Coussy O., Bourbie T.: Propagation des ondes acoustiques dans les milieux poreux sature s. Rev. Inst. Fr. Pet. 39, 47-66 (1984)

Coussy O. Dormieux L. Detournay E.: From mixture theory to Biots approach for porous media. Int. J. Solids Struct. 35, 46194635 (1998)

Coussy O.: Poromechanics, John Wiley Sons, Chichester (2004)

Cowin SC: Bone Mechanics Handbook, Boca Raton, FL: CRC Press (2001)

Debergue P. et al.: Boundary conditions for the weak formulation of the mixed (u,p) poroelasticity prob- lem. J. Acoust. Soc. Am. 106:5, 2383-2390 (1999)

de Boer R.: Highlights in the historical development of the porous media theory: toward a consistent macroscopic theory. Appl. Mech. Rev. 49:4, 201-262 (1996)

de Boer R.: Contemporary progress in porous media theory. Appl. Mech. Rev. 53, 323370 (2000)

de Boer R.: Theoretical poroelasticity a new approach. Chaos Solitons Fractals 25, 861-878 (2005)

de Buhan P., Dormieux L., Chateau X.: A micro-macro approach to the constitutive formulation of large strain poroelasticity. Poromechanics a tribute to Maurice A. Biot. Thimus et al., Rotterdam (1998) 21 CISM Course C-1006 Udine, July 12-16 2010

de Buhan P., Chateau X., Dormieux L.: The constitutive equations of finite strain poroelasticity in the light of a micro-macro approach. Eur. J. Mech. A/Solids. 17, 909-921 (1998)

Cryer C.W.: A comparison of the three dimensional consolidation theories of Biot and Terzaghi. Q. J. Mech. Appl. Math. 16:4, 401-412 (1963) de la Cruz V., Hube J., Spanos T.J.T.: Reflection and transmission of seismic waves at the boundaries of porous media. Wave Motion 16, 323-338 (1992)

dellIsola F., Rosa L., Woz niak C.: A micro-structured continuum modelling compacting fluid-saturated grounds: the effects of pore-size scale parameter. Acta Mech. 127, 1:4, 165-182 (1998)

dellIsola, F., Guarascio, M., Hutter, K.: A variational approach for the deformation of a saturated porous solid. A second gradient theory extending Terzaghis effective stress principle. Arch. Appl. Mech. 70, 323- 337 (2000).

dellIsola, F., Sciarra, G., Batra, R.C.: Static Deformations of a Linear Elastic Porous Body filled with an Inviscid Fluid. J. Elasticity 72, 99-120 (2003).

Deresiewicz H.: The effect of boundaries on wave propagation in a liquidfilled porous solid: I. Reflection of plane waves at a free plane boundary (non-dissipative case). Bull. Seismol. Soc. Am. 50, 599607 (1960)

Deresiewicz H.: The effect of boundaries on wave propagation in a liquid-filled porous solid: III. Reflec- tion of plane waves at a free plane boundary (general case) Bull. Seismol. Soc. Am. 52, 595625 (1962)

Deresiewicz H.: A note on Love waves in a homogeneous crust overlying an inhomogeneous substratum. Bull. Seismol. Soc. Am. 52, 639-645 (1962)

Deresiewicz H.: The effect of boundaries on wave propagation in a liquidfilled porous solid: IV. surface in a half-space. Bull. Seismol. Soc. Am. 50, 627-638 (1962)

Deresiewicz H.: On uniqueness in dynamic poroelasticity. Bull. Seismol. Soc. Am. 53, 783-788 (1963)

Deresiewicz H.: The effect of boundaries on wave propagation in a liquidfilled porous solid: V. Trans- mission across a plane interface. Bull. Seismol. Soc. Am. 54, 409-416 (1964)

Deresiewicz H.: The effect of boundaries on wave propagation in a liquidfilled porous solid: VII. Surface waves in a half-space in the presence of a liquid layer. Bull. Seismol. Soc. Am. 54, 425430 (1964)

Dormieux L.,Coussy O., de Buhan P.: Mode lisation me canique dun milieu polyphasique par la me thode des puissances virtuelles. C.R. Acad. Sci. Paris. 313, Se rie IIb, 863-868 (1991)

Dormieux L., Stolz C.: Approche variationelle en poroe lasticite . C.R. Acad. Sci. Paris, 315, Se rie IIb, 407-412 (1992)

Dormieux L., Kondo D. and Ulm F.-J. Microporomechanics Wiley (2006).

Fillunger P.: Erbdaumechanik. Selbst Verlag des Verfassers, Wien (1936).

Gavrilyuk S.L., Gouin H., Perepechko Yu.V.: A variational principle for two-fluid models. C.R. Acad. Sci. Paris, 324, Se rie IIb, 483-490 (1997)

Gavrilyuk S.L., Gouin H., Perepechko Yu.V.: Hyperbolic Models of Homogeneous Two-Fluid Mixtures, Meccanica 33, 161-175(1998)

Gavrilyuk S., Perepechko Yu. V.: Variational approach to constructing hyperbolic models of two-velocity media. J. Appl. Mech. Technical Phys. 39:5, 684-698 (1998)

GavrilyukS.L.,GouinH.:A new form of governing equations of fluids arising from Hamiltons principle. Int. J. Eng. Sci. 37, 1495-1520 (1999)

Gavrilyuk S., Saurel R.: Rankine-Hugoniot relations for shocks in heterogeneous mixtures J. Fluid Mech. 575, 495-507 (2007)

Germain P.: Cours de me canique des milieux continus tome 1 the orie ge ne rale, Masson (1973)

Gouin H.: Variational theory of mixtures in continuum mechanics. Eur. J. Mech., B/Fluids 9, 469-491 (1990)

Gouin H., Gavrilyuk S.L.: Hamiltons principle and Rankine-Hugoniot conditions for general motions of mixtures. Meccanica 34, 39-47 (1998) 22 CISM Course C-1006 Udine, July 12-16 2010 Goyeau B. et al.: Momentum transport at a fluidporous interface. Int. J. Heat Mass Transf. 46, 4071-4081 (2003)

Haber S., Mauri R.: Boundary conditions for Darcys flow through porous media. Int. J. Multiphase Flow 9, 561574 (1983)

Hassanizadeh S. M. Gray W.: Boundary and interface conditions in porous media. Water Resources Res. 25, 17051715 (1989)

Hornung U.: Homogenization and Porous Media. Interdiscip. Appl. Math., vol. 6, Springer, Berlin (1997)

HoulsbyaG.T.,PuzrinA.M.:Rate-dependent plasticity models derived from potential functions. J. Rheol. 46(1), 113-126 (2002)

Jager W., Mikelic A.: On the interface boundary condition of Beavers, Joseph, and Saffman. SIAM J. Appl. Math. 60, 1111-1127 (2000)

Kaasschieter E.F., Frijns A.J.H.: Squeezing a sponge: a three-dimensional solution in poroelasticity Computational Geosciences 7: 49-59, (2003).

Kosin ski W.: Field Singularities and Wave Analysis in Continuum Mechanics. PWN-Polish Scientific Publishers, Warszawa (1986)

Kubik J., Cieszko M.: Analysis of matching conditions at the boundary surface of a fluid-saturated porous solid and a bulk fluid: the use of Lagrange multipliers. Continuum Mech. Thermodyn. 17:4, 351-359 (2005)

Kuznetsov A.V.: Influence of the stress jump condition at the porous medium/clear fluid interface on a flow at porous wall. Int. Comm. Heat Mass Transf. 24, 401-410. (1997)

Le Bars M., Grae Worster M.: Interfacial conditions between a pure fluid and a porous medium: impli- cations for binary alloy solidification. J. Fluid Mech. 550, 149-173 (2006)

Lee C.K.: Flow and deformation in poroelastic media with moderate load and weak inertia. Proc. R. Soc. Lond. A 460, 20512087 (2004)

Levy T., Sanchez-Palencia E.: . On boundary conditions for fluid flow in porous media Int. J. Eng. Sci. 13, 923-940 (1975)

Madeo A., dellIsola F., Ianiro N., Sciarra G.: A variational deduction of second gradient poroelasticity II: an application to the Consolidation Problem. J. Mech. Mater. Struct. 3:4 , 607-625 (2008)

Mandel J.: Consolidation des sols (e tude mathe matique). Geotechnique 3, 287-299 (1953)

Marle C.M.: On macroscopic equations governing multiphase flow with diffusion and chemical reactions in porous media. Int. J. Eng. Sci. 20:5, 643662 (1982)

Mobbs S.D.: Variational principles for perfect and dissipative fluid flows. Proc. R. Soc. Lond. A 381, 457468 (1982)

Neale G., Nader W.: Practical significance of Brinkmans extension of Darcys law: coupled parallel flows within a channel and a bounding porous medium. Can. J. Chem. Eng. 52, 475-478 (1974)

Ochoa-Tapia J.A., Whitaker S.: Momentum transfer at the boundary between a porous; medium and a homogeneous fluid I. Theoretical development. Int. J. Heat Mass Transf. 38, 2635-2646 (1995)

Ochoa-Tapia J. A., Whitaker S.: Momentum transfer at the boundary between a porous medium and a homogeneous fluid II. Comparison with experiment. Int. J. Heat Mass Transf. 38, 2647-2655 (1995).

Ochoa-Tapia J.A., Whitaker S.: Momentum jump condition at the boundary between a porous medium and a homogeneous fluid: inertial effects. J. Porous Media 1, 201-217 (1998).

Orsi G.,Petrazzuoli S.M.,Wohletz K.:Mechanical and thermo-fluid behaviour during unrest at the Campi Flegrei caldera (Italy) J. Vulcanol. Geotherm. Res. 91, 453-470 (1999)

Pan Y., Horne R.N.: Generalized Macroscopic Models for Fluid Flow in Deformable Porous Media. Theories Transport in Porous Media 45: 1-27, 2001.

Prat M.: On the boundary conditions at the macroscopic level‘ Transp. Porous Med. 4, 259-280 (1988) 23 CISM Course C-1006 Udine, July 12-16 2010

Poulikakos D., Kazmierczak M.: Forced convection in a duct partially filled with a porous material. J. Heat Transf. 109, 653-662 (1987)

Quiroga-Goode G., Carcione J.M.: Heterogeneous modelling behaviour at an interface in porous media. Comput. Geosciences 1, 109125 (1997)

Rajagopal K.R. et al.: On boundary conditions for a certain class of problems in mixture theory. Int. J. Eng. Sci. 24, 14531463 (1986)

Rajagopal K.R., Tao L.: Mechanics of mixtures. Ser. Adva. Math.

Appl. Sci. Vol. 35 (1995)

Rasolofosaon N.J.P., Coussy O.: Propagation des ondes acoustiques dans les milieux poreux sature s: effets dinterface I. Rev. Inst. Fr. Pet. 40, 581-594 (1985)

Rasolofosaon N.J.P., Coussy O.: Propagation des ondes acoustiques dans les milieux poreux sature s: effets dinterface II. Rev. Inst. Fr. Pet. 40, 785802 (1985)

Rasolofosaon N.J.P., Coussy O.: Propagation des ondes acoustiques dans les milieux poreux sature s: effets dinterface III. Rev. Inst. Fr. Pet. 41, 91103 (1986)

Saffman P. G.: On the boundary conditions at the surface of a porous medium. Stud. Appl. Math. L, 93-101 (1971)

Sciarra G., dellIsola F., Ianiro N., Madeo A.: A variational deduction of second gradient poroelasticity I: general theory. J. Mech. Mater. Struct. 3:3, 507-526 (2008)

Seliger R.L., Whitham G.B.: Variational principles in continuum mechanics. Proc. R. Soc. Lond. A 305, 1-25 (1968)

Sharma M.D.: Wave propagation across the boundary between two dissimilar poroelastic solids. J. Sound Vib. 314, 657-671 (2008)

SonnetA.M., Maffettone P.L. and Virga E.G.: Continuum theory for nematic liquid crystals with tensorial order. J. Non-Newtonian Fluid Mech. 119 51-59 (2004)

V. Terzaghi K.: Theoretical soil mechanics. John Wiley Sons, Chichester (1943)

Savare G., Tomarelli F.: Superposition and chain rule for bounded Hessian functions. Adv. Math. 140, 237-281 (1998)

Vafai K., Thiyagaraja R.: Analysis of flow and heat transfer at the interface region of a porous medium. Int. J. Heat Mass Transf. 30, 1391-1405 (1987)

Vafai K., Kim S.J.: Analysis of surface enhancement by a porous substrate. J. Heat Transf. 112, 700-706 (1990)

Valde s-Paradaa F.J., Goyeaub B., Ochoa-Tapiaa J.A.: Diffusive mass transfer between a microporous medium and a homogeneous fluid: Jump boundary conditions. Chem. Eng. Sci. 61 1692-1704 (2006)

WilmanskiK.:Waves in porous and granular materials. In: Kinetic and Continuum Theories of Granular and Porous Media, 131-185, K. Hutter and K. Wilmanski(eds.), Springer Wien New York (1999)

Wilmanski K.: A few remarks on Biots model and linear acoustics of poroelastic saturated materials. Soil Dyn. Earthq. Eng. 26, 509-536 (2006)

Yang J.: Importance of flow condition on seismic waves at a saturated porous solid boundary. J. Sound Vib. 221:3, 391413 (1999)

FSC
www.fsc.org
MIX
Papier | Fördert gute Waldnutzung
FSC® C083411

Kolibri 360 GmbH
Ferdinand-Jühlke-Straße 7
99095 Erfurt, Deutschland
produktsicherheit@kolibri360.de